SIGMUND FREUD

OUT OF YOUR VULNERABILITIES WILL COME YOUR STRENGTH.

The Interpretation of Dreams

CLASSICS

Published 2023

FiNGERPRINT! CLASSiCS
An imprint of Prakash Books India Pvt. Ltd

113/A, Darya Ganj,
New Delhi-110 002
Email: info@prakashbooks.com/sales@prakashbooks.com

facebook www.facebook.com/fingerprintpublishing
twitter www.twitter.com/FingerprintP
www.fingerprintpublishing.com

ISBN: 978 93 5856 084 8

Processed & printed in India

"Success remains a matter of ingenious conjecture, of direct intuition, and for this reason dream interpretation has naturally been elevated to an art, which seems to depend upon extraordinary gifts."

The founder of psychoanalysis, a clinical method for treating mental disorders through dialogue between a psychoanalyst and a patient, Sigmund Freud was born as Sigismund Schlomo Freud on May 6, 1856, in the Moravian town of Freiberg in the Austro-Hungarian Empire. He was the son of Jakob Freud, a wool merchant, and Amalia Nathansohn, Jakob's third wife.

An outstanding student in high school, Freud graduated with honours from the Matura in the year 1873 and entered the University of Vienna, where he joined the medical faculty. He studied philosophy under the influential German philosopher, Franz Brentano; physiology under Ernst Brücke; and zoology under the German zoologist, Carl Claus; and graduated in 1881 with an MD. Freud's medical career began in 1882, when he entered the Vienna General hospital. He worked in various departments of the hospital and spent time as a clinical assistant at Theodor Meynert's psychiatric clinic. His interest in clinical work increased when he worked as a locum in a local asylum. In 1884, he published a seminal paper on the palliative effects of cocaine. It was a result of his research work in cerebral anatomy. In 1885, Freud was appointed as a university lecturer in neuropathology.

Freud went to Paris in the winter of 1885–86, where he continued his studies in neuropathology under the guidance of Jean-Martin Charcot at the Salpêtrière clinic. This time turned out to be a turning point in his career. Through Charcot's work with patients classified as "hysterics," Freud was introduced to the possibility of psychological disorders originating in the subconscious rather than the conscious. In 1886, he returned to Vienna.

Freud married Martha Bernays in the summer of 1886. The granddaughter of Isaac Bernays, a chief rabbi in Hamburg, she was to bear him six children. Anna Freud, their sixth child, was to follow the path of her father and become a distinguished psychoanalyst. Freud set up his private practice in the same year. He used hypnosis in his clinical work and adopted the approach of Josef Breuer, his friend and collaborator. Breuer's work with his patient, known as Anna O, proved to be transformative for Freud's clinical practice. Anna O was invited to talk about her symptoms while under hypnosis, in the course of which her symptoms reduced.

Freud found that the patient's dreams could be analysed to reveal the complex structuring of the unconscious domain and to demonstrate the psychic action of repression. He used the term "psychoanalysis" to refer to his new clinical method.

During the development of these new theories, Freud was going through a period in which he experienced depression, disturbing dreams, and heart irregularities. He self-analysed his dreams and memories of childhood. Freud co-authored *Studies on Hysteria* with Josef Breuer. Published in 1895, it described the evolution of his clinical methods. *The Interpretation of Dreams* was published in 1899. Considered to be his masterwork, it introduces Freud's theory of the unconscious with respect to dream interpretation and provides interpretations of Freud's own dreams and the dreams of his patients in terms of wish-fulfilments. According to his theory, dreams represent the fulfilment of our unconscious wishes. Though the first edition of the book did not sell out for almost eight years, seven more editions were published during Freud's lifetime that gained popularity.

On Dream, an abridged version of *The Interpretation of Dreams* was published in 1901. His other significant works include *The Psychopathology of Everyday Life* (1901), *Jokes and their Relation to the Unconscious* (1905), *Three Essays on the Theory of Sexuality* (1905), *Introduction to Psychoanalysis* (1916–17), and *Moses and Monotheism* (1939).

Freud died in September 1939 after suffering from cancer of the jaw. He was cremated at the Golders Green Crematorium, three days after his death.

Contents

Translator's Preface

Since the appearance of the author's *Selected Papers on Hysteria and other Psychoneuroses,* and *Three Contributions to the Sexual Theory,*[1] much has been said and written about Freud's works. Some of our readers have made an honest endeavour to test and utilise the author's theories, but they have been handicapped by their inability to read fluently very difficult German, for only two of Freud's works have hitherto been accessible to English readers. For them this work will be of invaluable assistance. To be sure, numerous articles on the Freudian psychology have of late made their appearance in our literature;[2] but these scattered papers, read by those unacquainted with the original work, often serve to confuse rather than enlighten. For Freud cannot be mastered from the reading of a few pamphlets, or even one or two of his original works. Let me repeat what I have so often said: No one is really qualified to use or to judge Freud's psychoanalytic method who has not thoroughly mastered his theory of the neuroses—*The Interpretation of Dreams, Three Contributions to the Sexual Theory, Psychopathology of Everyday Life,* and *Wit and Its Relation to the Unconscious,* and who has not had considerable experience in analysing the dreams and psychopathological actions of himself and others. That there is required also a thorough training in normal and abnormal psychology goes without saying.

The Interpretation of Dreams is the author's greatest and most important work; it is here that he develops his psychoanalytic technique, a thorough knowledge of which is absolutely indispensable for every worker in this field. The difficult task of making a translation of this work has, therefore, been undertaken primarily for the purpose of assisting those who are actively engaged in treating patients by Freud's psychoanalytic method. Considered apart from its practical aim, the book presents much that is of interest to the psychologist and the general reader. For, notwithstanding the fact that dreams have of late years been the subject of investigation at the hands of many competent observers, only few have contributed anything tangible towards their solution; it was Freud who divested the dream of its mystery, and solved its riddles. He not only showed us that the dream is full of meaning, but amply demonstrated that it is intimately connected with normal and abnormal mental life. It is in the treatment of the abnormal mental states that we must recognise the most important value of dream interpretation. The dream does not only reveal to us the cryptic mechanisms of hallucinations, delusions, phobias, obsessions, and other psychopathological conditions, but it is also the most potent instrument in the removal of these.[3]

I take this opportunity of expressing my indebtedness to Professor F. C. Prescott for reading the manuscript and for helping me overcome the almost insurmountable difficulties in the translation.

A. A. BRILL.
NEW YORK CITY.

Notes:

1. Translated by A. A. Brill (*Journal of Nervous and Mental Disease* Publishing Company).
2. *Cf.* the works of Ernest Jones, James J. Putnam, the present writer, and others.
3. For examples demonstrating these facts, *cf.* my work, *Psychoanalysis; its Theories and Practical Application,* W. B. Saunders' Publishing Company, Philadelphia & London.

"Flectere si nequeo superos, Acheronta movebo"

['If I can't bend those above, I'll stir the lower regions'
(Virgil, *Aeneid*, VII, 312)]

1

The Scientific Literature
on the Problems of the Dream

In[1] the following pages I shall prove that there exists a psychological technique by which dreams may be interpreted, and that upon the application of this method every dream will show itself to be a senseful psychological structure which may be introduced into an assignable place in the psychic activity of the waking state. I shall furthermore endeavour to explain the processes which give rise to the strangeness and obscurity of the dream, and to discover through them the nature of the psychic forces which operate, whether in combination or in opposition, to produce the dream. This accomplished, my investigation will terminate, as it will have reached the point where the problem of the dream meets with broader problems, the solution of which must be attempted through other material.

I must presuppose that the reader is acquainted with the work done by earlier authors as well as with the present status of the dream problem in science, since in the course of this treatise I shall not often have occasion to return to them. For, notwithstanding the effort of several thousand years, little progress has been made in the scientific understanding of dreams. This has been so universally acknowledged by the authors that it seems unnecessary to quote individual opinions. One will find in the writings indexed at the end of this book many stimulating observations and plenty of interesting material for our subject, but little or nothing that concerns the true nature of the dream or that solves definitively any of its enigmas. Still less of course has been transmitted to the knowledge of the educated laity.

The first book in which the dream is treated as an object of psychology seems to be that of Aristotle (*Concerning Dreams and their Interpretation*). Aristotle asserts that the dream is of demoniacal, though not of divine nature, which indeed contains deep meaning, if it be correctly interpreted. He was also acquainted with some of the characteristics of dream life, *e.g.,* he knew that the dream turns slight sensations perceived during sleep into great ones ("one imagines that one walks through fire and feels hot, if this or that part of the body becomes slightly warmed"), which led him to conclude that dreams might easily betray to the physician the first indications of an incipient change in the body passing unnoticed during the day. I have been unable to go more deeply into the Aristotelian treatise, because of insufficient preparation and lack of skilled assistance.

As every one knows, the ancients before Aristotle did not consider the dream a product of the dreaming mind, but a divine inspiration, and in ancient times the two antagonistic streams, which one finds throughout in the estimates of dream life, were already noticeable. They distinguished between true and valuable dreams, sent to the dreamer to warn him or to foretell the future, and vain, fraudulent, and empty dreams, the object of which was to misguide or lead him to destruction.[2] This pre-scientific conception of the dream among the ancients was certainly in perfect keeping with their general view of life, which was wont to project as reality in the outer world that which possessed reality only within the mind. It, moreover, accounted for the main impression made upon the waking life by the memory left from the dream in the morning, for in this memory the dream, as compared with the rest of the psychic content, seems something strange, coming, as it were, from another world. It would likewise be wrong to suppose that the theory of the supernatural origin of dreams lacks followers in our own day; for leaving out of consideration all bigoted and mystical authors—who are perfectly justified in adhering to the remnants of the once extensive realm of the supernatural until they have been swept away by scientific explanation—one meets even sagacious men averse to anything adventurous, who go so far as to base their religious belief in the existence and co-operation of superhuman forces on the inexplicableness of the dream manifestations (Haffner). The validity ascribed to the dream life by some schools of philosophy, *e.g.,* the school of Schelling, is a distinct echo of the undisputed divinity of dreams in antiquity, nor is discussion closed on the subject of the mantic or prophetic power of dreams. This is due to the fact that the attempted psychological explanations are too inadequate to overcome the accumulated material, however strongly all those who devote themselves to a scientific mode of thought may feel that such assertions should be repudiated.

To write a history of our scientific knowledge of dream problems is so difficult because, however valuable some parts of this knowledge may have been, no progress in definite directions has been discernible. There has been no construction of a foundation of assured results upon which future investigators could continue to build, but every new author takes up the same problems afresh and from the very beginning. Were I to follow the authors in chronological order, and give a review of the opinions each has held concerning the problems of the dream, I should be prevented from drawing a clear and complete picture of the present state of knowledge on the subject. I have therefore preferred to base the treatment upon themes rather than upon the authors, and I shall cite for each problem of the dream the material found in the literature for its solution.

But as I have not succeeded in mastering the entire literature, which is widely disseminated and interwoven with that on other subjects, I must ask my readers to rest content provided no fundamental fact or important viewpoint be lost in my description.

Until recently most authors have been led to treat the subjects of sleep and dream in the same connection, and with them they have also regularly treated analogous states of psychopathology, and other dreamlike states like hallucinations, visions, &c. In the more recent works, on the other hand, there has been a tendency to keep more closely to the theme, and to take as the subject one single question of the dream life. This change, I believe, is an expression of the conviction that enlightenment and agreement in such obscure matters can only be brought about by a series of detailed investigations. It is such a detailed investigation and one of a special psychological nature, that I would offer here. I have little occasion to study the problem of sleep, as it is essentially a psychological problem, although the change of functional determinations for the mental apparatus must be included in the character of sleep. The literature of sleep will therefore not be considered here.

A scientific interest in the phenomena of dreams as such leads to the following in part interdependent inquiries:

(A) The Relation of the Dream to the Waking State

The naïve judgment of a person on awakening assumes that the dream—if indeed it does not originate in another world—at any rate has taken the dreamer into another world. The old physiologist, Burdach, to whom we are indebted for a careful and discriminating description of the phenomena of dreams, expressed this conviction in an often-quoted passage, p. 474: "The waking life never repeats itself with its trials and joys, its pleasures and pains, but, on the contrary, the dream aims to relieve us of these. Even when our whole mind is filled with one subject, when profound sorrow has torn our hearts or when a task has claimed the whole power of our mentality, the dream either gives us something entirely strange, or it takes for its combinations only a few elements from reality, or it only enters into the strain of our mood and symbolises reality."

L. Strümpell expresses himself to the same effect in his *Nature and Origin of Dreams* (p. 16), a study which is everywhere justly held in high respect: "He who dreams turns his back upon the world of waking consciousness" (p. 17). "In the dream the memory of the orderly content of the waking consciousness and its normal behaviour is as good as entirely lost" (p. 19). "The almost complete isolation of the mind in the dream from the regular normal content and course of the waking state. . . . "

But the overwhelming majority of the authors have assumed a contrary view of the relation of the dream to waking life. Thus Haffner (p. 19): "First of all the dream is the continuation of the waking state. Our dreams always unite themselves with those ideas which have shortly before been in our consciousness. Careful examination will nearly always find a thread by which the dream has connected itself with the experience of the previous day." Weygandt (p. 6), flatly contradicts the above-cited statement of Burdach: "For it may often be observed, apparently in the great majority of dreams, that they lead us directly back into everyday life, instead of releasing us from it." Maury (p. 56), says in a concise formula: "Nous rêvons de ce que nous avons vu, dit, desiré ou fait." Jessen, in his *Psychology,* published in 1855

(p. 530), is somewhat more explicit: "The content of dreams is more or less determined by the individual personality, by age, sex, station in life, education, habits, and by events and experiences of the whole past life."

The ancients had the same idea about the dependence of the dream content upon life. I cite Radestock (p. 139): "When Xerxes, before his march against Greece, was dissuaded from this resolution by good counsel, but was again and again incited by dreams to undertake it, one of the old rational dream-interpreters of the Persians, Artabanus, told him very appropriately that dream pictures mostly contain that of which one has been thinking while awake."

In the didactic poem of Lucretius, *De Rerum Natura* (IV, v. 959), occurs this passage:—

> "Et quo quisque fere studio devinctus adhaeret,
> aut quibus in rebus multum sumus ante morati
> atque in ea ratione fuit contenta magis mens,
> in somnis eadem plerumque videmur obire;
> causidici causas agere et componere leges,
> induperatores pugnare ac proelia obire," &c., &c.

Cicero (*De Divinatione,* II) says quite similarly, as does also Maury much later:—

"Maximeque reliquiae earum rerum moventur in animis et agitantur, de quibus vigilantes aut cogitavimus aut egimus."

The contradiction expressed in these two views as to the relation between dream life and waking life seems indeed insoluble. It will therefore not be out of place to mention the description of F. W. Hildebrandt (1875), who believes that the peculiarities of the dream can generally be described only by calling them a "series of contrasts which apparently shade off into contradictions" (p. 8). "The first of these contrasts is formed on the one hand by the strict isolation or seclusion of the dream from true and actual life, and on the other hand by the continuous encroachment of the one upon the other, and the constant dependency of one upon the other. The dream is something absolutely separated from the reality experienced during the waking state; one may call it an existence hermetically sealed up and separated from real life by an unsurmountable chasm. It frees us from reality, extinguishes normal recollection of reality, and places us in another world and in a totally different life, which at bottom has nothing in common with reality. . . . " Hildebrandt then asserts that in falling asleep our whole being, with all its forms of existence, disappears "as through an invisible trap door." In the dream one is perhaps making a voyage to St. Helena in order to offer the imprisoned Napoleon something exquisite in the way of Moselle wine. One is most amicably received by the ex-emperor, and feels almost sorry when the interesting illusion is destroyed on awakening. But let us now compare the situation of the dream with reality. The dreamer has never been a wine merchant, and has no desire to become one. He has never made a sea voyage, and St. Helena is the last place he would take as destination for such a voyage. The dreamer entertains no sympathetic feeling for Napoleon, but on the contrary a strong patriotic hatred. And finally the dreamer was not yet among the living when Napoleon died on the island; so that it was beyond the reach of possibility for him to have had any personal relations with Napoleon. The dream experience thus appears as something strange, inserted between two perfectly harmonising and succeeding periods.

"Nevertheless," continues Hildebrandt, "the opposite is seemingly just as true and correct. I believe that hand in hand with this seclusion and isolation there can still exist the most intimate relation and connection. We may justly say that no matter what the dream offers, it finds its material in reality and in the psychic life arrayed around this reality. However strange the dream may seem, it can never detach itself from reality, and its most sublime as well as its most farcical structures must always borrow their elementary material either from what we have seen with our eyes in the outer world, or from what has previously found a place somewhere in our waking thoughts; in other words, it must be taken from what we had already experienced either objectively or subjectively."

(B) The Material of the Dream—Memory in the Dream

That all the material composing the content of the dream in some way originates in experience, that it is reproduced in the dream, or recalled,—this at least may be taken as an indisputable truth. Yet it would be wrong to assume that such connection between dream content and reality will be readily disclosed as an obvious product of the instituted comparison. On the contrary, the connection must be carefully sought, and in many cases it succeeds in eluding discovery for a long time. The reason for this is to be

16

found in a number of peculiarities evinced by the memory in dreams, which, though universally known, have hitherto entirely eluded explanation. It will be worth while to investigate exhaustively these characteristics.

It often happens that matter appears in the dream content which one cannot recognise later in the waking state as belonging to one's knowledge and experience. One remembers well enough having dreamed about the subject in question, but cannot recall the fact or time of the experience. The dreamer is therefore in the dark as to the source from which the dream has been drawing, and is even tempted to believe an independently productive activity on the part of the dream, until, often long afterwards, a new episode brings back to recollection a former experience given up as lost, and thus reveals the source of the dream. One is thus forced to admit that something has been known and remembered in the dream that has been withdrawn from memory during the waking state.

Delboeuf narrates from his own experience an especially impressive example of this kind. He saw in his dream the courtyard of his house covered with snow, and found two little lizards half-frozen and buried in the snow. Being a lover of animals, he picked them up, warmed them, and put them back into a crevice in the wall which was reserved for them. He also gave them some small fern leaves that had been growing on the wall, which he knew they were fond of. In the dream he knew the name of the plant: *Asplenium ruta muralis*. The dream then continued, returning after a digression to the lizards, and to his astonishment Delboeuf saw two other little animals falling upon what was left of the ferns. On turning his eyes to the open field he saw a fifth and a sixth lizard running into the hole in the wall, and finally the street was covered with a procession of lizards, all wandering in the same direction, &c.

In his waking state Delboeuf knew only a few Latin names of plants, and nothing of the Asplenium. To his great surprise he became convinced that a fern of this name really existed and that the correct name was *Asplenium ruta muraria*, which the dream had slightly disfigured. An accidental coincidence could hardly be considered, but it remained a mystery for Delboeuf whence he got his knowledge of the name Asplenium in the dream.

The dream occurred in 1862. Sixteen years later, while at the house of one of his friends, the philosopher noticed a small album containing dried plants resembling the albums that are sold as souvenirs to visitors in many parts of Switzerland. A sudden recollection occurred to him; he opened the herbarium, and discovered therein the Asplenium of his dream, and recognised his own handwriting in the accompanying Latin name. The connection could now be traced. While on her wedding trip, a sister of this friend visited Delboeuf in 1860—two years prior to the lizard dream. She had with her at the time this album, which was intended for her brother, and Delboeuf took the trouble to write, at the dictation of a botanist, under each of the dried plants the Latin name.

The favourable accident which made possible the report of this valuable example also permitted Delboeuf to trace another portion of this dream to its forgotten source. One day in 1877 he came upon an old volume of an illustrated journal, in which he found pictured the whole procession of lizards just as he had dreamed it in 1862. The volume bore the date of 1861, and Delboeuf could recall that he had subscribed to the journal from its first appearance.

That the dream has at its disposal recollections which are inaccessible to the waking state is such a remarkable and theoretically important fact that I should like to urge more attention to it by reporting several other "Hypermnesic Dreams." Maury relates that for some time the word Mussidan used to occur to his mind during the day. He knew it to be the name of a French city, but nothing else. One night he dreamed of a conversation with a certain person who told him that she came from Mussidan, and, in answer to his question where the city was, she replied: "Mussidan is a principal country town in the Département de La Dordogne." On waking, Maury put no faith in the information received in his dream; the geographical lexicon, however, showed it to be perfectly correct. In this case the superior knowledge of the dream is confirmed, but the forgotten source of this knowledge has not been traced.

Jessen tells (p. 55) of a quite similar dream occurrence, from more remote times. "Among others we may here mention the dream of the elder Scaliger (Hennings, *l.c.,* p. 300), who wrote a poem in praise of celebrated men of Verona, and to whom a man, named Brugnolus, appeared in a dream, complaining that he had been neglected. Though Scaliger did not recall ever having heard of him, he wrote some verses in his honour, and his son later discovered at Verona that a Brugnolus had formerly been famous there as a critic.

Myers is said to have published a whole collection of such hypermnesic dreams in the *Proceedings of the Society for Psychical Research,* which are unfortunately inaccessible to me. I believe every one who occupies himself with dreams will recognise as a very common phenomenon the fact that the dream gives proof of knowing and recollecting matters unknown to the waking person. In my psychoanalytic investigations of nervous patients, of which I shall speak later, I am every week more than once in position to convince my patients from their dreams that they are well acquainted with quotations, obscene expressions, &c., and that they make use of these in their dreams, although they have forgotten them in the waking state. I shall cite here a simple case of dream hypermnesia because it was easy to trace the source which made the knowledge accessible to the dream.

A patient dreamed in a lengthy connection that he ordered a "Kontuszówka" in a café, and after reporting this inquired what it might mean, as he never heard the name before. I was able to answer that Kontuszówka was a Polish liquor which he could not have invented in his dream, as the name had long been familiar to me in advertisements. The patient would not at first believe me, but some days later, after he had realised his dream of the café, he noticed the name on a signboard at the street corner, which he had been obliged to pass for months at least twice a day.

I have learned from my own dreams how largely the discovery of the origin of some of the dream elements depends on accident. Thus, for years before writing this book, I was haunted by the picture of a very simply formed church tower which I could not recall having seen. I then suddenly recognised it with absolute certainty at a small station between Salzburg and Reichenhall. This was in the later nineties, and I had travelled over the road for the first time in the year 1886. In later years, when I was already busily engaged in the study of dreams, I was quite annoyed at the frequent recurrence of the dream picture of a certain peculiar locality. I saw it in definite local relation to my person—to my left, a dark space from which many grotesque sandstone figures stood out. A glimmer of recollection, which I did not quite credit, told me it was the entrance to a beer-cellar, but I could explain neither the meaning nor the origin of this dream picture. In 1907 I came by chance to Padua, which, to my regret, I had been unable to visit since 1895. My first visit to this beautiful university city was unsatisfactory; I was unable to see Giotto's frescoes in the church of the Madonna dell' Arena, and on my way there turned back on being informed that the little church was closed on the day. On my second visit, twelve years later, I thought of compensating myself for this, and before everything else I started out for Madonna dell' Arena. On the street leading to it, on my left, probably at the place where I had turned in 1895, I discovered the locality which I had so often seen in the dream, with its sandstone figures. It was in fact the entrance to a restaurant garden.

One of the sources from which the dream draws material for reproduction—material which in part is not recalled or employed in waking thought—is to be found in childhood. I shall merely cite some of the authors who have observed and emphasized this.

Hildebrandt (p. 23): "It has already been expressly admitted that the dream sometimes brings back to the mind with wonderful reproductive ability remote and even forgotten experiences from the earliest periods."

Strümpell (p. 40): "The subject becomes more interesting when we remember how the dream sometimes brings forth, as it were, from among the deepest and heaviest strata which later years have piled upon the earliest childhood experiences, the pictures of certain places, things, and persons, quite uninjured and with their original freshness. This is not limited merely to such impressions as have gained vivid consciousness during their origin or have become impressed with strong psychic validity, and then later return in the dream as actual reminiscences, causing pleasure to the awakened consciousness. On the contrary, the depths of the dream memory comprise also such pictures of persons, things, places, and early experiences as either possessed but little consciousness and no psychic value at all, or have long ago lost both, and therefore appear totally strange and unknown both in the dream and in the waking state, until their former origin is revealed."

Volkelt (p. 119): "It is essentially noteworthy how easily infantile and youthful reminiscences enter into the dream. What we have long ceased to think about, what has long since lost for us all importance, is constantly recalled by the dream."

The sway of the dream over the infantile material, which, as is well known, mostly occupies the gaps in the conscious memory, causes the origin of interesting hypermnestic dreams, a few of which I shall here report.

Maury relates (p. 92) that as a child he often went from his native city, Meaux, to the neighbouring Trilport, where his father superintended the construction of a bridge. On a certain night a dream transported him to Trilport, and he was again playing in the city streets. A man approached him wearing some sort of uniform. Maury asked him his name, and he introduced himself, saying that his name was C—, and that he was a bridge guard. On waking, Maury, who still doubted the reality of the reminiscence, asked his old servant, who had been with him in his childhood, whether she remembered a man of this name. "Certainly," was the answer, "he used to be watchman on the bridge which your father was building at that time."

Maury reports another example demonstrating just as nicely the reliability of infantile reminiscences appearing in dreams. Mr. F—, who had lived as a child in Montbrison, decided to visit his home and old friends of his family after an absence of twenty-five years. The night before his departure he dreamt that he had reached his destination, and that he met near Montbrison a man, whom he did not know by sight, who told him he was Mr. F., a friend of his father. The dreamer remembered that as a child he had known a gentleman of this name, but on waking he could no longer recall his features. Several days later, having really arrived at Montbrison, he found the supposedly unknown locality of his dream, and there met a man whom he at once recognised as the Mr. F. of his dream. The real person was only older than the one in the dream picture.

I may here relate one of my own dreams in which the remembered impression is replaced by an association. In my dream I saw a person whom I recognised, while dreaming, as the physician of my native town. The features were indistinct and confused with the picture of one of my colleague teachers, whom I still see occasionally. What association there was between the two persons I could not discover on awakening. But upon questioning my mother about the physician of my early childhood, I discovered that he was a one-eyed man. My teacher, whose figure concealed that of the physician in the dream, was also one-eyed. I have not seen the physician for thirty-eight years, and I have not to my knowledge thought of him in my waking state, although a scar on my chin might have reminded me of his help.

As if to counterbalance the immense rôle ascribed to the infantile impressions in the dream, many authors assert that the majority of dreams show elements from the most recent time. Thus Robert (p. 46) declares that the normal dream generally occupies itself only with the impressions of the recent days. We learn indeed that the theory of the dream advanced by Robert imperatively demands that the old impressions should be pushed back, and the recent ones brought to the front. Nevertheless the fact claimed by Robert really exists; I can confirm this from my own investigations. Nelson, an American author, thinks that the impressions most frequently found in the dream date from two or three days before, as if the impressions of the day immediately preceding the dream were not sufficiently weakened and remote.

Many authors who are convinced of the intimate connection between the dream content and the waking state are impressed by the fact that impressions which have intensely occupied the waking mind appear in the dream only after they have been to some extent pushed aside from the elaboration of the waking thought. Thus, as a rule, we do not dream of a dead beloved person while we are still overwhelmed with sorrow. Still Miss Hallam, one of the latest observers, has collected examples showing the very opposite behaviour, and claims for the point the right of individual psychology.

The third and the most remarkable and incomprehensible peculiarity of the memory in dreams, is shown in the selection of the reproduced material, for stress is laid not only on the most significant, but also on the most indifferent and superficial reminiscences. On this point I shall quote those authors who have expressed their surprise in the most emphatic manner.

Hildebrandt (p. 11): "For it is a remarkable fact that dreams do not, as a rule, take their elements from great and deep-rooted events or from the powerful and urgent interests of the preceding day, but from unimportant matters, from the most worthless fragments of recent experience or of a more remote past. The most shocking death in our family, the impressions of which keep us awake long into the night, becomes obliterated from our memories, until the first moment of awakening brings it back to us with depressing force. On the other hand, the wart on the forehead of a passing stranger, of whom we did not think for a second after he was out of sight, plays its part in our dreams."

Strümpell (p. 39): " . . . such cases where the analysis of a dream brings to light elements which, although derived from events of the previous day or the day before the last, yet prove to be so unimportant and worthless for the waking state that they merge into forgetfulness shortly after coming to light. Such occurrences may be statements of others heard accidentally or actions superficially observed, or fleeting perceptions of things or persons, or single phrases from books, &c."

Havelock Ellis (p. 727): "The profound emotions of waking life, the questions and problems on which we spread our chief voluntary mental energy, are not those which usually present themselves at once to dream-consciousness. It is, so far as the immediate past is concerned, mostly the trifling, the incidental, the "forgotten" impressions of daily life which reappear in our dreams. The psychic activities that are awake most intensely are those that sleep most profoundly."

Binz (p. 45) takes occasion from the above-mentioned characteristics of the memory in dreams to express his dissatisfaction with explanations of dreams which he himself has approved of: "And the normal dream raises similar questions. Why do we not always dream of memory impressions from the preceding days, instead of going back to the almost forgotten past lying far behind us without any perceptible reason? Why in a dream does consciousness so often revive the impression of indifferent memory pictures while the cerebral cells bearing the most sensitive records of experience remain for the most part inert and numb, unless an acute revival during the waking state has shortly before excited them?"

We can readily understand how the strange preference of the dream memory for the indifferent and hence the unnoticed details of daily experience must usually lead us to overlook altogether the dependence of the dream on the waking state, or at least make it difficult to prove this dependence in any individual case. It thus happened that in the statistical treatment of her own and her friend's dreams, Miss Whiton Calkins found 11 per cent. of the entire number that showed no relation to the waking state. Hildebrandt was certainly correct in his assertion that all our dream pictures could be genetically explained if we devoted enough time and material to the tracing of their origin. To be sure, he calls this "a most tedious and thankless job." For it would at most lead us to ferret out all kinds of quite worthless psychic material from the most remote corners of the memory chamber, and to bring to light some very indifferent moments from the remote past which were perhaps buried the next hour after their appearance." I must, however, express my regret that this discerning author refrained from following the road whose beginning looked so unpromising; it would have led him directly to the centre of the dream problem.

The behaviour of the memory in dreams is surely most significant for every theory of memory in general. It teaches us that "nothing which we have once psychically possessed is ever entirely lost" (Scholz); or as Delboeuf puts it, "que toute impression même la plus insignifiante, laisse une trace inaltérable, indéfiniment susceptible de reparaître au jour," a conclusion to which we are urged by so many of the other pathological manifestations of the psychic life. Let us now bear in mind this extraordinary capability of the memory in the dream, in order to perceive vividly the contradictions which must be advanced in certain dream theories to be mentioned later, when they endeavour to explain the absurdities and incoherence of dreams through a partial forgetting of what we have known during the day.

One might even think of reducing the phenomenon of dreaming to that of memory, and of regarding the dream as the manifestation of an activity of reproduction which does not rest even at night, and which is an end in itself. Views like those expressed by Pilcz would corroborate this, according to which intimate relations are demonstrable between the time of dreaming and the contents of the dream from the fact that the impressions reproduced by the dream in sound sleep belong to the remotest past while those reproduced towards morning are of recent origin. But such a conception is rendered improbable from the outset by the manner of the dream's behaviour towards the material to be remembered. Strümpell justly calls our attention to the fact that repetitions of experiences do not occur in the dream. To be sure the dream makes an effort in that direction, but the next link is wanting, or appears in changed form, or it is replaced by something entirely novel. The dream shows only fragments of reproduction; this is so often the rule that it admits of theoretical application. Still there are exceptions in which the dream repeats an episode as thoroughly as our memory would in its waking state. Delboeuf tells of one of his university colleagues who in his dream repeated, with all its details, a dangerous wagon ride in which he escaped accident as if by miracle. Miss Calkins mentions two dreams, the contents of which exactly reproduced incidents from the day before, and I shall later take occasion to report an example which came to my notice, showing a childish experience which returned unchanged in a dream.[3]

(C)   Dream Stimuli and Dream Sources

What is meant by dream stimuli and dream sources may be explained by referring to the popular saying, "Dreams come from the stomach." This notion conceals a theory which conceives the dream as a result of a disturbance of sleep. We should not have dreamed if some disturbing element had not arisen in sleep, and the dream is the reaction from this disturbance.

The discussion of the exciting causes of dreams takes up the most space in the descriptions of the authors. That this problem could appear only after the dream had become an object of biological investigation is self-evident. The ancients who conceived the dream as a divine inspiration had no need of looking for its exciting source; to them the dream resulted from the will of the divine or demoniacal powers, and its content was the product of their knowledge or intention. Science, however, soon raised the question whether the stimulus to the dream is always the same, or whether it might be manifold, and thus led to the question whether the causal explanation of the dream belongs to psychology or rather to physiology. Most authors seem to assume that the causes of the disturbance of sleep, and hence the sources of the dream, might be of various natures, and that physical as well as mental irritations might assume the rôle of dream inciters. Opinions differ greatly in preferring this or that one of the dream sources, in ranking them, and indeed as to their importance for the origin of dreams.

Wherever the enumeration of dream sources is complete we ultimately find four forms, which are also utilised for the division of dreams:—

I.   External (objective) sensory stimuli.
II. Internal (subjective) sensory stimuli.
III. Internal (organic) physical excitations.
IV. Purely psychical exciting sources.

I. The External Sensory Stimuli

The younger Strümpell, son of the philosopher whose writings on the subject have already more than once served us as a guide in the problem of dreams, has, as is well known, reported his observations on a patient who was afflicted with general anaesthesia of the skin and with paralysis of several of the higher sensory organs. This man merged into sleep when his few remaining sensory paths from the outer world were shut off. When we wish to sleep we are wont to strive for a situation resembling the one in Strümpell's experiment. We close the most important sensory paths, the eyes, and we endeavour to keep away from the other senses every stimulus and every change of the stimuli acting upon them. We then fall asleep, although we are never perfectly successful in our preparations. We can neither keep the stimuli away from the sensory organs altogether, nor can we fully extinguish the irritability of the sensory organs. That we may at any time be awakened by stronger stimuli should prove to us "that the mind has remained in constant communication with the material world even during sleep." The sensory stimuli which reach us during sleep may easily become the source of dreams.

There are a great many stimuli of such nature, ranging from those that are unavoidable, being brought on by the sleeping state or at least occasionally induced by it, to the accidental waking stimuli which are adapted or calculated to put an end to sleep. Thus a strong light may force itself into the eyes, a noise may become perceptible, or some odoriferous matter may irritate the mucous membrane of the nose. In the spontaneous movements of sleep we may lay bare parts of the body and thus expose them to a sensation of cold, or through change of position we may produce sensations of pressure and touch. A fly may bite us, or a slight accident at night may simultaneously attack more than one sense. Observers have called attention to a whole series of dreams in which the stimulus verified on waking, and a part of the dream content corresponded to such a degree that the stimulus could be recognised as the source of the dream.

I shall here cite a number of such dreams collected by Jessen (p. 527), traceable to more or less accidental objective sensory stimuli. "Every indistinctly perceived noise gives rise to corresponding dream pictures; the rolling of thunder takes us into the thick of battle, the crowing of a cock may be transformed into human shrieks of terror, and the creaking of a door may conjure up dreams of burglars breaking into the house. When one of our blankets slips off at night we may dream that we are walking about naked or falling into the water. If we lie diagonally across the bed with our feet extending beyond the edge, we may dream of standing on the brink of a terrifying precipice, or of falling from a steep height. Should our head accidentally get under the pillow we may then imagine a big rock hanging over us and about to crush us under its weight. Accumulation of semen produces voluptuous dreams, and local pain the idea of suffering ill treatment, of hostile attacks, or of accidental bodily injuries."

"Meier (*Versuch einer Erklärung des Nachtwandelns,* Halle, 1758, p. 33), once dreamed of being assaulted by several persons who threw him flat on the ground and drove a stake into the ground between his big and second toes. While imagining this in his dream he suddenly awoke and felt a blade of straw sticking between his toes. The same author, according to Hemmings (*Von den Traumen und Nachtwandeln,* Weimar, 1784, p. 258) dreamed on another occasion that he was being hanged when his shirt was pinned somewhat tight around his neck. Hauffbauer dreamed in his youth of having fallen from a high wall and found upon waking that the bedstead had come apart, and that he had actually fallen to the floor. . . . Gregory relates that he once applied a hot-water bottle to his feet, and dreamed of taking a trip to the summit of Mount Aetna, where he found the heat on the ground almost unbearable. After having applied a blistering plaster to his head, a second man dreamed of being scalped by Indians; a third, whose shirt was damp, dreamed of being dragged through a stream. An attack of gout caused the patient to believe that he was in the hands of the Inquisition, and suffering pains of torture (Macnish)."

The argument based upon the resemblance between stimulus and dream content is reinforced if through a systematic induction of stimuli we succeed in producing dreams corresponding to the stimuli. According to Macnish such experiments have already been made by Giron de Buzareingues. "He left his knee exposed and dreamed of travelling in a mail coach at night. He remarked in this connection that travellers would well know how cold the knees become in a coach at night. Another time he left the back of his head uncovered, and dreamed of taking part in a religious ceremony in the open air. In the country where he lived it was customary to keep the head always covered except on such occasions."

Maury reports new observations on dreams produced in himself. (A number of other attempts produced no results.)

1. He was tickled with a feather on his lips and on the tip of his nose. He dreamed of awful torture, viz. that a mask of pitch was stuck to his face and then forcibly torn off, taking the skin with it.
2. Scissors were sharpened on pincers. He heard bells ringing, then sounds of alarm which took him back to the June days of 1848.
3. Cologne water was put on his nose. He found himself in Cairo in the shop of John Maria Farina. This was followed by mad adventures which he was unable to reproduce.
4. His neck was lightly pinched. He dreamed that a blistering plaster was put on him, and thought of a doctor who treated him in his childhood.
5. A hot iron was brought near his face. He dreamed that *chauffeurs*[4] broke into the house and forced the occupants to give up their money by

sticking their feet into burning coals. The Duchess of Abrantés, whose secretary he imagined himself in the dream, then entered.

6. A drop of water was let fall on his forehead. He imagined himself in Italy perspiring heavily and drinking white wine of Orvieto.

7. When a burning candle was repeatedly focussed on him through red paper, he dreamed of the weather, of heat, and of a storm at sea which he once experienced in the English Channel.

D'Hervey, Weygandt, and others have made other attempts to produce dreams experimentally.

Many have observed the striking skill of the dream in interweaving into its structure sudden impressions from the outer world in such a manner as to present a gradually prepared and initiated catastrophe (Hildebrandt). "In former years," this author relates, "I occasionally made use of an alarm clock in order to wake regularly at a certain hour in the morning. It probably happened hundreds of times that the sound of this instrument fitted into an apparently very long and connected dream, as if the entire dream had been especially designed for it, as if it found in this sound its appropriate and logically indispensable point, its inevitable issue."

I shall cite three of these alarm-clock dreams for another purpose.

Volkelt (p. 68) relates: "A composer once dreamed that he was teaching school, and was just explaining something to his pupils. He had almost finished when he turned to one of the boys with the question: 'Did you understand me?' The boy cried out like one possessed 'Ya.' Annoyed at this, he reprimanded him for shouting. But now the entire class was screaming 'Orya,' then 'Euryo,' and finally 'Feueryo.' He was now aroused by an actual alarm of fire in the street."

Garnier (*Traité des Facultés de l'Âme*, 1865), reported by Radestock, relates that Napoleon I., while sleeping in a carriage, was awakened from a dream by an explosion which brought back to him the crossing of the Tagliamento and the bombarding of the Austrians, so that he started up crying, "We are undermined!"

The following dream of Maury has become celebrated. He was sick, and remained in bed; his mother sat beside him. He then dreamed of the reign of terror at the time of the Revolution. He took part in terrible scenes of murder, and finally he himself was summoned before the Tribunal. There he saw Robespierre, Marat, Fouquier-Tinville, and all the sorry heroes of that cruel epoch; he had to give an account of himself, and, after all sort of incidents which did not fix themselves in his memory, he was sentenced to death. Accompanied by an enormous crowd, he was led to the place of execution. He mounted the scaffold, the executioner tied him to the board, it tipped, and the knife of the guillotine fell. He felt his head severed from the trunk, and awakened in terrible anxiety, only to find that the top piece of the bed had fallen down, and had actually struck his cervical vertebra in the same manner as the knife of a guillotine.

This dream gave rise to an interesting discussion introduced by Le Lorrain and Egger in the *Revue Philosophique*. The question was whether, and how, it was possible for the dreamer to crowd together an amount of dream content apparently so large in the short space of time elapsing between the perception of the waking stimulus and the awakening.

Examples of this nature make it appear that the objective stimuli during sleep are the most firmly established of all the dream sources; indeed, it is the only stimulus which plays any part in the layman's knowledge. If we ask an educated person, who is, however, unacquainted with the literature of dreams, how dreams originate, he is sure to answer by referring to a case familiar to him in which a dream has been explained after waking by a recognised objective stimulus. Scientific investigation cannot, however, stop here, but is incited to further research by the observation that the stimulus influencing the senses during sleep does not appear in the dream at all in its true form, but is replaced by some other presentation which is in some way related to it. But the relation existing between the stimulus and the result of the dream is, according to Maury, "une affinité quelconque mais qui n'est pas unique et exclusive" (p. 72). If we read, *e.g.,* three of Hildebrandt's "Alarm Clock Dreams," we will then have to inquire why the same stimulus evoked so many different results, and why just these results and no others.(P. 37).

"I am taking a walk on a beautiful spring morning. I saunter through the green fields to a neighbouring village, where I see the natives going to church in great numbers, wearing their holiday attire and carrying their hymn-books under their arms. I remember that it is Sunday, and that the morning service will soon begin. I decide to attend it, but as I am somewhat overheated I also decide to cool off in the cemetery surrounding the church. While reading the various epitaphs, I hear the sexton ascend the tower and see the small village bell in the cupola which is about to give signal for the beginning of the devotions. For another short while it hangs motionless, then it begins to swing, and suddenly its notes resound so clearly and penetratingly that my sleep comes to an end. But the sound of bells comes from the alarm clock."

"A second combination. It is a clear day, the streets are covered with deep snow. I have promised to take part in a sleigh-ride, but have had to wait for some time before it was announced that the sleigh is in front of my house. The preparations for getting into the sleigh are now made. I put on my furs and adjust my muff, and at last I am in my place. But the departure is still delayed, until the reins give the impatient horses the perceptible sign. They start, and the sleigh bells, now forcibly shaken, begin their familiar janizary music with a force that instantly tears the gossamer of my dream. Again it is only the shrill sound of my alarm clock."

Still a third example. "I see the kitchen-maid walk along the corridor to the dining-room with several dozen plates piled up. The porcelain column in her arms seems to me to be in danger of losing its equilibrium. 'Take care,' I exclaim, 'you will drop the whole pile.' The usual retort is naturally not wanting—that she is used to such things. Meanwhile I continue to follow her with my worried glance, and behold! at the door-step the fragile dishes fall, tumble, and roll across the floor in hundreds of pieces. But I soon notice that the noise continuing endlessly is not really a rattling but a true ringing, and with this ringing the dreamer now becomes aware that the alarm clock has done its duty."

The question why the dreaming mind misjudges the nature of the objective sensory stimulus has been answered by Strümpell, and almost identically by Wundt, to the effect that the reaction of the mind to the attacking stimuli in sleep is determined by the formation of illusions. A sensory impression is recognised by us and correctly interpreted, *i.e.* it is classed with the memory group to which it belongs according to all previous experience, if the impression is strong, clear, and long enough, and if we have the necessary time at our disposal for this reflection. If these conditions are not fulfilled, we mistake the objects which give rise to the impression, and on its basis we form an illusion. "If one takes a walk in an open field and perceives indistinctly a distant object, it may happen that he will at first take it for a horse." On closer inspection the image of a cow resting may obtrude itself, and the presentation may finally resolve itself with certainty into a group of people sitting. The impressions which the mind receives during sleep through outer stimuli are of a similar indistinct nature; they give rise to illusions because the impression evokes a greater or lesser number of memory pictures through which the impression receives its psychic value. In which of the many spheres of memory to be taken into consideration the corresponding pictures are aroused, and which of the possible association connections thereby come into force, this, even according to Strümpell, remains indeterminable, and is left, as it were, to the caprice of the psychic life.

We may here take our choice. We may admit that the laws of the dream formation cannot really be traced any further, and therefore refrain from asking whether or not the interpretation of the illusion evoked by the sensory impression depends upon still other conditions; or we may suppose that the objective sensory stimulus encroaching upon sleep plays only a modest part as a dream source, and that other factors determine the choice of the memory picture to be evoked. Indeed, on carefully examining Maury's experimentally produced dreams, which I have purposely reported in detail, one is apt to think that the experiment really explains the origin of only one of the dream elements, and that the rest of the dream content appears in fact too independent, too much determined in detail, to be explained by the one demand, viz. that it must agree with the element experimentally introduced. Indeed one even begins to doubt the illusion theory, and the power of the objective impression to form the dream, when one learns that this impression at times experiences the most peculiar and far-fetched interpretations during the sleeping state. Thus B. M. Simon tells of a dream in which he saw persons of gigantic stature[5] seated at a table, and heard distinctly the awful rattling produced by the impact of their jaws while chewing. On waking he heard the clacking of the hoofs of a horse galloping past his window. If the noise of the horse's hoofs had recalled ideas from the memory sphere of "Gulliver's Travels," the sojourn with the giants of Brobdingnag and the virtuous horse-creatures—as I should perhaps interpret it without any assistance on the author's part—should not the choice of a memory sphere so uncommon for the stimulus have some further illumination from other motives?

II. Internal (Subjective) Sensory Stimuli

Notwithstanding all objections to the contrary, we must admit that the rôle of the objective sensory stimuli as a producer of dreams has been indisputably established, and if these stimuli seem perhaps insufficient in

31

their nature and frequency to explain all dream pictures, we are then directed to look for other dream sources acting in an analogous manner. I do not know where the idea originated that along with the outer sensory stimuli the inner (subjective) stimuli should also be considered, but as a matter of fact this is done more or less fully in all the more recent descriptions of the etiology of dreams. "An important part is played in dream illusions," says Wundt (p. 363), "by those subjective sensations of seeing and hearing which are familiar to us in the waking state as a luminous chaos in the dark field of vision, ringing, buzzing, &c., of the ears, and especially irritation of the retina. This explains the remarkable tendency of the dream to delude the eyes with numbers of similar or identical objects. Thus we see spread before our eyes numberless birds, butterflies, fishes, coloured beads, flowers, &c. Here the luminous dust in the dark field of vision has taken on phantastic figures, and the many luminous points of which it consists are embodied by the dream in as many single pictures, which are looked upon as moving objects owing to the mobility of the luminous chaos. This is also the root of the great fondness of the dream for the most complex animal figures, the multiplicity of forms readily following the form of the subjective light pictures."

The subjective sensory stimuli as a source of the dream have the obvious advantage that unlike the objective stimuli they are independent of external accidents. They are, so to speak, at the disposal of the explanation as often as it needs them. They are, however, in so far inferior to the objective sensory stimuli that the rôle of dream inciter, which observation and experiment have proven for the latter, can be verified in their case only with difficulty or not at all. The main proof for the dream-inciting power of subjective sensory excitements is offered by the so-called hypnogogic hallucinations, which have been described by John Müller as "phantastic visual manifestations." They are those very vivid and changeable pictures which occur regularly in many people during the period of falling asleep, and which may remain for awhile even after the eyes have been opened. Maury, who was considerably troubled by them, subjected them to a thorough study, and maintained that they are related to or rather identical with dream pictures—this has already been asserted by John Müller. Maury states that a certain psychic passivity is necessary for their origin; it requires a relaxation of the tension of attention (p. 59). But in any ordinary disposition a hypnogogic hallucination may be produced by merging for a second into such lethargy, after which one perhaps awakens until this oft-repeated process terminates in sleep. According to Maury, if one awakens shortly thereafter, it is often possible to demonstrate the same pictures in the dream which one has perceived as hypnogogic hallucinations before falling asleep (p. 134). Thus it once happened to Maury with a group of pictures of grotesque figures, with distorted features and strange headdresses, which obtruded themselves upon him with incredible importunity during the period of falling asleep, and which he recalled having dreamed upon awakening. On another occasion, while suffering from hunger, because he kept himself on a rather strict diet, he saw hypnogogically a plate and a hand armed with a fork taking some food from the plate. In his dream he found himself at a table abundantly supplied with food, and heard the rattle made by the diners with their forks. On still another occasion, after falling asleep with irritated and painful eyes, he had the hypnogogic hallucination of seeing microscopically small characters which he was forced to decipher one by one with great exertion; having been awakened from his sleep an hour later, he recalled a dream in which there was an open book with very small letters, which he was obliged to read through with laborious effort.

Just as in the case of these pictures, auditory hallucinations of words, names, &c., may also appear hypnogogically, and then repeat themselves in the dream, like an overture announcing the principal motive of the opera which is to follow.

A more recent observer of hypnogogic hallucinations, G. Trumbull Ladd, takes the same path pursued by John Müller and Maury. By dint of practice he succeeded in acquiring the faculty of suddenly arousing himself, without opening his eyes, two to five minutes after having gradually fallen asleep, which gave him opportunity to compare the sensations of the retina just vanishing with the dream pictures remaining in his memory. He assures us that an intimate relation between the two can always be recognised, in the sense that the luminous dots and lines of the spontaneous light of the retina produced, so to speak, the sketched outline or scheme for the psychically perceived dream figures. A dream, *e.g.,* in which he saw in front of him clearly printed lines which he read and studied, corresponded to an arrangement of the luminous dots and lines in the retina in parallel lines, or, to express it in his own words: "The clearly printed page, which he was reading in the dream, resolved itself into an object which appeared to his waking perception like part of an actual printed sheet looked at through a little hole in a piece of paper, from too great a distance to be made out distinctly." Without in any way under-estimating the central part of the phenomenon, Ladd believes that hardly any visual dream occurs in our minds that is not based on material furnished by this inner condition of stimulation in the retina. This is particularly true of dreams occurring shortly after falling asleep in a dark room, while dreams occurring in the morning near the period of awakening receive their stimulation from the objective light penetrating the eye from the lightened room. The shifting and endlessly variable character of the spontaneous luminous excitation of the retina corresponds exactly to the fitful succession of pictures presented to us in our dreams. If we attach any importance to Ladd's observations, we cannot underrate the productiveness of this subjective source of excitation for the dream; for visual pictures apparently form the principal constituent of our dreams. The share furnished from the spheres of the other senses, beside the sense of hearing, is more insignificant and inconstant.

III. Internal (Organic) Physical Excitation

If we are disposed to seek dream sources not outside, but inside, the organism, we must remember that almost all our internal organs, which in their healthy state hardly remind us of their existence, may, in states of excitation—as we call them—or in disease, become for us a source of the most painful sensations, which must be put on an equality with the external excitants of the pain and sensory stimuli. It is on the strength of very old experience that, *e.g.*, Strümpell declares that "during sleep the mind becomes far more deeply and broadly conscious of its connection with the body than in the waking state, and it is compelled to receive and be influenced by stimulating impressions originating in parts and changes of the body of which it is unconscious in the waking state." Even Aristotle declares it quite possible that the dream should draw our attention to incipient morbid conditions which we have not noticed at all in the waking state (owing to the exaggeration given by the dream to the impressions; and some medical authors, who were certainly far from believing in any prophetic power of the dream, have admitted this significance of the dream at least for the foretelling of disease. (Compare M. Simon, p. 31, and many older authors.)

Even in our times there seems to be no lack of authenticated examples of such diagnostic performances on the part of the dream. Thus Tissié cites from Artigues (*Essai sur la Valeur séméiologique des Rêves*), the history of a woman of forty-three years, who, during several years of apparently perfect health, was troubled with anxiety dreams, and in whom medical examination later disclosed an incipient affection of the heart to which she soon succumbed.

Serious disturbances of the internal organs apparently act as inciters of dreams in a considerable number of persons. Attention is quite generally called to the frequency of anxiety dreams in the diseases of the heart and lungs; indeed this relation of the dream life is placed so conspicuously in the foreground by many authors that I shall here content myself with a mere reference to the literature. (Radestock, Spitta, Maury, M. Simon, Tissié.) Tissié even assumes that the diseased organs impress upon the dream content their characteristic features. The dreams of persons suffering from diseases of the heart are generally very brief and terminate in a terrified awakening; the situation of death under terrible circumstances almost always plays a part in their content. Those suffering from diseases of the lungs dream of suffocation, of being crowded, and of flight, and a great many of them are subject to the well-known nightmare, which, by the way, Boerner has succeeded in producing experimentally by lying on the face and closing up the openings of the respiratory organs. In digestive disturbances the dream contains ideas from the sphere of enjoyment and disgust. Finally, the influence of sexual excitement on the dream content is perceptible enough in every one's experience, and lends the strongest support to the entire theory of the dream excitation through organic sensation.

Moreover, as we go through the literature of the dream, it becomes quite obvious that some of the authors (Maury, Weygandt) have been led to the study of dream problems by the influence of their own pathological state on the content of their dreams.

The addition to dream sources from these undoubtedly established facts is, however, not as important as one might be led to suppose; for the dream is a phenomenon which occurs in healthy persons—perhaps in all persons, and every night—and a pathological state of the organs is apparently not one of its indispensable conditions. For us, however, the question is not whence particular dreams originate, but what may be the exciting source for the ordinary dreams of normal persons.

But we need go only a step further to find a dream source which is more prolific than any of those mentioned above, which indeed promises to be inexhaustible in every case. If it is established that the bodily organs become in sickness an exciting source of dreams, and if we admit that the mind, diverted during sleep from the outer world, can devote more attention to the interior of the body, we may readily assume that the organs need not necessarily become diseased in order to permit stimuli, which in some way or other grow into dream pictures, to reach the sleeping mind. What in the waking state we broadly perceive as general sensation, distinguishable by its quality alone, to which, in the opinion of the physicians, all the organic systems contribute their shares—this general sensation at night attaining powerful efficiency and becoming active with its individual components— would naturally furnish the most powerful as well as the most common source for the production of the dream presentations. It still remains, however, to examine according to what rule the organic sensations become transformed into dream presentations.

The theory of the origin of dreams just stated has been the favourite with all medical authors. The obscurity which conceals the essence of our being—the *"moi splanchnique,"* as Tissié terms it—from our knowledge and the obscurity of the origin of the dream correspond too well not to be brought into relation with each other. The train of thought which makes organic sensation the inciter of the dream has besides another attraction for the physician, inasmuch as it favours the etiological union of the dream and mental diseases, which show so many agreements in their manifestations, for alterations in the organic sensations and excitations emanating from the inner organs are both of wide significance in the origin of the psychoses. It is therefore not surprising that the theory of bodily sensation can be traced to more than one originator who has propounded it independently.

A number of authors have been influenced by the train of ideas developed by the philosopher Schopenhauer in 1851. Our conception of the universe originates through the fact that our intellect recasts the impressions coming to it from without in the moulds of time, space, and causality. The sensations from the interior of the organism, proceeding from the sympathetic nervous system, exert in the day-time an influence on our mood for the most part unconscious. At night, however, when the overwhelming influence of the day's impressions is no longer felt, the impressions pressing upward from the interior are able to gain attention—just as in the night we hear the rippling of the spring that was rendered inaudible by the noise of the day. In what other way, then, could the intellect react upon these stimuli than by performing its characteristic function? It will transform the stimuli into figures, filling space and time, which move at the beginning of causality; and thus the dream originates. Scherner, and after him Volkelt, attempted to penetrate into closer relations between physical sensations and dream pictures; but we shall reserve the discussion of these attempts for the chapter on the theory of the dream.

In a study particularly logical in its development, the psychiatrist Krauss found the origin of the dream as well as of deliria and delusions in the same element, viz. the organically determined sensation. According to this author there is hardly a place in the organism which might not become the starting point of a dream or of a delusion. Now organically determined sensations "may be divided into two classes: (1) those of the total feeling (general sensations), (2) specific sensations which are inherent in the principal systems of the vegetative organism, which may be divided into five groups: (*a*) the muscular, (*b*) the pneumatic, (*c*) the gastric, (*d*) the sexual, (*e*) the peripheral sensations (p. 33 of the second article)."

The origin of the dream picture on the basis of the physical sensations is conceived by Krauss as follows: The awakened sensation evokes a presentation related to it in accordance with some law of association, and combines with this, thus forming an organic structure, towards which, however, consciousness does not maintain its normal attitude. For it does not bestow any attention on the sensation itself, but concerns itself entirely with the accompanying presentation; this is likewise the reason why the state of affairs in question should have been so long misunderstood (p. 11, &c.). Krauss finds for this process the specific term of "transubstantiation of the feeling into dream pictures" (p. 24).

That the organic bodily sensations exert some influence on the formation of the dream is nowadays almost universally acknowledged, but the question as to the law underlying the relation between the two is answered in various ways and often in obscure terms. On the basis of the theory of bodily excitation the special task of dream interpretation is to trace back the content of a dream to the causative organic stimulus, and if we do not recognise the rules of interpretation advanced by Scherner, we frequently find ourselves confronted with the awkward fact that the organic exciting source reveals itself in the content of the dream only.

A certain agreement, however, is manifested in the interpretation of the various forms of dreams which have been designated as "typical" because they recur in so many persons with almost the same contents. Among these are the well-known dreams of falling from heights, of the falling out of teeth, of flying, and of embarrassment because of being naked or barely clad. This last dream is said to be caused simply by the perception felt in sleep that one has thrown off the bedcover and is exposed. The dream of the falling out of teeth is explained by "dental irritation," which does not, however, of necessity imply a morbid state of excitation in the teeth. According to Strümpell, the flying dream is the adequate picture used by the mind to interpret the sum of excitation emanating from the rising and sinking of the pulmonary lobes after the cutaneous sensation of the thorax has been reduced to insensibility. It is this latter circumstance that causes a sensation related to the conception of flying. Falling from a height in a dream is said to have its cause in the fact that when unconsciousness of the sensation of cutaneous pressure has set in, either an arm falls away from the body or a flexed knee is suddenly stretched out, causing the feeling of cutaneous pressure to return to consciousness, and the transition to consciousness embodies itself psychically as a dream of falling. (Strümpell, p. 118). The weakness of these plausible attempts at explanation evidently lies in the fact that without any further elucidation they allow this or that group of organic sensations to disappear from psychic perception or to obtrude themselves upon it until the constellation favourable for the explanation has been established. I shall, however, later have occasion to recur to typical dreams and to their origin.

From comparison of a series of similar dreams, M. Simon endeavoured to formulate certain rules for the influence of the organic sensations on the determination of the resulting dream. He says (p. 34): "If any organic apparatus, which during sleep normally participates in the expression of an affect, for any reason merges into the state of excitation to which it is usually aroused by that affect, the dream thus produced will contain presentations which fit the affect."

Another rule reads as follows (p. 35): "If an organic apparatus is in a state of activity, excitation, or disturbance during sleep, the dream will bring ideas which are related to the exercise of the organic function which is performed by that apparatus."

Mourly Vold has undertaken to prove experimentally the influence assumed by the theory of bodily sensation for a single territory. He has made experiments in altering the positions of the sleeper's limbs, and has compared the resulting dream with his alterations. As a result he reports the following theories:—

1. The position of a limb in a dream corresponds approximately to that of reality, *i.e.* we dream of a static condition of the limb which corresponds to the real condition.
2. When one dreams of a moving limb it always happens that one of the positions occurring in the execution of this movement corresponds to the real position.
3. The position of one's own limb may be attributed in the dream to another person.
4. One may dream further that the movement in question is impeded.
5. The limb in any particular position may appear in the dream as an animal or monster, in which case a certain analogy between the two is established.
6. The position of a limb may incite in the dream ideas which bear some relation or other to this limb. Thus, *e.g.,* if we are employed with the fingers we dream of numerals.

Such results would lead me to conclude that even the theory of bodily sensation cannot fully extinguish the apparent freedom in the determination of the dream picture to be awakened.[6]

IV. Psychic Exciting Sources

In treating the relations of the dream to the waking life and the origin of the dream material, we learned that the earliest as well as the latest investigators agreed that men dream of what they are doing in the day-time, and of what they are interested in during the waking state. This interest continuing from waking life into sleep, besides being a psychic tie joining the dream to life, also furnishes us a dream source not to be under-estimated, which, taken with those stimuli which become interesting and active during sleep, suffices to explain the origin of all dream pictures. But we have also heard the opposite of the above assertion, viz. that the dream takes the sleeper away from the interests of the day, and that in most cases we do not dream of things that have occupied our attention during the day until after they have lost for the waking life the stimulus of actuality. Hence in the analysis of the dream life we are reminded at every step that it is inadmissible to frame general rules without making provision for qualifications expressed by such terms as "frequently," "as a rule," "in most cases," and without preparing for the validity of the exceptions.

If the conscious interest, together with the inner and outer sleep stimuli, sufficed to cover the etiology of the dreams, we ought to be in a position to give a satisfactory account of the origin of all the elements of a dream; the riddle of the dream sources would thus be solved, leaving only the task of separating the part played by the psychic and the somatic dream stimuli in individual dreams. But as a matter of fact no such complete solution of a dream has ever been accomplished in any case, and, what is more, every one attempting such solution has found that in most cases there have remained a great many components of the dream, the source of which he was unable to explain. The daily interest as a psychic source of dreams is evidently not far-reaching enough to justify the confident assertions to the effect that we all continue our waking affairs in the dream.

Other psychic sources of dreams are unknown. Hence, with the exception perhaps of the explanation of dreams given by Scherner, which will be referred to later, all explanations found in the literature show a large gap when we come to the derivation of the material for the presentation pictures, which is most characteristic for the dream. In this dilemma the majority of authors have developed a tendency to depreciate as much as possible the psychic factor in the excitations of dreams which is so difficult to approach. To be sure, they distinguish as a main division of dreams the nerve-exciting and the association dreams, and assert that the latter has its source exclusively in reproduction (Wundt, p. 365), but they cannot yet dismiss the doubt whether "they do not appear without being impelled by the psychical stimulus" (Volkelt, p. 127). The characteristic quality of the pure association dream is also found wanting. To quote Volkelt (p. 118): "In the association dreams proper we can no longer speak of such a firm nucleus. Here the loose grouping penetrates also into the centre of the dream. The ideation which is already set free from reason and intellect is here no longer held together by the more important psychical and mental stimuli, but is left to its own aimless shifting and complete confusion." Wundt, too, attempts to depreciate the psychic factor in the stimulation of dreams by declaring that the "phantasms of the dream certainly are unjustly regarded as pure hallucinations, and that probably most dream presentations are really illusions, inasmuch as they emanate from slight sensory impressions which are never extinguished during sleep" (p. 338,

&c.). Weygandt agrees with this view, but generalises it. He asserts that "the first source of all dream presentations is a sensory stimulus to which reproductive associations are then joined" (p. 17). Tissié goes still further in repressing the psychic exciting sources (p. 183): "Les rêves d'origine absolument psychique n'existent pas"; and elsewhere (p. 6), "Les pensées de nos rêves nous viennent de dehors. . . ."

Those authors who, like the influential philosopher Wundt, adopt a middle course do not fail to remark that in most dreams there is a co-operation of the somatic stimuli with the psychic instigators of the dream, the latter being either unknown or recognised as day interests.

We shall learn later that the riddle of the dream formation can be solved by the disclosure of an unsuspected psychic source of excitement. For the present we shall not be surprised at the over-estimation of those stimuli for the formation of the dream which do not originate from psychic life. It is not merely because they alone can easily be found and even confirmed by experiment, but the somatic conception of the origin of dreams thoroughly corresponds to the mode of thinking in vogue nowadays in psychiatry. Indeed, the mastery of the brain over the organism is particularly emphasized; but everything that might prove an independence of the psychic life from the demonstrable organic changes, or a spontaneity in its manifestations, is alarming to the psychiatrist nowadays, as if an acknowledgment of the same were bound to bring back the times of natural philosophy and the metaphysical conception of the psychic essence. The distrust of the psychiatrist has placed the psyche under a guardian, so to speak, and now demands that none of its feelings shall divulge any of its own faculties; but this attitude shows slight confidence in the stability of the causal concatenation which extends between the material and the psychic. Even where on investigation the psychic can be recognised as the primary course of a phenomenon, a more profound penetration will some day succeed in finding a continuation of the path to the organic determination of the psychic. But where the psychic must be taken as the terminus for our present knowledge, it should not be denied on that account.

(D) Why the Dream Is Forgotten after Awakening

That the dream "fades away" in the morning is proverbial. To be sure, it is capable of recollection. For we know the dream only by recalling it after awakening; but very often we believe that we remember it only incompletely, and that during the night there was more of it; we can observe how the memory of a dream which has been still vivid in the morning vanishes in the course of the day, leaving only a few small fragments; we often know that we have been dreaming, but we do not know what; and we are so well used to the fact that the dream is liable to be forgotten that we do not reject as absurd the possibility that one may have been dreaming even when one knows nothing in the morning of either the contents or the fact of dreaming. On the other hand, it happens that dreams manifest an extraordinary retentiveness in the memory. I have had occasion to analyse with my patients dreams which had occurred to them twenty-five years or more previously, and I can remember a dream of my own which is separated from the present day by at least thirty-seven years, and yet has lost nothing of its freshness in my memory. All this is very remarkable, and for the present incomprehensible.

The forgetting of dreams is treated in the most detailed manner by Strümpell. This forgetting is evidently a complex phenomenon; for Strümpell does not explain it by a single reason, but by a considerable number of reasons.

In the first place, all those factors which produce forgetfulness in the waking state are also determinant for the forgetting of dreams. When awake we are wont soon to forget a large number of sensations and perceptions because they are too feeble, and because they are connected with a slight amount of emotional feeling. This is also the case with many dream pictures; they are forgotten because they are too weak, while stronger pictures in proximity will be remembered. Moreover, the factor of intensity in itself is not the only determinant for the preservation of the dream pictures; Strümpell, as well as other authors (Calkins), admits that dream pictures are often rapidly forgotten, although they are known to have been vivid, whereas among those that are retained in memory there are many that are very shadowy and hazy. Besides, in the waking state one is wont to forget easily what happened only once, and to note more easily things of repeated occurrence. But most dream pictures are single experiences,[7] and this peculiarity equally contributes towards the forgetting of all dreams. Of greater significance is a third motive for forgetting. In order that feelings, presentations, thoughts and the like, should attain a certain degree of memory, it is important that they should not remain isolated, but that they should enter into connections and associations of a suitable kind. If the words of a short verse are taken and mixed together, it will be very difficult to remember them. "When well arranged in suitable sequence one word will help another, and the whole remains as sense easily and firmly in the memory for a long time. Contradictions we usually retain with just as much difficulty and rarity as things confused and disarranged." Now dreams in most cases lack sense and order. Dream compositions are by their very nature incapable of being remembered, and they are forgotten because they usually crumble together the very next moment. To be sure, these conclusions are not in full accord with the observation of Radestock (p. 168), that we retain best just those dreams which are most peculiar.

According to Strümpell, there are still other factors effective in the forgetting of dreams which are derived from the relation of the dream to the waking state. The forgetfulness of the waking consciousness for dreams is evidently only the counterpart of the fact already mentioned, that the dream (almost) never takes over successive memories from the waking state, but only certain details of these memories which it tears away from the habitual psychic connections in which they are recalled while we are awake. The dream composition, therefore, has no place in the company of psychic successions which fill the mind. It lacks all the aids of memory. "In this manner the dream structure rises, as it were, from the soil of our psychic life, and floats in psychic space like a cloud in the sky, which the next breath of air soon dispels" (p. 87). This is also aided by the fact that, upon awakening, the attention is immediately seized by the inrushing sensory world, and only very few dream pictures can withstand this power. They fade away before the impressions of the new day like the glow of the stars before the sunlight.

As a last factor favouring the forgetting of dreams, we may mention the fact that most people generally take little interest in their dreams. One who investigates dreams for a time, and takes a special interest in them, usually dreams more during that time than at any other; that is, he remembers his dreams more easily and more frequently.

Two other reasons for the forgetting of dreams added by Bonatelli (given by Benini) to those of Strümpell have already been included in the latter; namely, (1) that the change of the general feeling between the sleeping and waking states is unfavourable to the mutual reproductions, and (2) that the different arrangement of the presentation material in the dream makes the dream untranslatable, so to speak, for the waking consciousness.

It is the more remarkable, as Strümpell observes, that, in spite of all these reasons for forgetting the dream, so many dreams are retained in memory. The continued efforts of the authors to formulate laws for the remembering of dreams amounts to an admission that here too there is something puzzling and unsolved. Certain peculiarities relating to the memory of dreams have been particularly noticed of late, *e.g.,* that a dream which is considered forgotten in the morning may be recalled in the course of the day through a perception which accidentally touches the forgotten content of the dream (Radestock, Tissié). The entire memory of the dream is open to an objection calculated to depreciate its value very markedly in critical eyes. One may doubt whether our memory, which omits so much from the dream, does not falsify what it retained.

Such doubts relating to the exactness of the reproduction of the dream are expressed by Strümpell when he says: "It therefore easily happens that the active consciousness involuntarily inserts much in recollection of the dream; one imagines one has dreamt all sorts of things which the actual dream did not contain."

Jessen (p. 547) expresses himself very decidedly: "Moreover we must not lose sight of the fact, hitherto little heeded, that in the investigation and interpretation of orderly and logical dreams we almost always play with the truth when we recall a dream to memory. Unconsciously and unwittingly we fill up the gaps and supplement the dream pictures. Rarely, and perhaps never, has a connected dream been as connected as it appears to us in memory. Even the most truth-loving person can hardly relate a dream without exaggerating and embellishing it. The tendency of the human mind to conceive everything in connection is so great that it unwittingly supplies the deficiencies of connection if the dream is recalled somewhat disconnectedly."

The observations of V. Egger, though surely independently conceived, sound almost like a translation of Jessen's words: " . . . L'observation des rêves a ses difficultés spéciales et le seul moyen d'éviter toute erreur en pareille matière est de confier au papier sans le moindre retard ce que l'on vient d'éprouver et de remarquer; sinon, l'oubli vient vite ou total ou partiel; l'oubli total est sans gravité; mais l'oubli partiel est perfide; car si l'on se met ensuite à raconter ce que l'on n'a pas oublié, on est exposé à compléter par imagination les fragments incohérents et disjoints fourni par la mémoire . . . ; on devient artiste à son insu, et le récit, periodiquement répété s'impose à la créance de son auteur, qui, de bonne foi, le présente comme un fait authentique, dûment établi selon les bonnes méthodes. . . . "

Similarly Spitta, who seems to think that it is only in our attempt to reproduce the dream that we put in order the loosely associated dream elements: "To make connection out of disconnection, that is, to add the process of logical connection which is absent in the dream."

As we do not at present possess any other objective control for the reliability of our memory, and as indeed such a control is impossible in examining the dream which is our own experience, and for which our memory is the only source, it is a question what value we may attach to our recollections of dreams.

(E) The Psychological Peculiarities of Dreams

In the scientific investigation of the dream we start with the assumption that the dream is an occurrence of our own psychic activity; nevertheless the finished dream appears to us as something strange, the authorship of which we are so little forced to recognise that we can just as easily say "a dream appeared to me," as "I have dreamt." Whence this "psychic strangeness" of the dream? According to our discussion of the sources of dreams we may suppose that it does not depend on the material reaching the dream content; because this is for the most part common to the dream life and waking life. One may ask whether in the dream it is not changes in the psychic processes which call forth this impression, and may so put to test a psychological characteristic of the dream.

No one has more strongly emphasized the essential difference between dream and waking life, and utilised this difference for more far-reaching conclusions, than G. Th. Fechner in some observations in his *Elements of Psychophysic* (p. 520, part 11). He believes that "neither the simple depression of conscious psychic life under the main threshold," nor the distraction of attention from the influences of the outer world, suffices to explain the peculiarities of the dream life as compared with the waking life. He rather believes that the scene of dreams is laid elsewhere than in the waking presentation life. "If the scene of the psychophysical activity were the same during the sleeping and the waking states, the dream, in my opinion, could only be a continuation of the waking ideation maintaining itself at a lower degree of intensity, and must moreover share with the latter its material and form. But the state of affairs is quite different."

What Fechner really meant has never been made clear, nor has anybody else, to my knowledge, followed further the road, the clue to which he indicated in this remark. An anatomical interpretation in the sense of physiological brain localisations, or even in reference to histological sections of the cerebral cortex, will surely have to be excluded. The thought may, however, prove ingenious and fruitful if it can be referred to a psychic apparatus which is constructed out of many instances placed one behind another.

Other authors have been content to render prominent one or another of the tangible psychological peculiarities of the dream life, and perhaps to take these as a starting point for more far-reaching attempts at explanation.

It has been justly remarked that one of the main peculiarities of the dream life appears even in the state of falling asleep, and is to be designated as the phenomenon inducing sleep. According to Schleiermacher (p. 351), the characteristic part of the waking state is the fact that the psychic activity occurs in ideas rather than in pictures. But the dream thinks in pictures, and one may observe that with the approach of sleep the voluntary activities become difficult in the same measure as the involuntary appear, the latter belonging wholly to the class of pictures. The inability for such presentation work as we perceive to be intentionally desired, and the appearance of pictures which is regularly connected with this distraction, these are two qualities which are constant in the dream, and which in its psychological analysis we must recognise as essential characters of the dream life. Concerning the pictures—the hypnogogic hallucinations—we have discovered that even in their content they are identical with the dream pictures.

The dream therefore thinks preponderately, but not exclusively, in visual pictures. It also makes use of auditory pictures, and to a lesser extent of the impressions of the other senses. Much is also simply thought or imagined (probably represented by remnants of word presentations), just as in the waking state. But still what is characteristic for the dream is only those elements of the content which act like pictures, *i.e.* which resemble more the perceptions than the memory presentations. Disregarding all the discussions concerning the nature of hallucinations, familiar to every psychiatrist, we can say, with all well-versed authors, that the dream hallucinates, that is, replaces thoughts through hallucinations. In this

46

respect there is no difference between visual and acoustic presentations; it has been noticed that the memory of a succession of sounds with which one falls asleep becomes transformed while shaking into sleep into an hallucination of the same melody, so as to make room again on awakening, which may repeatedly alternate with falling into a slumber, for the softer memory presentations which are differently formed in quality.

The transformation of an idea into an hallucination is not the only deviation of the dream from a waking thought which perhaps corresponds to it. From these pictures the dream forms a situation, it presents something in the present, it dramatises an idea, as Spitta (p. 145) puts it.[8] But the characteristic of this side of the dream life becomes complete only when it is remembered that while dreaming we do not—as a rule; the exceptions require a special explanation—imagine that we are thinking, but that we are living through an experience, *i.e.*, we accept the hallucination with full belief. The criticism that this has not been experienced but only thought in a peculiar manner—dreamt—comes to us only on awakening. This character distinguishes the genuine sleeping dream from day dreaming, which is never confused with reality.

The characteristics of the dream life thus far considered have been summed up by Burdach (p. 476) in the following sentences: "As characteristic features of the dream we may add (*a*) that the subjective activity of our mind appears as objective, inasmuch as our faculty of perception perceives the products of phantasy as if they were sensory activities . . . (*b*) sleep abrogates one's self-command, hence falling asleep necessitates a certain amount of passivity. . . . The slumber pictures are conditioned by the relaxation of one's self-command."

It is a question now of attempting to explain the credulity of the mind in reference to the dream hallucinations, which can only appear after the suspension of a certain arbitrary activity. Strümpell asserts that the mind behaves in this respect correctly, and in conformity with its mechanism. The dream elements are by no means mere presentations, but true and real experiences of the mind, similar to those that appear in the waking state as a result of the senses (p. 34). Whereas in the waking state the mind represents and thinks in word pictures and language, in the dream it represents and thinks in real tangible pictures (p. 35). Besides, the dream manifests a consciousness of space by transferring the sensations and pictures, just as in the waking state, into an outer space (p. 36). It must therefore be admitted that the mind in the dream is in the same relation to its pictures and perceptions as in the waking state (p. 43). If, however, it is thereby led astray, this is due to the fact that it lacks in sleep the criticism which alone can distinguish between the sensory perceptions emanating from within or from without. It cannot subject its pictures to the tests which alone can prove their objective reality. It furthermore neglects to differentiate between pictures that are arbitrarily interchanged and others where there is no free choice. It errs because it cannot apply to its content the law of causality (p. 58). In brief, its alienation from the outer world contains also the reason for its belief in the subjective dream world.

Delboeuf reaches the same conclusion through a somewhat different line of argument. We give to the dream pictures the credence of reality because in sleep we have no other impressions to compare them with, because we are cut off from the outer world. But it is not perhaps because we are unable to make tests in our sleep, that we believe in the truth of our hallucinations. The dream may delude us with all these tests, it may make us believe that we may touch the rose that we see in the dream, and still we only dream. According to Delboeuf there is no valid criterion to show whether something is a dream or a conscious reality, except—and that only in practical generality—the fact of awakening. "I declare delusional everything that is experienced between the period of falling asleep and awakening, if I notice on awakening that I lie in my bed undressed" (p. 84). "I have considered the dream pictures real during sleep in consequence of the mental habit, which cannot be put to sleep, of perceiving an outer world with which I can contrast my ego."[9]

As the deviation from the outer world is taken as the stamp for the most striking characteristics of the dream, it will be worth while mentioning some ingenious observations of old Burdach which will throw light on the relation of the sleeping mind to the outer world and at the same time serve to prevent us from over-estimating the above deductions. "Sleep results only under the condition," says Burdach, "that the mind is not excited by sensory stimuli . . . but it is not the lack of sensory stimuli that conditions sleep, but rather a lack of interest for the same; some sensory impressions are even necessary in so far as they serve to calm the mind; thus the miller can fall asleep only when he hears the rattling of his mill, and he who finds it necessary to burn a light at night, as a matter of precaution, cannot fall asleep in the dark" (p. 457).

"The psyche isolates itself during sleep from the outer world, and withdraws from the periphery. . . . Nevertheless, the connection is not entirely interrupted; if one did not hear and feel even during sleep, but only after awakening, he would certainly never awake. The continuance of sensation is even more plainly shown by the fact that we are not always awakened by the mere sensory force of the impression, but by the psychic relation of the same; an indifferent word does not arouse the sleeper, but if called by name he awakens . . . : hence the psyche differentiates sensations during sleep. . . . It is for this reason that we may be awakened by the lack of a sensory stimulus if it relates to the presentation of an important thing; thus one awakens when the light is extinguished, and the miller when the mill comes to a standstill; that is, the awakening is due to the cessation of a sensory activity, which presupposes that it has been perceived, and that it has not disturbed the mind, being indifferent or rather gratifying" (p. 460, &c.).

If we are willing to disregard these objections, which are not to be taken lightly, we still must admit that the qualities of the dream life thus far considered, which originate by withdrawing from the outer world, cannot fully explain the strangeness of the dream. For otherwise it would be possible to change back the hallucinations of the dream into presentations and the situations of the dream into thoughts, and thus to perform the task of dream interpretation. Now this is what we do when we reproduce the dream from memory after awakening, and whether we are fully or only partially successful in this back translation the dream still retains its mysteriousness undiminished.

Furthermore all the authors assume unhesitatingly that still other more far-reaching alterations take place in the presentation material of waking life. One of them, Strümpell expresses himself as follows (p. 17): "With the cessation of the objectively active outlook and of the normal consciousness, the psyche loses the foundation in which were rooted the feelings, desires, interests, and actions. Those psychic states, feelings, interests, estimates which cling in the waking state to the memory pictures also succumb to . . . an obscure pressure, in consequence of which their connection with the pictures becomes severed; the perception pictures of things, persons, localities, events, and actions of the waking state are singly very abundantly reproduced, but none of these brings along its psychic value. The latter is removed from them, and hence they float about in the mind dependent upon their own resources. . . . "

This deprivation the picture suffers of its psychic value, which again goes back to the derivation from the outer world, is according to Strümpell mainly responsible for the impression of strangeness with which the dream is confronted in our memory.

We have heard that even falling asleep carries with it the abandonment of one of the psychic activities—namely, the voluntary conduct of the presentation course. Thus the supposition, suggested also by other grounds, obtrudes itself, that the sleeping state may extend its influence also over the psychic functions. One or the other of these functions is perhaps entirely suspended; whether the remaining ones continue to work undisturbed, whether they can furnish normal work under the circumstances, is the next question. The idea occurs to us that the peculiarities of the dream may be explained through the inferior psychic activity during the sleeping state, but now comes the impression made by the dream upon our waking judgment which is contrary to such a conception. The dream is disconnected, it unites without hesitation the worst contradictions, it allows impossibilities, it disregards our authoritative knowledge from the day, and evinces ethical and moral dulness. He who would behave in the waking state as the dream does in its situations would be considered insane. He who in the waking state would speak in such manner or report such things as occur in the dream content, would impress us as confused and weak-minded. Thus we believe that we are only finding words for the fact when we place but little value on the psychic activity in the dream, and especially when we declare that the higher intellectual activities are suspended or at least much impaired in the dream.

With unusual unanimity—the exceptions will be dealt with elsewhere—the authors have pronounced their judgments on the dream—such judgments as lead immediately to a definite theory or explanation of the dream life. It is time that I should supplement the *résumé* which I have just given with a collection of the utterances of different authors—philosophers and physicians—on the psychological character of the dream.

According to Lemoine, the incoherence of the dream picture is the only essential character of the dream.

Maury agrees with him; he says (p. 163): "Il n'y a pas des rêves absolument raisonnables et qui ne contiennent quelque incohérence, quelque anachronisme, quelque absurdité."

According to Hegel, quoted by Spitta, the dream lacks all objective and comprehensible connection.

Dugas says: "Le rêve, c'est l'anarchie psychique, affective et mentale, c'est le jeu des fonctions livrées à elles-mêmes et s'exerçant sans contrôle et sans but; dans le rêve l'esprit est un automate spirituel."

"The relaxation, solution, and confusion of the presentation life which is held together through the logical force of the central ego" is conceded even by Volkelt (p. 14), according to whose theory the psychic activity during sleep seems in no way aimless.

The absurdity of the presentation connections appearing in the dream can hardly be more strongly condemned than it was by Cicero (*De Divin.* II.): "Nihil tam praepostere, tam incondite, tam monstruose cogitari potest, quod non possimus somniare."

Fechner says (p. 522): "It is as if the psychological activity were transferred from the brain of a reasonable being into the brain of a fool."

Radestock (p. 145) says: "It seems indeed impossible to recognise in this absurd action any firm law. Having withdrawn itself from the strict police of the rational will guiding the waking presentation life, and of the attention, the dream whirls everything about kaleidoscopically in mad play."

Hildebrandt (p. 45) says: "What wonderful jumps the dreamer allows himself, *e.g.,* in his chain of reasoning! With what unconcern he sees the most familiar laws of experience turned upside down! What ridiculous contradictions he can tolerate in the orders of nature and society before things go too far, as we say, and the overstraining of the nonsense brings an awakening! We often multiply quite unconcernedly: three times three make twenty; we are not at all surprised when a dog recites poetry for us, when a dead person walks to his grave, and when a rock swims on the water; we go in all earnestness by high command to the duchy of Bernburg or the principality of Lichtenstein in order to observe the navy of the country, or we allow ourselves to be recruited as a volunteer by Charles XII. shortly before the battle of Poltawa."

Binz (p. 33) points to a dream theory resulting from the impressions. "Among ten dreams nine at least have an absurd content. We unite in them persons or things which do not bear the slightest relation to one another. In the next moment, as in a kaleidoscope, the grouping changes, if possible to one more nonsensical and irrational than before; thus the changing play of the imperfectly sleeping brain continues until we awaken, and put our hand to our forehead and ask ourselves whether we really still possess the faculty of rational imagination and thought."

Maury (p. 50) finds for the relation of the dream picture to the waking thoughts, a comparison most impressive for the physician: "La production de ces images que chez l'homme éveillé fait le plus souvent naître la volonté, correspond, pour l'intelligence, à ce que cont pour la motilité certains mouvements que nous offrent la chorée et les affections paralytiques. . . . " For the rest, he considers the dream "toute une série de dégradation de la faculté pensant et raisonant" (p. 27).

It is hardly necessary to mention the utterances of the authors which repeat Maury's assertion for the individual higher psychic activities.

According to Strümpell, some logical mental operations based on relations and connections disappear in the dream—naturally also at points where the nonsense is not obvious (p. 26). According to Spitta (p. 148), the presentations in the dream are entirely withdrawn from the laws of causality. Radestock and others emphasize the weakness of judgment and decision in the dream. According to Jodl (p. 123), there is no critique in the dream, and no correcting of a series of perceptions through the content of the sum of consciousness. The same author states that "all forms of conscious activity occur in the dream, but they are imperfect, inhibited, and isolated from one another." The contradictions manifested in the dream towards our conscious knowledge are explained by Stricker (and many others), on the ground that facts are forgotten in the dream and logical relations between presentations are lost (p. 98), &c., &c.

The authors who in general speak thus unfavourably about the psychic capacities in the dream, nevertheless admit that the dream retains a certain remnant of psychic activity. Wundt, whose teaching has influenced so many other workers in the dream problems, positively admits this. One might inquire as to the kind and behaviour of the remnants of the psychic life which manifest themselves in the dream. It is now quite universally acknowledged that the reproductive capacity, the memory in the dream, seems to have been least affected; indeed it may show a certain superiority over the same function in the waking life (*vid. supra,* p. 10), although a part of the absurdities of the dream are to be explained by just this forgetfulness of the dream life. According to Spitta, it is the emotional life of the psyche that is not overtaken by sleep and that then directs the dream. "By emotion

["Gemüth"] we understand the constant comprehension of the feelings as the inmost subjective essence of man" (p. 84).

Scholz (p. 37) sees a psychic activity manifested in the dream in the "allegorising interpretation" to which the dream material is subjected. Siebeck verifies also in the dream the "supplementary interpretative activity" (p. 11) which the mind exerts on all that is perceived and viewed. The judgment of the apparently highest psychic function, the consciousness, presents for the dream a special difficulty. As we can know anything only through consciousness, there can be no doubt as to its retention; Spitta, however, believes that only consciousness is retained in the dream, and not self-consciousness. Delboeuf confesses that he is unable to conceive this differentiation.

The laws of association which govern the connection of ideas hold true also for the dream pictures; indeed, their domination evinces itself in a purer and stronger expression in the dream than elsewhere. Strümpell (p. 70) says: "The dream follows either the laws of undisguised presentations as it seems exclusively or organic stimuli along with such presentations, that is, without being influenced by reflection and reason, aesthetic sense, and moral judgment." The authors whose views I reproduce here conceive the formation of the dream in about the following manner: The sum of sensation stimuli affecting sleep from the various sources, discussed elsewhere, at first awaken in the mind a sum of presentations which represent themselves as hallucinations (according to Wundt, it is more correct to say as illusions, because of their origin from outer and inner stimuli). These unite with one another according to the known laws of association, and, following the same rules, in turn evoke a new series of presentations (pictures). This entire material is then elaborated as well as possible by the still active remnant of the organising and thinking mental faculties (cf. Wundt and Weygandt). But thus far no one has been successful in finding the motive which would decide that the awakening of pictures which do not originate objectively follow this or that law of association.

But it has been repeatedly observed that the associations which connect the dream presentations with one another are of a particular kind, and different from those found in the waking mental activity. Thus Volkelt says: "In the dream, the ideas chase and hunt each other on the strength of accidental similarities and barely perceptible connections. All dreams are pervaded by such loose and free associations." Maury attaches great value to this characteristic of connection between presentations, which allows him to bring the dream life in closer analogy to certain mental disturbances. He recognises two main characters of the *délire:* "(1) une action spontanée et comme automatique de l'esprit; (2) une association vicieuse et irregulière des idées" (p. 126). Maury gives us two excellent examples from his own dreams, in which the mere similarity of sound forms the connection of the dream presentations. He dreamed once that he undertook a pilgrimage (*pélerinage*) to Jerusalem or Mecca. After many adventures he was with the chemist Pelletier; the latter after some talk gave him a zinc shovel (*pelle*) which became his long battle sword in the dream fragment which followed (p. 137). On another occasion he walked in a dream on the highway and read the kilometres on the milestones; presently he was with a spice merchant who had large scales with which to weigh Maury; the spice merchant then said to him: "You are not in Paris; but on the island Gilolo." This was followed by many pictures, in which he saw the flower Lobelia, then the General Lopez, of whose demise he had read shortly before. He finally awoke while playing a game of lotto.

We are, however, quite prepared to hear that this depreciation of the psychic activities of the dream has not remained without contradiction from the other side. To be sure, contradiction seems difficult here. Nor is it of much significance that one of the deprecators of dream life, Spitta (p. 118), assures us that the same psychological laws which govern the waking state rule the dream also, or that another (Dugas) states: "Le rêve n'est pas déraison ni même irraison pure," as long as neither of them has made any effort to bring this estimation into harmony with the psychic anarchy and dissolution of all functions in the dream described by them. Upon others, however, the possibility seems to have dawned that the madness of the dream is perhaps not without its method—that it is perhaps only a sham, like that of the Danish prince, to whose madness the intelligent judgment here cited refers. These authors must have refrained from judging by appearances, or the appearance which the dream showed to them was quite different.

Without wishing to linger at its apparent absurdity, Havelock Ellis considers the dream as "an archaic world of vast emotions and imperfect thoughts," the study of which may make us acquainted with primitive stages of development of the psychic life. A thinker like Delboeuf asserts— to be sure without adducing proof against the contradictory material, and hence indeed unjustly: "Dans le sommeil, hormis la perception, toutes les facultés de l'esprit, intelligence, imagination, mémoire, volonté, moralité, restant intactes dans leur essence; seulement, elles s'appliquent à des objets imaginaires et mobiles. Le songeur est un acteur qui joue à volonté les fous et les sages, les bourreaux et les victimes, les nains et les géants, les démons et les anges" (p. 222). The Marquis of Hervey, who is sharply controverted by Maury, and whose work I could not obtain despite all effort, seems to combat most energetically the under-estimation of the psychic capacity in the dream. Maury speaks of him as follows (p. 19):

"M. le Marquis d'Hervey prête à l'intelligence, durant le sommeil toute sa liberté d'action et d'attention et il ne semble faire consister le sommeil que dans l'occlusion des sens, dans leur fermeture au monde extérieur; en sorte que l'homme qui dort ne se distingué guère, selon sa manière de voir, de l'homme qui laisse vaguer sa pensée en se bouchant les sens; toute la différence qui séparé alors la pensée ordinaire du celle du dormeur c'est que, chez celui-ci, l'idée prend une forme visible, objective et ressemble, à s'y meprendre, à la sensation déterminée par les objets extérieurs; le souvenir revêt l'apparence du fait présent."

Maury adds, however; "Qu'il y a une différence de plus et capitale à savoir que les facultés intellectuelles de l'homme endormi n'offrent pas l'équilibre qu'elles gardent chez l'homme l'éveillé."

The scale of the estimation of the dream as a psychic product has a great range in the literature; it reaches from the lowest under-estimation, the expression of which we have come to know, through the idea of a value not yet revealed to the over-estimation which places the dream far above the capacities of the waking life. Hildebrandt, who, as we know, sketches the psychological characteristics into three antinomies, sums up in the third of these contradistinctions the extreme points of this series as follows (p. 19):

"It is between a climax, often an involution which raises itself to virtuosity, and on the other hand a decided diminution and weakening of the psychic life often leading below the human niveau."

"As for the first, who could not confirm from his own experience that, in the creations and weavings of the genius of the dream, there sometimes comes to light a profundity and sincerity of emotion, a tenderness of feeling, a clearness of view, a fineness of observation, and a readiness of wit, all which we should modestly have to deny that we possess as a constant property during the waking life? The dream has a wonderful poetry, an excellent allegory, an incomparable humour, and a charming irony. It views the world under the guise of a peculiar idealisation, and often raises the effect of its manifestations into the most ingenious understanding of the essence lying at its basis. It represents for us earthly beauty in true heavenly radiance, the sublime in the highest majesty, the actually frightful in the most gruesome figure, and the ridiculous in the indescribably drastic comical; and at times we are so full of one of these impressions after awakening that we imagine that such a thing has never been offered to us by the real world."

One may ask, is it really the same object that the depreciating remarks and these inspired praises are meant for? Have the latter overlooked the stupid dreams and the former the thoughtful and ingenious dreams? And if both kinds do occur—that is, dreams that merit to be judged in this or that manner—does it not seem idle to seek the psychological character of the dream? would it not suffice to state that everything is possible in the dream, from the lowest depreciation of the psychic life to a raising of the same which is unusual in the waking state? As convenient as this solution would be it has this against it, that behind the efforts of all dream investigators, it seems to be presupposed that there is such a definable character of the dream, which is universally valid in its essential features and which must eliminate these contradictions.

It is unquestionable that the psychic capacities of the dream have found quicker and warmer recognition in that intellectual period which now lies behind us, when philosophy rather than exact natural science ruled intelligent minds. Utterances like those of Schubert, that the dream frees the mind from the power of outer nature, that it liberates the soul from the chains of the sensual, and similar opinions expressed by the younger Fichte,[10] and others, who represent the dream as a soaring up of the psychic life to a higher stage, hardly seem conceivable to us to-day; they are only repeated at present by mystics and devotees. With the advance of the scientific mode of thinking, a reaction took place in the estimation of the dream. It is really the medical authors who are most prone to underrate the psychic activity in the dream, as being insignificant and invaluable, whereas, philosophers and unprofessional observers—amateur psychologists— whose contributions in this realm can surely not be overlooked, in better agreement with the popular ideas, have mostly adhered to the psychic value of the dream. He who is inclined to underrate the psychic capacity in the dream prefers, as a matter of course, the somatic exciting sources in the etiology of the dream; he who leaves to the dreaming mind the greater part of its capacities, naturally has no reason for not also admitting independent stimuli for dreaming.

Among the superior activities which, even on sober comparison, one is tempted to ascribe to the dream life, memory is the most striking; we have fully discussed the frequent experiences which prove this fact. Another superiority of the dream life, frequently extolled by the old authors, viz. that it can regard itself supreme in reference to distance of time and space, can be readily recognised as an illusion. This superiority, as observed by Hildebrandt, is only illusional; the dream takes as much heed of time and space as the waking thought, and this because it is only a form of thinking. The dream is supposed to enjoy still another advantage in reference to time; that is, it is independent in still another sense of the passage of time. Dreams like the guillotine dream of Maury, reported above, seem to show that the dream can crowd together more perception content in a very short space of time than can be controlled by our psychic activity in the waking mind. These conclusions have been controverted, however, by many arguments; the essays of Le Lorrain and Egger "Concerning the apparent duration of dreams" gave rise to a long and interesting discussion which has probably not said the last word upon this delicate and far-reaching question.

That the dream has the ability to take up the intellectual work of the day and bring to a conclusion what has not been settled during the day, that it can solve doubt and problems, and that it may become the source of new inspiration in poets and composers, seems to be indisputable, as is shown by many reports and by the collection compiled by Chabaneix. But even if there be no dispute as to the facts, nevertheless their interpretation is open in principle to a great many doubts.

Finally the asserted divinatory power of the dream forms an object of contention in which hard unsurmountable reflection encounters obstinate and continued faith. It is indeed just that we should refrain from denying all that is based on fact in this subject, as there is a possibility that a number of such cases may perhaps be explained on a natural psychological basis.

(F)   The Ethical Feelings in the Dream

For reasons which will be understood only after cognisance has been taken of my own investigations of the dream, I have separated from the psychology of the dream the partial problem whether and to what extent the moral dispositions and feelings of the waking life extend into the dreams. The same contradictions which we were surprised to observe in the authors' descriptions of all the other psychic capacities strike us again here. Some affirm decidedly that the dream knows nothing of moral obligations; others as decidedly that the moral nature of man remains even in his dream life.

A reference to our dream experience of every night seems to raise the correctness of the first assertion beyond doubt. Jessen says (p. 553): "Nor does one become better or more virtuous in the dream; on the contrary, it seems that conscience is silent in the dream, inasmuch as one feels no compassion and can commit the worst crimes, such as theft, murder, and assassination, with perfect indifference and without subsequent remorse."

Radestock (p. 146) says: "It is to be noticed that in the dream the associations terminate and the ideas unite without being influenced by reflection and reason, aesthetic taste, and moral judgment; the judgment is extremely weak, and ethical indifference reigns supreme."

Volkelt (p. 23) expresses himself as follows: "As every one knows, the sexual relationship in the dream is especially unbridled. Just as the dreamer himself is shameless in the extreme, and wholly lacking moral feeling and judgment, so also he sees others, even the most honoured persons, engaged in actions which even in thought he would blush to associate with them in his waking state."

Utterances like those of Schopenhauer, that in the dream every person acts and talks in accordance with his character, form the sharpest contrast to those mentioned above. R. P. Fischer[11] maintains that the subjective feelings and desires or affects and passions manifest themselves in the wilfulness of the dream life, and that the moral characteristics of a person are mirrored in his dream.

Haffner (p. 25): "With rare exceptions . . . a virtuous person will be virtuous also in his dreams; he will resist temptation, and show no sympathy for hatred, envy, anger, and all other vices; while the sinful person will, as a rule, also find in his dreams the pictures which he has before him while awake."

Scholz (p. 36): "In the dream there is truth; despite all masking in pride or humility, we still recognise our own self. . . . The honest man does not commit any dishonourable offence even in the dream, or, if this does occur, he is terrified over it as if over something foreign to his nature. The Roman emperor who ordered one of his subjects to be executed because he dreamed that he cut off the emperor's head, was not wrong in justifying his action on the ground that he who has such dreams must have similar thoughts while awake. About a thing that can have no place in our mind we therefore say significantly: 'I would never dream of such a thing.'"

Pfaff,[12] varying a familiar proverb, says: "Tell me for a tune your dreams, and I will tell you what you are within."

The short work of Hildebrandt, from which I have already taken so many quotations, a contribution to the dream problem as complete and as rich in thought as I found in the literature, places the problem of morality in the dream as the central point of its interest. For Hildebrandt, too, it is a strict rule that the purer the life, the purer the dream; the impurer the former, the impurer the latter.

The moral nature of man remains even in the dream:

"But while we are not offended nor made suspicious by an arithmetical error no matter how obvious, by a reversal of science no matter how romantic, or by an anachronism no matter how witty, we nevertheless do not lose sight of the difference between good and evil, right and wrong, virtue and vice. No matter how much of what follows us during the day may vanish in our hours of sleep—Kant's categorical imperative sticks to our heels as an inseparable companion from whom we cannot rid ourselves even in slumber. . . . This can be explained, however, only by the fact that the fundamental in human nature, the moral essence, is too firmly fixed to take part in the activity of the kaleidoscopic shaking up to which phantasy, reason, memory, and other faculties of the same rank succumb in the dream" (p. 45, &c.).

In the further discussion of the subject we find remarkable distortion and inconsequence in both groups of authors. Strictly speaking, interest in immoral dreams would cease for all those who assert that the moral personality of the person crumbles away in the dream. They could just as calmly reject the attempt to hold the dreamer responsible for his dreams, and to draw inferences from the badness of his dreams as to an evil strain in his nature, as they rejected the apparently similar attempt to demonstrate the insignificance of his intellectual life in the waking state from the absurdity of his dreams. The others for whom "the categorical imperative" extends also into the dream, would have to accept full responsibility for the immoral dreams; it would only be desirable for their own sake that their own objectionable dreams should not lead them to abandon the otherwise firmly held estimation of their own morality.

Still it seems that no one knows exactly about himself how good or how bad he is, and that no one can deny the recollection of his own immoral dreams. For besides the opposition already mentioned in the criticism of the morality of the dream, both groups of authors display an effort to explain the origin of the immoral dream and a new opposition is developed, depending on whether their origin is sought in the functions of the psychic life or in the somatically determined injuries to this life. The urgent force of the facts then permits the representatives of the responsibility, as well as of the irresponsibility of the dream life, to agree in the recognition of a special psychic source for the immorality of dreams.

All those who allow the continuance of the morality in the dream nevertheless guard against accepting full responsibility for their dreams. Haffner says (p. 24): "We are not responsible for dreams because the basis upon which alone our life has truth and reality is removed from our thoughts. . . . Hence there can be no dream wishing and dream acting, no virtue or sin." Still the person is responsible for the sinful dream in so far as he brings it about indirectly. Just as in the waking state, it is his duty to cleanse his moral mind, particularly so before retiring to sleep.

The analysis of this mixture of rejection and recognition of responsibility for the moral content of the dream is followed much further by Hildebrandt. After specifying that the dramatic maimer of representation in the dream, the crowding together of the most complicated processes of deliberation in the briefest period of time, and the depreciation and the confusion of the presentation elements in the dream admitted by him must be recognised as unfavourable to the immoral aspect of dreams; he nevertheless confesses that, yielding to the most earnest reflection, he is inclined simply to deny all responsibility for faults and dream sins.

(P. 49):

"If we wish to reject very decisively any unjust accusation, especially one that has reference to our intentions and convictions, we naturally make use of the expression: I should never have dreamed of such a thing. By this we mean to say, of course, that we consider the realm of the dream the last and remotest place in which we are to be held responsible for our thoughts, because there these thoughts are only loosely and incoherently connected with our real being, so that we should hardly still consider them as our own; but as we feel impelled expressly to deny the existence of such thoughts, even in this realm, we thus at the same time indirectly admit that our justification will not be complete if it does not reach to that point. And I believe that, though unconsciously, we here speak the language of truth."

(P. 52):

"No dream thought can be imagined whose first motive has not already moved through the mind while awake as some wish, desire, or impulse." Concerning this original impulse we must say that the dream has not discovered it—it has only imitated and extended it, it has only elaborated a bit of historical material which it has found in us, into dramatic form; it enacts the words of the apostle: He who hates his brother is a murderer. And whereas, after we awaken and become conscious of our moral strength, we may smile at the boldly executed structure of the depraved dream, the original formative material, nevertheless, has no ridiculous side. One feels responsible for the transgressions of the dreamer, not for the whole sum, but still for a certain percentage. "In this sense, which is difficult to impugn, we understand the words of Christ: Out of the heart come evil thoughts—for we can hardly help being convinced that every sin committed in the dream brings with it at least a vague minimum of guilt."

Hildebrandt thus finds the source of the immorality of dreams in the germs and indications of evil impulses which pass through our minds during the day as tempting thoughts, and he sees fit to add these immoral elements to the moral estimation of the personality. It is the same thoughts and the same estimation of these thoughts, which, as we know, have caused devout and holy men of all times to lament that they are evil sinners.

There is certainly no reason to doubt the general occurrence of these contrasting presentations—in most men and even also in other than ethical spheres. The judgment of these at times has not been very earnest. In Spitta we find the following relevant expression from A. Zeller (Article "Irre" in the *Allgemeinen Encyklopädie der Wissenchaften* of Ersch and Grüber, p. 144): "The mind is rarely so happily organised as to possess at all times power enough not to be disturbed, not only by unessential but also by perfectly ridiculous ideas running counter to the usual clear trend of thought; indeed, the greatest thinkers have had cause to complain of this dream-like disturbing and painful rabble of ideas, as it destroys their profoundest reflection and their most sacred and earnest mental work."

A clearer light is thrown on the psychological status of this idea of contrast by another observation of Hildebrandt, that the dream at times allows us to glance into the deep and inmost recesses of our being, which are generally closed to us in our waking state (p. 55). The same knowledge is revealed by Kant in his *Anthropology,* when he states that the dream exists in order to lay bare for us our hidden dispositions and to reveal to us not what we are, but what we might have been if we had a different education. Radestock (p. 84) says that the dream often only reveals to us what we do not wish to admit to ourselves, and that we therefore unjustly condemn it as a liar and deceiver. That the appearance of impulses which are foreign to our consciousness is merely analogous to the already familiar disposition which the dream makes of other material of the presentation, which is either absent or plays only an insignificant part in the waking state, has been called to our attention by observations like those of Benini, who says: "Certe nostre inclinazione che si credevano suffocate a spente da un pezzo, si ridestano; passioni vecchie e sepolte rivivono; cose e persone a cui non pensiamo mai, ci vengono dinanzi" (p. 149). Volkelt expresses himself in a similar way: "Even presentations which have entered into our consciousness almost unnoticed, and have never perhaps been brought out from oblivion, often announce through the dream their presence in the mind (p. 105). Finally, it is not out of place to mention here that, according to Schleiermacher, the state of falling asleep is accompanied by the appearance of undesirable presentations (pictures).

We may comprise under "undesirable presentations" this entire material of presentations, the occurrence of which excites our wonder in immoral as well as in absurd dreams. The only important difference consists in the fact that our undesirable presentations in the moral sphere exhibit an opposition to our other feelings, whereas the others simply appear strange to us. Nothing has been done so far to enable us to remove this difference through a more penetrating knowledge.

But what is the significance of the appearance of undesirable presentations in the dream? What inferences may be drawn for the psychology of the waking and dreaming mind from these nocturnal manifestations of contrasting ethical impulses? We may here note a new diversity of opinion, and once more a different grouping of the authors. The stream of thought followed by Hildebrandt, and by others who represent his fundamental view, cannot be continued in any other way than by ascribing to the immoral impulses a certain force even in the waking state, which, to be sure, is inhibited from advancing to action, and asserting that something falls off during sleep, which, having the effect of an inhibition, has kept us from noticing the existence of such an impulse. The dream thus shows the real, if not the entire nature of man, and is a means of making the hidden psychic life accessible to our understanding. It is only on such assumption that Hildebrandt can attribute to the dream the rôle of monitor who calls our attention to the moral ravages in the soul, just as in the opinion of physicians it can announce a hitherto unobserved physical ailment. Spitta, too, cannot be guided by any other conception when he refers to the stream of excitement which, *e.g.,* flows in upon the psyche during puberty, and consoles the dreamer by saying that he has done everything in his power when he has led a strictly virtuous life during his waking state, when he has made an effort to suppress the sinful thoughts as often as they arise, and has kept them from maturing and becoming actions. According to this conception, we might designate the "undesirable" presentations as those that are "suppressed" during the day, and must recognise in their appearance a real psychic phenomenon.

If we followed other authors we would have no right to the last inference. For Jessen the undesirable presentations in the dream as in the waking state, in fever and other deliria, merely have "the character of a voluntary activity put to rest and a somewhat mechanical process of pictures and presentations produced by inner impulses" (p. 360). An immoral dream proves nothing for the psychic life of the dreamer except that he has in some way become cognizant of the ideas in question; it is surely not a psychic impulse of his own. Another author, Maury, makes us question whether he, too, does not attribute to the dream state the capacity for dividing the psychic activity into its components instead of destroying it aimlessly. He speaks as follows about dreams in which one goes beyond the bounds of morality: "Ce sont nos penchants qui parlent et qui nous font agir, sans que la conscience nous retienne, bien que parfoit elle nous avertisse. J'ai mes défauts et mes penchants vicieux; à l'état de veille, je tache de lutter contre eux, et il m'arrive assez souvent de n'y pas succomber. Mais dans mes songes j'y succombe toujours ou pour mieux dire j'agis, par leur impulsion, sans crainte et sans remords. . . . . Evidement les visions qui se déroulent devant ma pensée et qui constituent le rêve, me sont suggérées par les incitations que je lessens et que ma volonté absente ne cherche pas à réfouler" (p. 113).

If one believes in the capacity of the dream to reveal an actually existing but repressed or concealed immoral disposition of the dreamer, he could not emphasize his opinion more strongly than with the words of Maury (p. 115): "En rêve l'homme se révèle donc tout entier à soi-même dans sa nudité et sa misère natives. Dès qu'il suspend l'exercice de sa volonté, il dévient le jouet de toutes les passions contre les-quelles, à l'état de veille, la conscience, le sentiment d'honneur, la crainte nous défendent." In another place he finds the following striking words (p. 462): "Dans le rêve, c'est surtout l'homme instinctif que se révèle. . . . . L'homme revient pour ainsi dire à l'état de nature quand il rêve; mais moins les idées acquises ont pénétre dans son esprit, plus les penchants en désaccord avec elles conservent encore sur lui d'influence dans le rêve." He then mentions as an example that his dreams often show him as a victim of just those superstitions which he most violently combats in his writing.

The value of all these ingenious observations for a psychological knowledge of the dream life, however, is marred by Maury through the fact that he refuses to recognise in the phenomena so correctly observed by him any proof of the "automatisme psychologique" which in his opinion dominates the dream life. He conceives this automatism as a perfect contrast to the psychic activity.

A passage in the studies on consciousness by Stricker reads: "The dream does not consist of delusions merely; if, *e.g.,* one is afraid of robbers in the dream, the robbers are, of course, imaginary, but the fear is real. One's attention is thus called to the fact that the effective development in the dream does not admit of the judgment which one bestows upon the rest of the dream content, and the problem arises what part of the psychic processes in the dream may be real, *i.e.* what part of them may demand to be enrolled among the psychic processes of the waking state?"

(G)   Dream Theories and Functions of the Dream

A statement concerning the dream which as far as possible attempts to explain from one point of view many of its noted characters, and which at the same time determines the relation of the dream to a more comprehensive sphere of manifestations, may be called a theory of dreams. Individual theories of the dream will be distinguished from one another through the fact that they raise to prominence this or that characteristic of the dream, and connect explanations and relations with it. It will not be absolutely necessary to derive from the theory a function, *i.e.* a use or any such activity of the dream, but our expectation, which is usually adjusted to teleology, will nevertheless welcome those theories which promise an understanding of the function of the dream.

We have already become acquainted with many conceptions of the dream which, more or less, merit the name of dream theories in this sense. The belief of the ancients that the dream was sent by the gods in order to guide the actions of man was a complete theory of the dream giving information concerning everything in the dream worth knowing. Since the dream has become an object of biological investigation we have a greater number of theories, of which, however, some are very incomplete.

If we waive completeness, we may attempt the following loose grouping of dream theories based on their fundamental conception of the degree and mode of the psychic activity in the dream:—

1. Theories, like those of Delboeuf, which allow the full psychic activity of the waking state to continue into the dream. Here the mind does not sleep; its apparatus remains intact, and, being placed under the conditions different from the waking state, it must in normal activity furnish results different from those of the waking state. In these theories it is a question whether they are in position to derive the distinctions between dreaming

and waking thought altogether from the determinations of the sleeping state. They moreover lack a possible access to a function of the dream; one cannot understand why one dreams, why the complicated mechanism of the psychic apparatus continues to play even when it is placed under conditions for which it is not apparently adapted. There remain only two expedient reactions—to sleep dreamlessly or to awake when approached by disturbing stimuli—instead of the third, that of dreaming.

2. Theories which, on the contrary, assume for the dream a diminution for the psychic activity, a loosening of the connections, and an impoverishment in available material. In accordance with these theories, one must assume for sleep a psychological character entirely different from the one given by Delboeuf. Sleep extends far beyond the mind—it does not consist merely in a shutting off of the mind from the outer world; on the contrary, it penetrates into its mechanism, causing it at times to become useless. If I may draw a comparison from psychiatrical material, I may say that the first theories construct the dream like a paranoia, while the second make it after the model of a dementia or an amentia.

The theory that only a fragment of the psychic activity paralysed by sleep comes to expression is by far the favourite among the medical writers and in the scientific world. As far as one may presuppose a more general interest in dream interpretation, it may well be designated as the ruling theory of the dream. It is to be emphasized with what facility this particular theory escapes the worst rock threatening every dream interpretation, that is to say, being shipwrecked upon one of the contrasts embodied in the dream. As this theory considers the dream the result of a partial waking (or as Herbart's *Psychology* of the dream says, "a gradual, partial, and at the same time very anomalous waking"), it succeeds in covering the entire series of inferior activities in the dream which reveal themselves in its absurdities, up to the full concentration of mental activity, by following a series of states which become more and more awake until they reach full awakening.

One who finds the psychological mode of expression indispensable, or who thinks more scientifically, will find this theory of the dream expressed in the discussion of Binz (p. 43):—

"This state [of numbness], however, gradually approaches its end in the early morning hours. The accumulated material of fatigue in the albumen of the brain gradually becomes less. It is gradually decomposed or carried away by the constantly flowing circulation. Here and there some masses of cells can be distinguished as awake, while all around everything still remains in a state of torpidity. *The isolated work of the individual groups* now appears before our clouded consciousness, which lacks the control of other parts of the brain governing the associations. Hence the pictures created, which mostly correspond to the objective impressions of the recent past, fit with each other in a wild and irregular manner. The number of the brain cells set free becomes constantly greater, the irrationality of the dream constantly less."

The conception of the dream as an incomplete, partial waking state, or traces of its influence, can surely be found among all modern physiologists and philosophers. It is most completely represented by Maury. It often seems as if this author represented to himself the state of being awake or asleep in anatomical regions; at any rate it appears to him that an anatomical province is connected with a definite psychic function. I may here merely mention that if the theory of partial waking could be confirmed, there would remain much to be accomplished in its elaboration.

Naturally a function of the dream cannot be found in this conception of the dream life. On the contrary, the criticism of the status and importance of the dream is consistently uttered in this statement of Binz (p. 357): "All the facts, as we see, urge us to characterise the dream as a physical process in all cases useless, in many cases even morbid."

The expression "physical" in reference to the dream, which owes its prominence to this author, points in more than one direction. In the first place, it refers to the etiology of the dream, which was especially clear to Binz, as he studied the experimental production of dreams by the administration of poisons. It is certainly in keeping with this kind of dream theory to ascribe the incitement of the dream exclusively to somatic origin whenever possible. Presented in the most extreme form, it reads as follows: After we have put ourselves to sleep by removing the stimuli, there would be no need and no occasion for dreaming until morning, when the gradual awakening through the incoming stimuli would be reflected in the phenomenon of dreaming. But as a matter of fact, it is not possible to keep sleep free from stimuli; just as Mephisto complains about the germs of life, so stimuli reach the sleeper from every side—from without, from within, and even from certain bodily regions which never give us any concern during the waking state. Thus sleep is disturbed; the mind is aroused, now by this, now by that little thing, and functionates for a while with the awakened part only to be glad to fall asleep again. The dream is a reaction to the stimulus causing a disturbance of sleep—to be sure, it is a purely superfluous reaction.

To designate the dream as a physical process, which for all that remains an activity of the mental organ, has still another sense. It is meant to dispute the dignity of a psychic process for the dream. The application to the dream of the very old comparison of the "ten fingers of a musically ignorant person running over the keyboard of an instrument," perhaps best illustrates in what estimation the dream activity has been held by the representatives of exact science. In this sense it becomes something entirely untranslatable, for how could the ten fingers of an unmusical player produce any music?

The theory of partial wakefulness has not passed without objection even in early times. Thus Burdach, in 1830, says: "If we say that the dream is a partial wakefulness, in the first place, we explain thereby neither the waking nor the sleeping state; secondly, this expresses nothing more than that certain forces of the mind are active in the dream while others are at rest. But such irregularities take place throughout life . . . " (p. 483).

Among extant dream theories which consider the dream a "physical" process, there is one very interesting conception of the dream, first propounded by Robert in 1866, which is attractive because it assigns to the dream a function or a useful end. As a basis for this theory, Robert takes from observation two facts which we have already discussed in our consideration of the dream material (see p. 21-22). These facts are: that one very often dreams about the insignificant impressions of the day, and that one rarely carries over into the dream the absorbing interests of the day. Robert asserts as exclusively correct, that things which have been fully settled never become dream inciters, but only such things as are incomplete in the mind or touch it fleetingly (p. 11). "We cannot usually explain our dreams because their causes are to be found *in sensory impressions of the preceding day which have not attained sufficient recognition by the dreamer.*" The conditions allowing an impression to reach the dream are therefore, either that this impression has been disturbed in its elaboration, or that being too insignificant it has no claim to such elaboration.

Robert therefore conceives the dream "as a physical process of elimination which has reached to cognition in the psychic manifestation of its reaction." *Dreams are eliminations of thoughts nipped in the bud.* "A man deprived of the capacity for dreaming would surely in time become mentally unbalanced, because an immense number of unfinished and unsolved thoughts and superficial impressions would accumulate in his brain, under the pressure of which there would be crushed all that should be incorporated as a finished whole into memory." The dream acts as a safety-valve for the overburdened brain. *Dreams possess healing and unburdening properties* (p. 32).

It would be a mistake to ask Robert how representation in the dream can bring about an unburdening of the mind. The author apparently concluded from those two peculiarities of the dream material that during sleep such ejection of worthless impressions is effected as a somatic process, and that dreaming is not a special psychic process but only the knowledge that we receive of such elimination. To be sure an elimination is not the only thing that takes place in the mind during sleep. Robert himself adds that the incitements of the day are also elaborated, and "what cannot be eliminated from the undigested thought material lying in the mind becomes connected *by threads of thought borrowed from the phantasy into a finished whole,* and thus enrolled in the memory as a harmless phantasy picture" (p. 23).

But it is in his criticism of the dream sources that Robert appears most bluntly opposed to the ruling theory. Whereas according to the existing theory there would be no dream if the outer and inner sensory stimuli did not repeatedly wake the mind, according to Robert the impulse to dream lies in the mind itself. It lies in the overcharging which demands discharge, and Robert judges with perfect consistency when he maintains that the causes determining the dream which depend on the physical state assume a subordinate rank, and could not incite dreams in a mind containing no material for dream formation taken from waking consciousness. It is admitted, however, that the phantasy pictures originating in the depths of the mind can be influenced by the nervous stimuli (p. 48). Thus, according to Robert, the dream is not quite so dependent on the somatic element. To be sure, it is not a psychic process, and has no place among the psychic processes of the waking state; it is a nocturnal somatic process in the apparatus devoted to mental activity, and has a function to perform, viz. to guard this apparatus against overstraining, or, if the comparison may be changed, to cleanse the mind.

Another author, Yves Delage, bases his theory on the same characteristics of the dream, which become clear in the selection of the dream material, and it is instructive to observe how a slight turn in the conception of the same things gives a final result of quite different bearing.

Delage, after having lost through death a person very dear to him, found from his own experience that we do not dream of what occupies us intently during the day, or that we begin to dream of it only after it is overshadowed by other interests of the day. His investigations among other persons corroborated the universality of this state of affairs. Delage makes a nice observation of this kind, if it turn out to be generally true, about the dreaming of newly married people: "S'ils ont été fortement épris, presque jamais ils n'ont rêve l'un de l'autre avant le mariage ou pendant la lune de miel; et s'ils ont rêve d'amour c'est pour être infidèles avec quelque personne indifferente ou odieuse." But what does one dream of? Delage recognises that the material occurring in our dreams consists of fragments and remnants of impressions from the days preceding and former times. All that appears in our dreams, what at first we may be inclined to consider creations of the dream life, proves on more thorough investigation to be unrecognised reproductions," souvenir inconscient." But this presentation material shows a common character; it originates from impressions which have probably affected our senses more forcibly than our mind, or from which the attention has been deflected soon after their appearance. The less conscious, and at the same time the stronger the impression, the more prospect it has of playing a part in the next dream.

These are essentially the same two categories of impressions, the insignificant and the unadjusted, which were emphasized by Robert, but Delage changes the connection by assuming that these impressions become the subject of dreams, not because they are indifferent, but because they are unadjusted. The insignificant impressions, too, are in a way not fully adjusted; they, too, are from then" nature as new impressions "autant de ressorts tendus," which will be relaxed during sleep. Still more entitled to a rôle in the dream than the weak and almost unnoticed impression is a strong impression which has been accidentally detained in its elaboration or intentionally repressed. The psychic energy accumulated during the day through inhibition or suppression becomes the main-spring of the dream at night.

Unfortunately Delage stops here in his train of thought; he can ascribe only the smallest part to an independent psychic activity in the dream, and thus in his dream theory reverts to the ruling doctrine of a partial sleep of the brain: "En somme le rêve est le produit de la pensée errante, sans but et sans direction, se fixant successivement sur les souvenirs, qui ont gardé assez d'intensité pour se placer sur sa route et l'arrêter au passage, établissant entre eux un lien tantôt faible et indécis, tantôt plus fort et plus serré, selon que l'activité actuelle du cerveau est plus ou moms abolie par le sommeil."

In a third group we may include those dream theories which ascribe to the dreaming mind the capacity and propensity for a special psychic activity, which in the waking state it can accomplish either not at all or only in an imperfect manner. From the activity of these capacities there usually results a useful function of the dream. The dignity bestowed upon the dream by older psychological authors falls chiefly in this category. I shall content myself, however, with quoting, in their place, the assertions of Burdach, by virtue of which the dream "is the natural activity of the mind, which is not limited by the force of the individuality, not disturbed by self-consciousness and not directed by self-determination, but is the state of life of the sensible central point indulging in free play" (p. 486).

Burdach and others apparently consider this revelling in the free use of one's own powers as a state in which the mind refreshes itself and takes on new strength for the day work, something after the manner of a vacation holiday. Burdach, therefore, cites with approval the admirable words in which the poet Novalis lauds the sway of the dream: "The dream is a bulwark against the regularity and commonness of life, a free recreation of the fettered phantasy, in which it mixes together all the pictures of life and interrupts the continued earnestness of grown-up men with a joyous children's play. Without the dream we should surely age earlier, and thus the dream may be considered perhaps not a gift directly from above, but a delightful task, a friendly companion, on our pilgrimage to the grave."

The refreshing and curative activity of the dream is even more impressively depicted by Purkinje. "The productive dreams in particular would perform these functions. They are easy plays of the imagination, which have no connection with the events of the day. The mind does not wish to continue the tension of the waking life, but to release it and recuperate from it. It produces, in the first place, conditions opposed to those of the waking state. It cures sadness through joy, worry through hope and cheerfully distracting pictures, hatred through love and friendliness, and fear through courage and confidence; it calms doubt through conviction and firm belief, and vain expectations through realisation. Many sore spots in the mind, which the day keeps continually open, sleep heals by covering them and guarding against fresh excitement. Upon this the curative effect of time is partially based." We all feel that sleep is beneficial to the psychic life, and the vague surmise of the popular consciousness apparently cannot be robbed of the notion that the dream is one of the ways in which sleep distributes its benefits.

The most original and most far-reaching attempt to explain the dream as a special activity of the mind, which can freely display itself only in the sleeping state, was the one undertaken by Scherner in 1861. Scherner's book, written in a heavy and bombastic style, inspired by an almost intoxicated enthusiasm for the subject, which must repel us unless it can carry us away with it, places so many difficulties in the way of an analysis that we gladly resort to the clearer and shorter description in which the philosopher Volkelt presents Scherner's theories: "From the mystic conglomerations and from all the gorgeous and magnificent billows there indeed flashes and irradiates an ominous light of sense, but the path of the philosopher does not thereby become clearer." Such is the criticism of Scherner's description from one of his own adherents.

Scherner does not belong to those authors who allow the mind to take along its undiminished capacities into the dream life. He indeed explains how in the dream the centrality and the spontaneous energy of the ego are enervated, how cognition, feeling, will, and imagination become changed through this decentralisation, and how no true mental character, but only the nature of a mechanism, belongs to the remnants of these psychic forces. But instead, the activity of the mind designated as phantasy, freed from all rational domination and hence completely uncontrolled, rises in the dream to absolute supremacy. To be sure, it takes the last building stones from the memory of the waking state, but it builds with them constructions as different from the structures of the waking state as day and night. It shows itself in the dream not only reproductive, but productive. Its peculiarities give to the dream life its strange character. It shows a preference for the unlimited, exaggerated, and prodigious, but because freed from the impeding thought categories, it gains a greater flexibility and agility and new pleasure; it is extremely sensitive to the delicate emotional stimuli of the mind and to the agitating affects, and it rapidly recasts the inner life into the outer plastic clearness. The dream phantasy lacks the language of ideas; what it wishes to say, it must clearly depict; and as the idea now acts strongly, it depicts it with the richness, force, and immensity of the mode in question. Its language, however simple it may be, thus becomes circumstantial, cumbersome, and heavy. Clearness of language is rendered especially difficult by the fact that it shows a dislike for expressing an object by its own picture, but prefers a strange picture, if the latter can only express that moment of the object which it wishes to describe. This is the symbolising activity of the phantasy. . . . It is, moreover, of great significance that the dream phantasy copies objects not in detail, but only in outline and even this in the broadest manner. Its paintings, therefore, appear ingeniously light and graceful. The dream phantasy, however, does not stop at the mere representation of the object, but is impelled from within to mingle with the object more or less of the dream ego, and in this way to produce an action. The visual dream, *e.g.,* depicts gold coins in the street; the dreamer picks them up, rejoices, and carries them away.

According to Scherner, the material upon which the dream phantasy exerts its artistic activity is preponderately that of the organic sensory stimuli which are so obscure during the day; hence the phantastic theory of Scherner, and the perhaps over-sober theories of Wundt and other physiologists, though otherwise diametrically opposed, agree perfectly in their assumption of the dream sources and dream excitants. But whereas, according to the physiological theory, the psychic reaction to the inner physical stimuli becomes exhausted with the awakening of any ideas suitable to these stimuli, these ideas then by way of association calling to their aid other ideas, and with this stage the chain of psychic processes seeming to terminate according to Scherner, the physical stimuli only supply the psychic force with a material which it may render subservient to its phantastic intentions. For Scherner the formation of the dream only commences where in the conception of others it comes to an end.

The treatment of the physical stimuli by the dream phantasy surely cannot be considered purposeful. The phantasy plays a tantalising game with them, and represents the organic source which gives origin to the stimuli in the correspondent dream, in any plastic symbolism. Indeed Scherner holds the opinion, not shared by Volkelt and others, that the dream phantasy has a certain favourite representation for the entire organism; this representation would be the *house.* Fortunately, however, it does not seem to limit itself in its presentation to this material; it may also conversely employ a whole series of houses to designate a single organ, *e.g.,* very long rows of houses for the intestinal excitation. On other occasions particular parts of the house actually represent particular parts of the body, as *e.g.,* in the headache-dream, the ceiling of the room (which the dream sees covered with disgusting reptile-like spiders) represents the head.

Quite irrespective of the house symbolism, any other suitable object may be employed for the representation of these parts of the body which excite the dream. "Thus the breathing lungs find their symbol in the flaming stove with its gaseous roaring, the heart in hollow boxes and baskets, the bladder in round, bag-shaped, or simply hollowed objects. The male dream of sexual excitement makes the dreamer find in the street the upper portion of a clarinette, next to it the same part of a tobacco pipe, and next to that a piece of fur. The clarinette and tobacco pipe represent the approximate shape of the male sexual organ, while the fur represents the pubic hair. In the female sexual dream the tightness of the closely approximated thighs may be symbolised by a narrow courtyard surrounded by houses, and the vagina by a very narrow, slippery and soft footpath, leading through the courtyard, upon which the dreamer is obliged to walk, in order perhaps to carry a letter to a gentleman" (Volkelt, p. 39). It is particularly noteworthy that at the end of such a physically exciting dream, the phantasy, as it were, unmasks by representing the exciting organ or its function unconcealed. Thus the "tooth-exciting dream" usually ends with the dreamer taking a tooth out of his mouth.

The dream phantasy may, however, not only direct its attention to the shape of the exciting organ, but it may also make the substance contained therein the object of the symbolisation. Thus the dream of intestinal excitement, *e.g.,* may lead us through muddy streets, the bladder-exciting dream to foaming water. Or the stimulus itself, the manner of its excitation, and the object it covets, are represented symbolically, or the dream ego enters into a concrete combination with the symbolisation of its own state, as *e.g.,* when, in the case of painful stimuli, we struggle desperately with vicious dogs or raging bulls, or when in the sexual dream the dreamer sees herself pursued by a naked man. Disregarding all the possible prolixity of elaboration, a symbolising phantastic activity remains as the central force of every dream. Volkelt, in his finely and fervently written book, next attempted to penetrate further into the character of this phantasy and to assign to the psychical activity thus recognised, its position in a system of philosophical ideas, which, however, remains altogether too difficult of comprehension for any one who is not prepared by previous schooling for the sympathetic comprehension of philosophical modes of thinking.

Scherner connects no useful function with the activity of the symbolising phantasy in dreams. In the dream the psyche plays with the stimuli at its disposal. One might presume that it plays in an improper manner. One might also ask us whether our thorough study of Scherner's dream theory, the arbitrariness and deviation of which from the rules of all investigation are only too obvious, can lead to any useful results. It would then be proper for us to forestall the rejection of Scherner's theory without examination by saying that this would be too arrogant. This theory is built up on the impression received from his dreams by a man who paid great attention to them, and who would appear to be personally very well fitted to trace obscure psychic occurrences. Furthermore it treats a subject which, for thousands of years, has appeared mysterious to humanity though rich in its contents and relations; and for the elucidation of which stern science, as it confesses itself, has contributed nothing beyond attempting, in entire opposition to popular sentiment, to deny the substance and significance of the object. Finally, let us frankly admit that apparently we cannot avoid the phantastical in our attempts to elucidate the dream. There are also phantastic ganglia cells; the passage cited on p. 73 from a sober and exact investigator like Binz, which depicts how the aurora of awakening flows along the dormant cell masses of the cerebrum, is not inferior in fancifulness and in improbability to Scherner's attempts at interpretation. I hope to be able to demonstrate that there is something actual underlying the latter, though it has only been indistinctly observed and does not possess the character of universality entitling it to the claim of a dream theory. For the present, Scherner's theory of the dream, in its contrast to the medical theory, may perhaps lead us to realise between what extremes the explanation of dream life is still unsteadily vacillating.

(H)   Relations between the Dream and Mental Diseases

When we speak of the relation of the dream to mental disturbances, we may think of three different things: (1) Etiological and clinical relations, as when a dream represents or initiates a psychotic condition, or when it leaves such a condition behind it. (2) Changes to which the dream life is subjected in mental diseases. (3) Inner relations between the dream and the psychoses, analogies indicating an intimate relationship. These manifold relations between the two series of phenomena have been a favourite theme of medical authors in the earlier periods of medical science—and again in recent times—as we learn from the literature on the subject gathered from Spitta, Radestock, Maury, and Tissié. Sante de Sanctis has lately directed his attention to this relationship. For the purposes of our discussion it will suffice merely to glance at this important subject.

In regard to the clinical and etiological relations between the dream and the psychoses, I will report the following observations as paradigms. Hohnbaum asserts, that the first attack of insanity frequently originates in an anxious and terrifying dream, and that the ruling idea has connection with this dream. Sante de Sanctis adduces similar observations in paranoiacs, and declares the dream to be, in some of them, the "vraie cause déterminante de la folie." The psychosis may come to life all of a sudden with the dream causing and containing the explanation for the mental disturbances, or it may slowly develop through further dreams that have yet to struggle against doubt. In one of de Sanctis's cases, the affecting dream was accompanied by light hysterical attacks, which in their turn were followed by an anxious, melancholic state. Féré (cited by Tissié) refers to a dream which caused an hysterical paralysis. Here the dream is offered us as an etiology of mental disturbance, though we equally consider the prevailing conditions when we declare that the mental disturbance shows its first manifestation in dream life, that it has its first outbreak in the dream. In other instances the dream life contained the morbid symptoms, or the psychosis was limited to the dream life. Thus Thomayer calls attention to anxiety dreams which must be conceived as equivalent to epileptic attacks. Allison has described nocturnal insanity (cited by Radestock), in which the subjects are apparently perfectly well in the day-time, while hallucinations, fits of frenzy, and the like regularly appear at night. De Sanctis and Tissié report similar observations (paranoiac dream-equivalent in an alcoholic, voices accusing a wife of infidelity). Tissié reports abundant observations from recent times in which actions of a pathological character (based on delusions, obsessive impulses) had their origin in dreams. Guislain describes a case in which sleep was replaced by an intermittent insanity.

There is hardly any doubt that along with the psychology of the dream, the physician will one day occupy himself with the psychopathology of the dream.

In cases of convalescence from insanity, it is often especially obvious that, while the functions of the day are normal, the dream life may still belong to the psychosis. Gregory is said first to have called attention to such cases (cited by Krauss). Macario (reported by Tissié) gives account of a maniac who, a week after his complete recovery again experienced in dreams the flight of ideas and the passionate impulses of his disease.

Concerning the changes to which the dream life is subjected in chronic psychotic persons, very few investigations have so far been made. On the other hand, timely attention has been called to the inner relationship between the dream and mental disturbance, which shows itself in an extensive agreement of the manifestations occurring to both. According to Maury, Cubanis, in his *Rapports du physique et du moral,* first called attention to this; following him came Lelut, J. Moreau, and more particularly the philosopher Maine de Biran. To be sure, the comparison is still older. Radestock begins the chapter dealing with this comparison, by giving a collection of expressions showing the analogy between the dream and insanity. Kant somewhere says: "The lunatic is a dreamer in the waking state." According to Krauss "Insanity is a dream with the senses awake." Schopenhauer terms the dream a short insanity, and insanity a long dream. Hagen describes the delirium as dream life which has not been caused by sleep but by disease. Wundt, in the *Physiological Psychology,* declares: "As a matter of fact we may in the dream ourselves live through almost all symptoms which we meet in the insane asylums."

The specific agreements, on the basis of which such an identification commends itself to the understanding, are enumerated by Spitta. And indeed, very similarly, by Maury in the following grouping: "(1) Suspension or at least retardation, of self-consciousness, consequent ignorance of the condition as such, and hence incapability of astonishment and lack of moral consciousness. (2) Modified perception of the sensory organs; that is, perception is diminished in the dream and generally enhanced in insanity. (3) Combination of ideas with each other exclusively in accordance with the laws of association and of reproduction, hence automatic formation of groups and for this reason disproportion in the relations between ideas (exaggerations, phantasms). And as a result of all this: (4) Changing or transformation of the personality and at times of the peculiarities of character (perversities)."

Radestock gives some additional features or analogies in the material: "Most hallucinations and illusions are found in the sphere of the senses of sight and hearing and general sensation. As in the dream, the smallest number of elements is supplied by the senses of smell and taste. The

77

fever patient, like the dreamer, is assaulted by reminiscences from the remote past; what the waking and healthy man seems to have forgotten is recollected in sleep and in disease." The analogy between the dream and the psychosis receives its full value only when, like a family resemblance, it is extended to the finer mimicry and to the individual peculiarities of facial expression.

"To him who is tortured by physical and mental sufferings the dream accords what has been denied him by reality, to wit, physical well-being and happiness; so the insane, too, see the bright pictures of happiness, greatness, sublimity, and riches. The supposed possession of estates and the imaginary fulfilment of wishes, the denial or destruction of which have just served as a psychic cause of the insanity, often form the main content of the delirium. The woman who has lost a dearly beloved child, in her delirium experiences maternal joys; the man who has suffered reverses of fortune deems himself immensely wealthy; and the jilted girl pictures herself in the bliss of tender love."

The above passage from Radestock, an abstract of a keen discussion of Griesinger (p. 111), reveals with the greatest clearness the wish fulfilment as a characteristic of the imagination, common to the dream and the psychosis. (My own investigations have taught me that here the key to a psychological theory of the dream and of the psychosis is to be found.)

"Absurd combinations of ideas and weakness of judgment are the main characteristics of the dream and of insanity." The over-estimation of one's own mental capacity, which appears absurd to sober judgment, is found alike both in one and the other, and the rapid course of ideas in the dream corresponds to the flight of ideas in the psychosis. Both are devoid of any measure of time. The dissociation of personality in the dream, which, for instance, distributes one's own knowledge between two persons, one of whom, the strange one, corrects in the dream one's own ego, fully corresponds to the well-known splitting of personality in hallucinatory paranoia; the dreamer, too, hears his own thoughts expressed by strange voices. Even the constant delusions find their analogy in the stereotyped recurring pathological dreams (*rêve obsédant*). After recovering from a delirium, patients not infrequently declare that the disease appeared to them like an uncomfortable dream; indeed, they inform us that occasionally, even during the course of their sickness, they have felt that they were only dreaming, just as it frequently happens in the sleeping dream.

Considering all this, it is not surprising that Radestock condenses his own opinion and that of many others into the following: "Insanity, an abnormal phenomenon of disease, is to be regarded as an enhancement of the periodically recurring normal dream states" (p. 228).

Krauss attempted to base the relationship between the dream and insanity upon the etiology (or rather upon the exciting sources), perhaps making the relationship even more intimate than was possible through the analogy of the phenomena they manifest. According to him, the fundamental element common to both is, as we have learned, the organically determined sensation, the sensation of physical stimuli, the general feeling produced by contributions from all the organs. *Cf.* Peise, cited by Maury (p. 60).

The incontestable agreement between the dream and mental disturbance, extending into characteristic details, constitutes one of the strongest supports of the medical theory of dream life, according to which the dream is represented as a useless and disturbing process and as the expression of a reduced psychic activity. One cannot expect, however, to derive the final explanation of the dream from the mental disturbances, as it is generally known in what unsatisfactory state our understanding of the origin of the latter remains. It is very probably, however, that a modified conception of the dream must also influence our views in regard to the inner mechanism of mental disturbances, and hence we may say that we are engaged in the elucidation of the psychosis when we endeavour to clear up the mystery of the dream.

I shall have to justify myself for not extending my summary of the literature of the dream problems over the period between the first appearance of this book and its second edition. If this justification may not seem very satisfactory to the reader, I was nevertheless influenced by it. The motives which mainly induced me to summarise the treatment of the dream in the literature have been exhausted with the foregoing introduction; to have continued with this work would have cost me extraordinary effort and would have afforded little advantage or knowledge. For the period of nine years referred to has yielded nothing new or valuable either for the conception of the dream in actual material or in points of view. In most of the publications that have since appeared my work has remained unmentioned and unregarded; naturally least attention has been bestowed upon it by the so-called "investigators of dreams," who have thus afforded a splendid example of the aversion characteristic of scientific men to learning something new. "Les savants ne sont pas curieux," said the scoffer Anatole France. If there were such a thing in science as right to revenge, I in turn should be justified in ignoring the literature since the appearance of this book. The few accounts that have appeared in scientific journals are so full of folly and misconception that my only possible answer to my critics would be to request them to read this book over again. Perhaps also the request should be that they read it as a whole.

In the works of those physicians who make use of the psychoanalytic method of treatment (Jung, Abraham, Riklin, Muthmann, Stekel, Rank, and others), an abundance of dreams have been reported and interpreted in accordance with my instructions. In so far as these works go beyond the confirmation of my assertions I have noted their results in the context of my discussion. A supplement to the bibliography at the end of this book brings together the most important of these new publications. The voluminous book on the dream by Sante de Sanctis, of which a German translation appeared soon after its publication, has, so to speak, crossed with mine, so that I could take as little notice of him as the Italian author could of me. Unfortunately, I am further obliged to declare that this laborious work is exceedingly poor in ideas, so poor that one could never divine from it the existence of the problems treated by me.

I have finally to mention two publications which show a near relation to my treatment of the dream problems. A younger philosopher, H. Swoboda, who has undertaken to extend W. Fliesse's discovery of biological periodicity (in groups of twenty-three and twenty-eight days) to the psychic field, has produced an imaginative work,[13] in which, among other things, he has used this key to solve the riddle of the dream. The interpretation of dreams would herein have fared badly; the material contained in dreams would be explained through the coincidence of all those memories which during the night complete one of the biological periods for the first or the n-th time. A personal statement from the author led me to assume that he himself no longer wished to advocate this theory earnestly. But it seems I was mistaken in this conclusion; I shall report in another place some observations in reference to Swoboda's assertion, concerning the conclusions of which I am, however, not convinced. It gave me far greater pleasure to find accidentally, in an unexpected place, a conception of the dream in essentials fully agreeing with my own. The circumstances of time preclude the possibility that this conception was influenced by a reading of my book; I must therefore greet this as the only demonstrable concurrence in the literature with the essence of my dream theory. The book which contains the passage concerning the dream which I have in mind was published as a second edition in 1900 by Lynkus under the title *Phantasien eines Realisten.*

Notes:

1. To the first publication of this book, 1900.

2. Compare, on the other hand, O. Gruppe, *Griechische Mythologie und Religionsgeschichte,* p. 390. "Dreams were divided into two classes; the first were influenced only by the present (or past), and were unimportant for the future: they embraced the [Greek], insomnia, which immediately produces the given idea or its opposite, *e.g.* hunger or its satiation, and the [Greek], which elaborates the given idea phantastically, as *e.g.* the nightmare, ephialtes. The second class was, on the other hand, determinant for the future. To this belong: (1) direct prophecies received in the dream ([Greek], oraculum); (2) the foretelling of a future event ([Greek]); (3) the symbolic or the dream requiring interpretation ([Greek], somnium). This theory has been preserved for many centuries."

3. From subsequent experience I am able to state that it is not at all rare to find in dreams repetitions of harmless or unimportant occupations of the waking state, such as packing trunks, preparing food, work in the kitchen, &c., but in such dreams the dreamer himself emphasizes not the character but the reality of the memory, "I have really done all this in the day time."

4. *Chauffeurs* were bands of robbers in the Vendée who resorted to this form of torture.

5. Gigantic persons in a dream justify the assumption that it deals with a scene from the dreamer's childhood.

6. The first volume of this Norwegian author, containing a complete description of dreams, has recently appeared in German. See bibliography.

7. Periodically recurrent dreams have been observed repeatedly. *Cf.* the collection of Chabaneix.

8. Silberer has shown by nice examples how in the state of sleepiness even abstract thoughts may be changed into illustrative plastic pictures which express the same thing (*Jahrbuch* von Bleuler-Freud, vol. i. 1900).

9. Haffner made an attempt similar to Delboeuf's to explain the dream activity on the basis of an alteration which must result in an introduction of an abnormal condition in the otherwise correct function of the intact psychic apparatus, but he described this condition in somewhat different words. He states that the first distinguishing mark of the dream is the absence of time and space, *i.e.* the emancipation of the presentation from the position in the order of time and space which is common to the individual. Allied to this is the second fundamental character of the dream, the mistaking of the hallucinations, imaginations, and phantasy-combinations for objective perceptions. The sum total of the higher psychic forces, especially formation of ideas, judgment, and argumentation on the one hand, and the free self-determination on the other hand, connect themselves with the sensory phantasy pictures and at all times have them as a substratum. These activities too, therefore, participate in the irregularity of the dream presentation. We say they participate, for our faculties of judgment and will power are in themselves in no way altered daring sleep. In reference to activity, we are just as keen and just as free as in the waking state. A man cannot act contrary to the laws of thought, even in the dream, *i.e.* he is unable to harmonise with that which represents itself as contrary to him, &c.; he can only desire in the dream that which he presents to himself as good (*sub ratione boni*). But in this application of the laws of thinking and willing the human mind is led astray in the dream through mistaking one presentation for another. It thus happens that we form and commit in the dream the greatest contradictions, while, on the other hand, we display the keenest judgments and the most consequential chains of reasoning, and can make the most virtuous and sacred resolutions. Lack of orientation is the whole secret of the flight by which our phantasy moves in the dream, and lack of critical reflection and mutual understanding with others is the main source of the reckless extravagances of our judgments, hopes, and wishes in the dream" (p. 18).
10. *Cf.* Haffner and Spitta.
11. *Grundzüge des Systems der Anthropologie.* Erlangen, 1850 (quoted by Spitta).
12. *Das Traumleben und seine Deutung,* 1868 (cited by Spitta, p. 192).
13. H. Swoboda, *Die Perioden des Menschlichen Organismus,* 1904.

2

Method of Dream Interpretation:
The Analysis of a Sample Dream

The title which I have given my treatise indicates the tradition which I wish to make the starting-point in my discussion of dreams. I have made it my task to show that dreams are capable of interpretation, and contributions to the solution of the dream problems that have just been treated can only be yielded as possible by-products of the settlement of my own particular problem. With the hypothesis that dreams are interpretable, I at once come into contradiction with the prevailing dream science, in fact with all dream theories except that of Scherner, for to "interpret a dream" means to declare its meaning, to replace it by something which takes its place in the concatenation of our psychic activities as a link of full importance and value. But, as we have learnt, the scientific theories of the dream leave no room for a problem of dream interpretation, for, in the first place, according to these, the dream is no psychic action, but a somatic process which makes itself known to the psychic apparatus by means of signs. The opinion of the masses has always been quite different. It asserts its privilege of proceeding illogically, and although it admits the dream to be incomprehensible and absurd, it cannot summon the resolution to deny the dream all significance. Led by a dim intuition, it seems rather to assume that the dream has a meaning, albeit a hidden one; that it is intended as a substitute for some other thought process, and that it is only a question of revealing this substitute correctly in order to reach the hidden signification of the dream.

The laity has, therefore, always endeavoured to "interpret" the dream, and in doing so has tried two essentially different methods. The first of these procedures regards the dream content as a whole and seeks to replace it by another content which is intelligible and in certain respects analogous. This is symbolic dream interpretation; it naturally goes to pieces at the outset in the case of those dreams which appear not only unintelligible but confused. The construction which the biblical Joseph places upon the dream of Pharaoh furnishes an example of its procedure. The seven fat kine, after which came seven lean ones which devour the former, furnish a symbolic substitute for a prediction of seven years of famine in the land of Egypt, which will consume all the excess which seven fruitful years have created. Most of the artificial dreams contrived by poets are intended for such symbolic interpretation, for they reproduce the thought conceived by the poet in a disguise found to be in accordance with the characteristics of our dreaming, as we know these from experience.[1] The idea that the dream concerns itself chiefly with future events whose course it surmises in advance—a relic of the prophetic significance with which dreams were once credited—now becomes the motive for transplanting the meaning of the dream, found by means of symbolic interpretation, into the future by means of an "it shall."

A demonstration of the way in which such symbolic interpretation is arrived at cannot, of course, be given. Success remains a matter of ingenious conjecture, of direct intuition, and for this reason dream interpretation has naturally been elevated to an art, which seems to depend upon extraordinary gifts.[2] The other of the two popular methods of dream interpretation entirely abandons such claims. It might be designated as the "cipher method," since it treats the dream as a kind of secret code, in which every sign is translated into another sign of known meaning, according to an established key. For example, I have dreamt of a letter, and also of a funeral or the like; I consult a "dream book," and find that "letter" is to be translated by "vexation," and "funeral" by "marriage, engagement." It now remains to establish a connection, which I again am to assume pertains to the future, by means of the rigmarole which I have deciphered.

An interesting variation of this cipher procedure, a variation by which its character of purely mechanical transference is to a certain extent corrected, is presented in the work on dream interpretation by Artemidoros of Daldis. Here not only the dream content, but also the personality and station in life of the dreamer, are taken into consideration, so that the same dream content has a significance for the rich man, the married man, or the orator, which is different from that for the poor man, the unmarried man, or, say, the merchant. The essential point, then, in this procedure is that the work of interpretation is not directed to the entirety of the dream, but to each portion of the dream content by itself, as though the dream were a conglomeration, in which each fragment demands a particular disposal. Incoherent and confused dreams are certainly the ones responsible for the invention of the cipher method.[3]

The worthlessness of both these popular interpretation procedures for the scientific treatment of the subject cannot be questioned for a moment. The symbolic method is limited in its application and is capable of no general demonstration. In the cipher method everything depends upon whether the key, the dream book, is reliable, and for that all guarantees are lacking. One might be tempted to grant the contention of the philosophers and psychiatrists and to dismiss the problem of dream interpretation as a fanciful one.

I have come, however, to think differently. I have been forced to admit that here once more we have one of those not infrequent cases where an ancient and stubbornly retained popular belief seems to have come nearer to the truth of the matter than the judgment of the science which prevails to-day. I must insist that the dream actually has significance, and that a scientific procedure in dream interpretation is possible. I have come upon the knowledge of this procedure in the following manner:—

For several years I have been occupied with the solution of certain psychopathological structures in hysterical phobias, compulsive ideas, and the like, for therapeutic purposes. I have been so occupied since becoming familiar with an important report of Joseph Breuer to the effect that in those structures, regarded as morbid symptoms, solution and treatment go hand in hand.[4] Where it has been possible to trace such a pathological idea back to the elements in the psychic life of the patient to which it owes its origin, this idea has crumbled away, and the patient has been relieved of it. In view of the failure of our other therapeutic efforts, and in the face of the mysteriousness of these conditions, it seems to me tempting, in spite of all difficulties, to press forward on the path taken by Breuer until the subject has been fully understood. We shall have elsewhere to make a detailed report upon the form which the technique of this procedure has finally assumed, and the results of the efforts which have been made. In the course of these psychoanalytical studies, I happened upon dream interpretation. My patients, after I had obliged them to inform me of all the ideas and thoughts which came to them in connection with the given theme, related their dreams, and thus taught me that a dream may be linked into the psychic concatenation which must be followed backwards into the memory from the pathological idea as a starting-point. The next step was to treat the dream as a symptom, and to apply to it the method of interpretation which had been worked out for such symptoms.

For this a certain psychic preparation of the patient is necessary. The double effort is made with him, to stimulate his attention for his psychic perceptions and to eliminate the critique with which he is ordinarily in the habit of viewing the thoughts which come to the surface in him. For the purpose of self-observation with concentrated attention, it is advantageous that the patient occupy a restful position and close his eyes; he must be explicitly commanded to resign the critique of the thought-formations which he perceives. He must be told further that the success of the psychoanalysis depends upon his noticing and telling everything that passes through his mind, and that he must not allow himself to suppress one idea because it seems to him unimportant or irrelevant to the subject, or another because it seems nonsensical. He must maintain impartiality towards his ideas; for it would be owing to just this critique if he were unsuccessful in finding the desired solution of the dream, the obsession, or the like.

I have noticed in the course of my psychoanalytic work that the state of mind of a man in contemplation is entirely different from that of a man who is observing his psychic processes. In contemplation there is a greater play of psychic action than in the most attentive self-observation; this is also shown by the tense attitude and wrinkled brow of contemplation, in contrast with the restful features of self-observation. In both cases, there must be concentration of attention, but, besides this, in contemplation one exercises a critique, in consequence of which he rejects some of the ideas which he has perceived, and cuts short others, so that he does not follow the trains of thought which they would open; toward still other thoughts he may act in such a manner that they do not become conscious at all—that is to say, they are suppressed before they are perceived. In self-observation, on the other hand, one has only the task of suppressing the critique; if he succeeds in this, an unlimited number of ideas, which otherwise would have been impossible for him to grasp, come to his consciousness. With the aid of this material, newly secured for the purpose of self-observation, the interpretation of pathological ideas, as well as of dream images, can be accomplished. As may be seen, the point is to bring about a psychic state to some extent analogous as regards the apportionment of psychic energy (transferable attention) to the state prior to falling asleep (and indeed also to the hypnotic state). In falling asleep, the "undesired ideas" come into prominence on account of the slackening of a certain arbitrary (and certainly also critical) action, which we allow to exert an influence upon the trend of our ideas; we are accustomed to assign "fatigue" as the reason for this slackening; the emerging undesired ideas as the reason are changed into visual and acoustic images. (*Cf.* the remarks of Schleiermacher) and others, p. 69.) In the condition which is used for the analysis of dreams and pathological ideas, this activity is purposely and arbitrarily dispensed with, and the psychic energy thus saved, or a part of it, is used for the attentive following of the undesired thoughts now coming to the surface, which retain their identity as ideas (this is the difference from the condition of falling asleep). "Undesired ideas" are thus changed into "desired" ones.

The suspension thus required of the critique for these apparently "freely rising" ideas, which is here demanded and which is usually exercised on them, is not easy for some persons. The "undesired ideas" are in the habit of starting the most violent resistance, which seeks to prevent them from coming to the surface. But if we may credit our great poet-philosopher Friedrich Schiller, a very similar tolerance must be the condition of poetic production. At a point in his correspondence with Koerner, for the noting of which we are indebted to Mr. Otto Rank, Schiller answers a friend who complains of his lack of creativeness in the following words:

"The reason for your complaint lies, it seems to me, in the constraint which your intelligence imposes upon your imagination. I must here make an observation and illustrate it by an allegory. It does not seem beneficial, and it is harmful for the creative work of the mind, if the intelligence inspects too closely the ideas already pouring in, as it were, at the gates. Regarded by itself, an idea may be very trifling and very adventurous, but it perhaps becomes important on account of one which follows it; perhaps in a certain connection with others, which may seem equally absurd, it is capable of forming a very useful construction. The intelligence cannot judge all these things if it does not hold them steadily long enough to see them in connection with the others. In the case of a creative mind, however, the intelligence has withdrawn its watchers from the gates, the ideas rush in pell-mell, and it is only then that the great heap is looked over and critically examined. Messrs. Critics, or whatever else you may call yourselves, you are ashamed or afraid of the momentary and transitory madness which is found in all creators, and whose longer or shorter duration distinguishes the thinking artist from the dreamer. Hence your complaints about barrenness, for you reject too soon and discriminate too severely" (Letter of December 1, 1788).

And yet, "such a withdrawal of the watchers from the gates of intelligence," as Schiller calls it, such a shifting into the condition of uncritical self-observation, is in no way difficult.

Most of my patients accomplish it after the first instructions; I myself can do it very perfectly, if I assist the operation by writing down my notions. The amount, in terms of psychic energy, by which the critical activity is in this manner reduced, and by which the intensity of the self-observation may be increased, varies widely according to the subject matter upon which the attention is to be fixed.

The first step in the application of this procedure now teaches us that not the dream as a whole, but only the parts of its contents separately, may be made the object of our attention. If I ask a patient who is as yet unpractised: "What occurs to you in connection with this dream?" as a rule he is unable to fix upon anything in his psychic field of vision. I must present the dream to him piece by piece, then for every fragment he gives me a series of notions, which may be designated as the "background thoughts" of this part of the dream. In this first and important condition, then, the method of dream interpretation which I employ avoids the popular, traditional method of interpretation by symbolism famous in the legends, and approaches the second, the "cipher method." Like this one it is an interpretation in detail, not *en masse;* like this it treats the dream from the beginning as something put together—as a conglomeration of psychic images.

In the course of my psychoanalysis of neurotics, I have indeed already subjected many thousand dreams to interpretation, but I do not now wish to use this material in the introduction to the technique and theory of dream interpretation. Quite apart from the consideration that I should expose myself to the objection that these are dreams of neuropathic subjects, the conclusions drawn from which would not admit of reapplication to the dreams of healthy persons, another reason forces me to reject them. The theme which is naturally always the subject of these dreams, is the history of the disease which is responsible for the neurosis. For this purpose there would be required a very long introduction and an investigation into the nature and logical conditions of psychoneuroses, things which are in themselves novel and unfamiliar in the highest degree, and which would thus distract attention from the dream problem. My purpose lies much more in the direction of preparing the ground for a solution of difficult problems in the psychology of the neuroses by means of the solution of dreams. But if I eliminate the dreams of neurotics, I must not treat the remainder too discriminatingly. Only those dreams still remain which have been occasionally related to me by healthy persons of my acquaintance, or which I find as examples in the literature of dream life. Unfortunately in all these dreams the analysis is lacking, without which I cannot find the meaning of the dream. My procedure is, of course, not as easy as that of the popular cipher method, which translates the given dream content according to an established key; I am much more prepared to find that the same dream may cover a different meaning in the case of different persons, and in a different connection I must then resort to my own dreams, as an abundant and convenient material, furnished by a person who is about normal, and having reference to many incidents of everyday life. I shall certainly be with doubts as to the trustworthiness of these "self-analyses." Arbitrariness is here in no way avoided. In my opinion, conditions are more likely to be favourable in self-observation than in the observation of others; in any case, it is permissible to see how much can be accomplished by means of self-analysis. I must overcome further difficulties arising from inner self. One has a readily understood aversion to exposing so many intimate things from one's own psychic life, and one does not feel safe from the misinterpretation of strangers. But one must be able to put one's self beyond this. "Toute psychologiste," writes Delboeuf, "est obligé de faire l'aveu memo de ses faiblesses s'il croit par là jeter du jour sur quelque problème obscure." And I may assume that in the case of the reader, the immediate interest in the indiscretions which I must commit will very soon give way to exclusive engrossment in the psychological problems which are illuminated by them.

I shall, therefore, select one of my own dreams and use it to elucidate my method of interpretation. Every such dream necessitates a preliminary statement. I must now beg the reader to make my interests his own for a considerable time, and to become absorbed with me in the most trifling details of my life, for an interest in the hidden significance of dreams imperatively demands such transference.

Preliminary statement: In the summer of 1895 I had psychoanalytically treated a young lady who stood in close friendship to me and those near to me. It is to be understood that such a complication of relations may be the source of manifold feelings for the physician, especially for the psychotherapist. The personal interest of the physician is greater, his authority is less. A failure threatens to undermine the friendship with the relatives of the patient. The cure ended with partial success, the patient got rid of her hysterical fear, but not of all her somatic symptoms. I was at that time not yet sure of the criteria marking the final settlement of a hysterical case, and expected her to accept a solution which did not seem acceptable to her. In this disagreement, we cut short the treatment on account of the summer season. One day a younger colleague, one of my best friends, who had visited the patient—Irma—and her family in their country resort, came to see me. I asked him how he found her, and received the answer: "She is better, but not altogether well." I realise that those words of my friend Otto, or the tone of voice in which they were spoken, made me angry. I thought I heard a reproach in the words, perhaps to the effect that I had promised the patient too much, and rightly or wrongly I traced Otto's supposed siding against me to the influence of the relatives of the patient, who, I assume, had never approved of my treatment. Moreover, my disagreeable impression did not become clear to me, nor did I give it expression. The very same evening, I wrote down the history of Irma's case, in order to hand it, as though for my justification, to Dr. M., a mutual friend, who was at that time a leading figure in our circle. During the night following this evening (perhaps rather in the morning) I had the following dream, which was registered immediately after waking:—

Dream of July 23–24, 1895

A great hall—many guests whom we are receiving—among them Irma, whom I immediately take aside, as though to answer her letter, to reproach her for not yet accepting the "solution." I say to her: "If you still have pains, it is really only your own fault." She answers: "If you only knew what pains I now have in the neck, stomach, and abdomen; I am drawn together." I am frightened and look at her. She looks pale and bloated; I think that after all I must be overlooking some organic affection. I take her to the window and look into her throat. She shows some resistance to this, like a woman who has a false set of teeth. I think anyway she does not need them. The mouth then really opens without difficulty and I find a large white spot to the right, and at another place I see extended grayish-white scabs attached to curious curling formations, which have obviously been formed like the turbinated bone—I quickly call Dr. M., who repeats the examination and confirms it. . . . Dr. M.'s looks are altogether unusual; he is very pale, limps, and has no beard on his chin. . . . My friend Otto is now also standing next to her, and my friend Leopold percusses her small body and says: "She has some dulness on the left below," and also calls attention to an infiltrated portion of the skin on the left shoulder (something which I feel as he does, in spite of the dress). . . . M. says: "No doubt it is an infection, but it does not matter; dysentery will develop too, and the poison will be excreted. . . . We also have immediate knowledge of the origin of the infection. My friend Otto has recently given her an injection with a propyl preparation when she felt ill, propyls. . . . Propionic acid . . . Trimethylamine (the formula of which I see printed before me in heavy type). . . . Such injections are not made so rashly. . . . Probably also the syringe was not clean.

This dream has an advantage over many others. It is at once clear with what events of the preceding day it is connected, and what subject it treats. The preliminary statement gives information on these points. The news about Irma's health which I have received from Otto, the history of the illness upon which I have written until late at night, have occupied my psychic activity even during sleep. In spite of all this, no one, who has read the preliminary report and has knowledge of the content of the dream, has been able to guess what the dream signifies. Nor do I myself know. I wonder about the morbid symptoms, of which Irma complains in the dream, for they are not the same ones for which I have treated her. I smile about the consultation with Dr. M. I smile at the nonsensical idea of an injection with propionic acid, and at the consolation attempted by Dr. M. Towards the end the dream seems more obscure and more terse than at the beginning. In order to learn the significance of all this, I am compelled to undertake a thorough analysis.

Analysis

The hall—many guests, whom we are receiving.

We were living this summer at the Bellevue, in an isolated house on one of the hills which lie close to the Kahlenberg. This house was once intended as a place of amusement, and on this account has unusually high, hall-like rooms. The dream also occurred at the Bellevue, a few days before the birthday of my wife. During the day, my wife had expressed the expectation that several friends, among them Irma, would come to us as guests for her birthday. My dream, then, anticipates this situation: It is the birthday of my wife, and many people, among them Irma, are received by us as guests in the great hall of the Bellevue.

I reproach Irma for not having accepted the solution. I say: "If you still have pains, it is your own fault."

I might have said this also, or did say it, while awake. At that time I had the opinion (recognised later to be incorrect) that my task was limited to informing patients of the hidden meaning of their symptoms. Whether they then accepted or did not accept the solution upon which success depended—for that I was not responsible. I am thankful to this error, which fortunately has now been overcome, for making life easier for me" at a time when, with all my unavoidable ignorance, I was to produce successful cures. But I see in the speech which I make to Irma in the dream, that above all things I do not want to be to blame for the pains which she still feels. If it is Irma's own fault, it cannot be mine. Should the purpose of the dream be looked for in this quarter?

Irma'a complaints; pains in the neck, abdomen, and stomach; she is drawn together.

Pains in the stomach belonged to the symptom-complex of my patient, but they were not very prominent; she complained rather of sensations of nausea and disgust. Pains in the neck and abdomen and constriction of the throat hardly played a part in her case. I wonder why I decided upon this choice of symptoms, nor can I for the moment find the reason.

She looks pale and bloated.

My patient was always ruddy. I suspect that another person is here being substituted for her.

I am frightened at the thought that I must have overlooked some organic affection.

This, as the reader will readily believe, is a constant fear with the specialist, who sees neurotics almost exclusively, and who is accustomed to ascribe so many manifestations, which other physicians treat as organic, to hysteria. On the other hand, I am haunted by a faint doubt—I know not whence it comes—as to whether my fear is altogether honest. If Irma's pains are indeed of organic origin, I am not bound to cure them. My treatment, of course, removes only hysterical pains. It seems to me, in fact, that I wish to find an error in the diagnosis; in that case the reproach of being unsuccessful would be removed.

I take her to the window in order to look into her throat. She resists a little, like a woman who has false teeth. I think she does not need them anyway.

I had never had occasion to inspect Irma's aural cavity. The incident in the dream reminds me of an examination, made some time before, of a governess who at first gave an impression of youthful beauty, but who upon opening her mouth took certain measures for concealing her teeth. Other memories of medical examinations and of little secrets which are discovered by them, unpleasantly for both examiner and examined, connect themselves with this case. "She does not need them anyway," is at first perhaps a compliment for Irma; but I suspect a different meaning. In careful analysis one feels whether or not the "background thoughts" which are to be expected have been exhausted. The way in which Irma stands at the window suddenly reminds me of another experience. Irma possesses an intimate woman friend, of whom I think very highly. One evening on paying her a visit I found her in the position at the window reproduced in the dream, and her physician, the same Dr. M., declared that she had a diphtheritic membrane. The person of Dr. M. and the membrane return in the course of the dream. Now it occurs to me that during the last few months, I have been given every reason to suppose that this lady is also hysterical. Yes, Irma herself has betrayed this to me. But what do I know about her condition? Only the one thing, that like Irma she suffers from hysterical choking in dreams. Thus in the dream I have replaced my patient by her friend. Now I remember that I have often trifled with the expectation that this lady might likewise engage me to relieve her of her symptoms. But even at the time I thought it improbable, for she is of a very shy nature.

She resists, as the dream shows. Another explanation might be that *she does not need it;* in fact, until now she has shown herself strong enough to master her condition without outside help. Now only a few features remain, which I can assign neither to Irma nor to her friend: *Pale, bloated, false teeth.* The false teeth lead me to the governess; I now feel inclined to be satisfied with bad teeth. Then another person, to whom these features may allude, occurs to me. She is not my patient, and I do not wish her to be my patient, for I have noticed that she is not at her ease with me, and I do not consider her a docile patient. She is generally pale, and once, when she had a particularly good spell, she was bloated.[5] I have thus compared my patient Irma with two others, who would likewise resist treatment. What can it mean that I have exchanged her for her friend in the dream? Perhaps that I wish to exchange her; either the other one arouses in me stronger sympathies or I have a higher opinion of her intelligence. For I consider Irma foolish because she does not accept my solution. The other one would be more sensible, and would thus be more likely to yield. *The mouth then really opens without difficulty;* she would tell more than Irma.[6]

What I see in the throat; a white spot and scabby nostrils.

The white spot recalls diphtheria, and thus Irma's friend, but besides this it recalls the grave illness of my eldest daughter two years before and all the anxiety of that unfortunate time. The scab on the nostrils reminds me of a concern about my own health. At that time I often used cocaine in order to suppress annoying swellings in the nose, and had heard a few days before that a lady patient who did likewise had contracted an extensive necrosis of the nasal mucous membrane. The recommendation of cocaine, which I had made in 1885, had also brought grave reproaches upon me. A dear friend, already dead in 1895, had hastened his end through the misuse of this remedy.

I quickly call Dr. M., who repeats the examination.

This would simply correspond to the position which M. occupied among us. But the word "quickly" is striking enough to demand a special explanation. It reminds me of a sad medical experience. By the continued prescription of a remedy (sulfonal) which was still at that time considered harmless, I had once caused the severe intoxication of a woman patient, and I had turned in great haste to an older, more experienced colleague for assistance. The fact that I really had this case in mind is confirmed by an accessory circumstance. The patient, who succumbed to the intoxication, bore the same name as my eldest daughter. I had never thought of this until now; now it seems to me almost like a retribution of fate—as though I ought to continue the replacement of the persons here in another sense; this Matilda for that Matilda; an eye for an eye, a tooth for a tooth. It is as though I were seeking every opportunity to reproach myself with lack of medical conscientiousness.

Dr. M. is pale, without a beard on his chin, and he limps.

Of this so much is correct, that his unhealthy appearance often awakens the concern of his friends. The other two characteristics must belong to another person. A brother living abroad occurs to me, who wears his chin clean-shaven, and to whom, if I remember aright, M. of the dream on the whole bears some resemblance. About him the news arrived some days before that he was lame on account of an arthritic disease in the hip. There must be a reason why I fuse the two persons into one in the dream. I remember that in fact I was on bad terms with both of them for similar reasons. Both of them had rejected a certain proposal which I had recently made to them.

My friend Otto is now standing next to the sick woman, and my friend Leopold examines her and calls attention to a dulness on the left below.

My friend Leopold is also a physician, a relative of Otto. Since the two practise the same specialty, fate has made them competitors, who are continually being compared with each other. Both of them assisted me for years, while I was still directing a public dispensary for nervous children. Scenes like the one reproduced in the dream have often taken place there. While I was debating with Otto about the diagnosis of a case, Leopold had examined the child anew and had made an unexpected contribution towards the decision. For there was a difference of character between the two similar to that between Inspector Brassig and his friend Charles. The one was distinguished for his brightness, the other was slow, thoughtful, but thorough. If I contrast Otto and the careful Leopold in the dream, I do it, apparently, in order to extol Leopold. It is a comparison similar to the one above between the disobedient patient Irma and her friend who is thought to be more sensible. I now become aware of one of the tracks along which the thought association of the dream progresses; from the sick child to the children's asylum. The dulness to the left, below, recalls a certain case corresponding to it, in every detail in which Leopold astonished me by his thoroughness. Besides this, I have a notion of something like a metastatic affection, but it might rather be a reference to the lady patient whom I should like to have instead of Irma. For this lady, as far as I can gather, resembles a woman suffering from tuberculosis.

An infiltrated portion of skin on the left shoulder.

I see at once that this is my own rheumatism of the shoulder, which I always feel when I have remained awake until late at night. The turn of phrase in the dream also sounds ambiguous; something which I feel . . . in spite of the dress. "Feel on my own body" is intended. Moreover, I am struck with the unusual sound of the term "infiltrated portion of skin." "An infiltration behind on the upper left" is what we are accustomed to; this would refer to the lung, and thus again to tuberculosis patients.

In spite of the dress.

This, to be sure, is only an interpolation. We, of course, examine the children in the clinic undressed; it is some sort of contradiction to the manner in which grown-up female patients must be examined. The story used to be told of a prominent clinician that he always examined his patients physically only through the clothes. The rest is obscure to me; I have, frankly, no inclination to follow the matter further.

Dr. M. says: "It is an infection, but it does not matter. Dysentery will develop, and the poison will be excreted.

This at first seems ridiculous to me; still it must be carefully analysed like everything else. Observed more closely, it seems, however, to have a kind of meaning. What I had found in the patient was local diphtheritis. I remember the discussion about diphtheritis and diphtheria at the time of my daughter's illness. The latter is the general infection which proceeds from local diphtheritis. Leopold proves the existence of such general infection by means of the dulness, which thus suggests a metastatic lesion. I believe, however, that just this kind of metastasis does not occur in the case of diphtheria. It rather recalls pyaemia.

It does not matter, is a consolation. I believe it fits in as follows: The last part of the dream has yielded a content to the effect that the pains of the patient are the result of a serious organic affection. I begin to suspect that with this I am only trying to shift the blame from myself. Psychic treatment cannot be held responsible for the continued presence of diphtheritic affection. But now, in turn, I am disturbed at inventing such serious suffering for Irma for the sole purpose of exculpating myself. It seems cruel. I need (accordingly) the assurance that the result will be happy, and it does not seem ill-advised that I should put the words of consolation into the mouth of Dr. M. But here I consider myself superior to the dream, a fact which needs explanation.

But why is this consolation so nonsensical?

Dysentery:

Some sort of far-fetched theoretical notion that pathological material may be removed through the intestines. Am I in this way trying to make fun of Dr. M.'s great store of farfetched explanations, his habit of finding curious pathological relationships? Dysentery suggests something else. A few months ago I had in charge a young man suffering from remarkable pains during evacuation of the bowels, a case which colleagues had treated as "anaemia with malnutrition." I realised that it was a question of hysteria; I was unwilling to use my psychotherapy on him, and sent him off on a sea voyage. Now a few days before I had received a despairing letter from him from Egypt, saying that while there he had suffered a new attack, which the physician had declared to be dysentery. I suspect, indeed, that the diagnosis was only an error of my ignorant colleague, who allows hysteria to make a fool of him; but still I cannot avoid reproaching myself for putting the invalid in a position where he might contract an organic affection of the bowels in addition to his hysteria. Furthermore, dysentery sounds like diphtheria, a word which does not occur in the dream.

Indeed it must be that, with the consoling prognosis: "Dysentery will develop, &c.," I am making fun of Dr. M., for I recollect that years ago he once jokingly told a very similar story of another colleague. He had been called to consult with this colleague in the case of a woman who was very seriously ill and had felt obliged to confront the other physician, who seemed very hopeful, with the fact that he found albumen in the patient's urine. The colleague, however, did not let this worry him, but answered calmly: "That does not matter, doctor; the albumen will without doubt be excreted." Thus I can no longer doubt that derision for those colleagues who are ignorant of hysteria is contained in this part of the dream. As though in confirmation, this question now arises in my mind: "Does Dr. M. know that the symptoms of his patient, of our friend Irma, which give cause for fearing tuberculosis, are also based on hysteria? Has he recognised this hysteria, or has he stupidly ignored it?"

But what can be my motive in treating this friend so badly? This is very simple: Dr. M. agrees with my solution as little as Irma herself. I have thus already in this dream taken revenge on two persons, on Irma in the words, "If you still have pains, it is your own fault," and on Dr. M. in the wording of the nonsensical consolation which has been put into his mouth.

We have immediate knowledge of the origin of the infection.

This immediate knowledge in the dream is very remarkable. Just before we did not know it, since the infection was first demonstrated by Leopold.

My friend Otto has recently given her an injection when she felt ill.

Otto had actually related that in the short time of his visit to Irma's family, he had been called to a neighbouring hotel in order to give an injection to some one who fell suddenly ill. Injections again recall the unfortunate friend who has poisoned himself with cocaine. I had recommended the remedy to him merely for internal use during the withdrawal of morphine, but he once gave himself injections of cocaine.

With a propyl preparation . . . propyls . . . propionic acid. How did this ever occur to me? On the same evening on which I had written part of the history of the disease before having the dream, my wife opened a bottle of cordial labelled "Ananas,"[7] (which was a present from our friend Otto. For he had a habit of making presents on every possible occasion; I hope he will some day be cured of this by a wife).[8] Such a smell of fusel oil arose from this cordial that I refused to taste it. My wife observed: "We will give this bottle to the servants," and I, still more prudent, forbade it, with the philanthropic remark: "They mustn't be poisoned either." The smell of fusel oil (amyl . . .) has now apparently awakened in my memory the whole series, propyl, methyl, &c., which has furnished the propyl preparation of the dream. In this, it is true, I have employed a substitution; I have dreamt of propyl, after smelling amyl, but substitutions of this kind are perhaps permissible, especially in organic chemistry.

Trimethylamin. I see the chemical formula of this substance in the dream, a fact which probably gives evidence of a great effort on the part of my memory, and, moreover, the formula is printed in heavy type, as if to lay special stress upon something of particular importance, as distinguished from the context. To what does this trimethylamin lead, which has been so forcibly called to my attention? It leads to a conversation with another friend who for years has known all my germinating activities, as I have his. At that time he had just informed me of some of his ideas about sexual chemistry, and had mentioned, among others, that he thought he recognised in trimethylamin one of the products of sexual metabolism. This substance thus leads me to sexuality, to that factor which I credit with the greatest significance for the origin of the nervous affections which I attempt to cure. My patient Irma is a young widow; if I am anxious to excuse the failure of her cure, I suppose I shall best do so by referring to this condition, which her admirers would be glad to change. How remarkably, too, such a dream is fashioned! The other woman, whom I take as my patient in the dream instead of Irma, is also a young widow.

I suspect why the formula of trimethylamin has made itself so prominent in the dream. So many important things are gathered up in this one word: Trimethylamin is not only an allusion to the overpowering factor of sexuality, but also to a person whose sympathy I remember with satisfaction when I feel myself forsaken in my opinions. Should not this friend, who plays such a large part in my life, occur again in the chain of thoughts of the dream? Of course, he must; he is particularly acquainted with the results which proceed from affections of the nose and its adjacent cavities, and has revealed to science several highly remarkable relations of the turbinated bones to the female sexual organs (the three curly formations in Irma's throat). I have had Irma examined by him to see whether the pains in her stomach might be of nasal origin. But he himself suffers from suppurative rhinitis, which worries him, and to this perhaps there is an allusion in pyaemia, which hovers before me in the metastases of the dream.

Such injections are not made so rashly. Here the reproach of carelessness is hurled directly at my friend Otto. I am under the impression that I had some thought of this sort in the afternoon, when he seemed to indicate his siding against me by word and look. It was perhaps: "How easily he can be influenced; how carelessly he pronounces judgment." Furthermore, the above sentence again points to my deceased friend, who so lightly took refuge in cocaine injections. As I have said, I had not intended injections of the remedy at all. I see that in reproaching Otto I again touch upon the story of the unfortunate Matilda, from which arises the same reproach against me. Obviously I am here collecting examples of my own conscientiousness, but also of the opposite.

Probably also the syringe was not clean. Another reproach directed at Otto, but originating elsewhere. The day before I happened to meet the son of a lady eighty-two years of age whom I am obliged to give daily two injections of morphine. At present she is in the country, and I have heard that she is suffering from an inflammation of the veins. I immediately thought that it was a case of infection due to contamination from the syringe. It is my pride that in two years I have not given her a single infection; I am constantly concerned, of course, to see that the syringe is perfectly clean. For I am conscientious. From the inflammation of the veins, I return to my wife, who had suffered from emboli during a period of pregnancy, and now three related situations come to the surface in my memory, involving my wife, Irma, and the deceased Matilda, the identity of which three persons plainly justifies my putting them in one another's place.

I have now completed the interpretation of the dream.[9] In the course of this interpretation I have taken great pains to get possession of all the notions to which a comparison between the dream content and the dream thoughts hidden behind it must have given rise. Meanwhile, the "meaning" of the dream has dawned upon me. I have become conscious of a purpose which is realised by means of the dream, and which must have been the motive for dreaming. The dream fulfils several wishes, which have been actuated in me by the events of the preceding evening (Otto's news, and the writing down of the history of the disease). For the result of the dream is that I am not to blame for the suffering which Irma still has, and that Otto is to blame for it. Now Otto has made me angry by his remark about Irma's imperfect cure; the dream avenges me upon him by turning the reproach back upon himself. The dream acquits me of responsibility for Irma's condition by referring it to other causes, which indeed furnish a great number of explanations. The dream represents a certain condition of affairs as I should wish it to be; *the content of the dream is thus the fulfilment of a wish; its motive is a wish.*

This much is apparent at first sight. But many things in the details of the dream become intelligible when regarded from the point of view of wish-fulfilment. I take revenge on Otto, not only for hastily taking part against me, in that I accuse him of a careless medical operation (the injection), but I am also avenged on him for the bad cordial which smells like fusel oil, and I find an expression in the dream which unites both reproaches; the injection with a preparation of propyl. Still I am not satisfied, but continue my revenge by comparing him to his more reliable competitor. I seem to say by this: "I like him better than you." But Otto is not the only one who must feel the force of my anger. I take revenge on the disobedient patient by exchanging her for a more sensible, more docile one. Nor do I leave the contradiction of Dr. M. unnoticed, but express my opinion of him in an obvious allusion, to the effect that his relation to the question is that of *an ignoramus* (*"dysentery will develop," &c.*).

It seems to me, indeed, as though I were appealing from him to some one better informed (my friend, who has told me about trimethylamin); just as I have turned from Irma to her friend, I turn from Otto to Leopold. Rid me of these three persons, replace them by three others of my own choice, and I shall be released from the reproaches which I do not wish to have deserved! The unreasonableness itself of these reproaches is proved to me in the dream in the most elaborate way. Irma's pains are not charged to me, because she herself is to blame for them, in that she refuses to accept my solution. Irma's pains are none of my business, for they are of an organic nature, quite impossible to be healed by a psychic cure. Irma's sufferings are satisfactorily explained by her widowhood (trimethylamin!); a fact which, of course, I cannot alter. Irma's illness has been caused by an incautious injection on the part of Otto, with an ill-suited substance—in a way I should never have made an injection. Irma's suffering is the result of an injection made with an unclean syringe, just like the inflammation of the veins in my old lady, while I never do any such mischief with my injections. I am aware, indeed, that these explanations of Irma's illness, which unite in acquitting me, do not agree with one another; they even exclude one another. The whole pleading—this dream is nothing else—recalls vividly the defensive argument of a man who was accused by his neighbour of having returned a kettle to him in a damaged condition. In the first place, he said, he had returned the kettle undamaged; in the second, it already had holes in it when he borrowed it; and thirdly, he had never borrowed the kettle from his neighbour at all. But so much the better; if even one of these three methods of defence is recognised as valid, the man must be acquitted.

Still other subjects mingle in the dream, whose relation to my release from responsibility for Irma's illness is not so transparent: the illness of my daughter and that of a patient of the same name, the harmfulness of cocaine, the illness of my patient travelling in Egypt, concern about the health of my wife, my brother, of Dr. M., my own bodily troubles, and concern about the absent friend who is suffering from suppurative rhinitis. But if I keep all these things in view, they combine into a single train of thought, labelled perhaps: Concern for the health of myself and others—professional conscientiousness. I recall an undefined disagreeable sensation as Otto brought me the news of Irma's condition. I should like to note finally the expression of this fleeting sensation, which is part of the train of thought that is mingled into the dream. It is as though Otto had said to me: "You do not take your physician's duties seriously enough, you are not conscientious, do not keep your promises." Thereupon this train of thought placed itself at my service in order that I might exhibit proof of the high degree in which I am conscientious, how intimately I am concerned with the health of my relatives, friends, and patients. Curiously enough, there are also in this thought material some painful memories, which correspond rather to the blame attributed to Otto than to the accusation against me. The material has the appearance of being impartial, but the connection between this broader material, upon which the dream depends, and the more limited theme of the dream which gives rise to the wish to be innocent of Irma's illness, is nevertheless unmistakable.

I do not wish to claim that I have revealed the meaning of the dream entirely, or that the interpretation is flawless.

I could still spend much time upon it; I could draw further explanations from it, and bring up new problems which it bids us consider. I even know the points from which further thought associations might be traced; but such considerations as are connected with every dream of one's own restrain me from the work of interpretation. Whoever is ready to condemn such reserve, may himself try to be more straightforward than I. I am content with the discovery which has been just made. If the method of dream interpretation here indicated is followed, it will be found that the dream really has meaning, and is by no means the expression of fragmentary brain activity, which the authors would have us believe. *When the work of interpretation has been completed the dream may be recognised as the fulfilment of a wish.*

Notes:

1. In a novel, *Gradiva,* of the poet W. Jensen, I accidentally discovered several artificial dreams which were formed with perfect correctness and which could be interpreted as though they had not been invented, but had been dreamt by actual persons. The poet declared, upon my inquiry, that he was unacquainted with my theory of dreams. I have made use of this correspondence between my investigation and the creative work of the poet as a proof of the correctness of my method of dream analysis ("Der Wahn und die Träume," in W. Jensen's

Gradiva, No. 1 of the *Schriften zur angewandten Seelenkunde,* 1906, edited by me). Dr. Alfred Robitsek has since shown that the dream of the hero in Goethe's *Egmont* may be interpreted as correctly as an actually experienced dream ("Die Analyse von Egmont's Träume," *Jahrbuch,* edited by Bleuler-Freud, vol. ii., 1910.)

2. After the completion of my manuscript, a paper by Stumpf came to my notice which agrees with my work in attempting to prove that the dream is full of meaning and capable of interpretation. But the interpretation is undertaken by means of an allegorising symbolism, without warrant for the universal applicability of the procedure.

3. Dr. Alfred Robitsiek calls my attention to the fact that Oriental dream books, of which ours are pitiful plagiarisms, undertake the interpretation of dream elements, mostly according to the assonance and similarity of the words. Since these relationships must be lost by translation into our language, the incomprehensibility of the substitutions in our popular "dream books" may have its origin in this fact. Information as to the extraordinary significance of puns and punning in ancient Oriental systems of culture may be found in the writings of Hugo Winckler. The nicest example of a dream interpretation which has come down to us from antiquity is based on a play upon words. Artemidoros relates the following (p. 225): "It seems to me that Aristandros gives a happy interpretation to Alexander of Macedon. When the latter held Tyros shut in and in a state of siege, and was angry and depressed over the great loss of time, he dreamed that he saw a Satyros dancing on his shield. It happened that Aristandros was near Tyros and in the convoy of the king, who was waging war on the Syrians. By disjoining the word Satyros into [Greek] and [Greek], he induced the king to become more aggressive in the siege, and thus he became master of the city. ([Greek]—thine is Tyros.) The dream, indeed, is so intimately connected with verbal expression that Ferenczi may justly remark that every tongue has its own dream language. Dreams are, as a rule, not translatable into other languages.

4. Breuer and Freud, *Studien über Hysterie,* Vienna, 1895; 2nd ed. 1909.

5. The complaint, as yet unexplained, of pains in the abdomen, may also be referred to this third person. It is my own wife, of course, who is in question; the abdominal pains remind me of one of the occasions upon which her shyness became evident to me. I must myself admit that I

do not treat Irma and my wife very gallantly in this dream, but let it be said for my excuse that I am judging both of them by the standard of the courageous, docile, female patient.

6. I suspect that the interpretation of this portion has not been carried far enough to follow every hidden meaning. If I were to continue the comparison of the three women, I would go far afield. Every dream has at least one point at which it is unfathomable, a central point, as it were, connecting it with the unknown.

7. "Ananas," moreover, has a remarkable assonance to the family name of my patient Irma.

8. In this the dream did not turn out to be prophetic. But in another sense, it proved correct, for the "unsolved" stomach pains, for which I did not want to be to blame, were the forerunners of a serious illness caused by gall stones.

9. Even if I have not, as may be understood, given account of everything which occurred to me in connection with the work of interpretation.

3

The Dream is the Fulfilment
of a Wish

When after passing a defile one has reached an eminence
where the ways part and where the view opens out
broadly in different directions, it is permissible to
stop for a moment and to consider where one is to turn next.
Something like this happens to us after we have mastered this
first dream interpretation. We find ourselves in the open light of
a sudden cognition. The dream is not comparable to the irregular
sounds of a musical instrument, which, instead of being touched
by the hand of the musician, is struck by some outside force; the
dream is not senseless, not absurd, does not presuppose that a
part of our store of ideas is dormant while another part begins
to awaken. It is a psychic phenomenon of full value, and indeed
the fulfilment of a wish; it takes its place in the concatenation of
the waking psychic actions which are intelligible to us, and it has
been built up by a highly complicated intellectual activity. But at
the very moment when we are inclined to rejoice in this discovery,
a crowd of questions overwhelms us. If the dream, according to
the interpretation, represents a wish fulfilled, what is the cause
of the peculiar and unfamiliar manner in which this fulfilment is
expressed? What changes have occurred in the dream thoughts

before they are transformed into the manifest dream which we remember upon awaking? In what manner has this transformation taken place? Whence comes the material which has been worked over into the dream? What causes the peculiarities which we observe in the dream thoughts, for example, that they may contradict one another? (The analogy of the kettle, p. 114). Is the dream capable of teaching us something new about our inner psychic processes, and can its content correct opinions which we have held during the day? I suggest that for the present all these questions be laid aside, and that a single path be pursued. We have found that the dream represents a wish as fulfilled. It will be our next interest to ascertain whether this is a universal characteristic of the dream, or only the accidental content of the dream ("of Irma's injection") with which we have begun our analysis, for even if we make up our minds that every dream has a meaning and psychic value, we must nevertheless allow for the possibility that this meaning is not the same in every dream. The first dream we have considered was the fulfilment of a wish; another may turn out to be a realised apprehension; a third may have a reflection as to its content; a fourth may simply reproduce a reminiscence. Are there then other wish dreams; or are there possibly nothing but wish dreams?

It is easy to show that the character of wish-fulfilment in dreams is often undisguised and recognisable, so that one may wonder why the language of dreams has not long since been understood. There is, for example, a dream which I can cause as often as I like, as it were experimentally. If in the evening I eat anchovies, olives, or other strongly salted foods, I become thirsty at night, whereupon I waken. The awakening, however, is preceded by a dream, which each time has the same content, namely, that I am drinking. I quaff water in long draughts, it tastes as sweet as only a cool drink can taste when one's throat is parched, and then I awake and have an actual desire to drink. The occasion for this dream is thirst, which I perceive when I awake. The wish to drink originates from this sensation, and the dream shows me this wish as fulfilled. It thereby serves a function the nature of which I soon guess. I sleep well, and am not accustomed to be awakened by a bodily need. If I succeed in assuaging my thirst by means of the dream that I am drinking, I need not wake up in order to satisfy it. It is thus a dream of convenience. The dream substitutes itself for action, as elsewhere in life. Unfortunately the need of water for quenching thirst cannot be satisfied with a dream, like my thirst for revenge upon Otto and

Dr. M., but the intention is the same. This same dream recently appeared in modified form. On this occasion I became thirsty before going to bed, and emptied the glass of water which stood on the little chest next to my bed. Several hours later in the night came a new attack of thirst, accompanied by discomfort. In order to obtain water, I should have had to get up and fetch the glass which stood on the night-chest of my wife. I thus quite appropriately dreamt that my wife was giving me a drink from a vase; this vase was an Etruscan cinerary urn which I had brought home from an Italian journey and had since given away. But the water in it tasted so salty (apparently from the ashes) that I had to wake. It may be seen how conveniently the dream is capable of arranging matters; since the fulfilment of a wish is its only purpose, it may be perfectly egotistic. Love of comfort is really not compatible with consideration for others. The introduction of the cinerary urn is probably again the fulfilment of a wish; I am sorry that I no longer possess this vase; it, like the glass of water at my wife's side, is inaccessible to me. The cinerary urn is also appropriate to the sensation of a salty taste which has now grown stronger, and which I know will force me to wake up.[1]

Such convenience dreams were very frequent with me in the years of my youth. Accustomed as I had always been to work until late at night, early awakening was always a matter of difficulty for me. I used then to dream that I was out of bed and was standing at the wash-stand. After a while I could not make myself admit that I have not yet got up, but meanwhile I had slept for a time. I am acquainted with the same dream of laziness as dreamt by a young colleague of mine, who seems to share my propensity for sleep. The lodging-house keeper with whom he was living in the neighbourhood of the hospital had strict orders to wake him on time every morning, but she certainly had a lot of trouble when she tried to carry out his orders. One morning sleep was particularly sweet. The woman called into the room: "Mr. Joe, get up; you must go to the hospital." Whereupon the sleeper dreamt of a room in the hospital, a bed in which he was lying, and a chart pinned over his head reading: "Joe H. . . . cand. med. 22 years old." He said to himself in the dream: "If I am already at the hospital, I don't have to go there," turned over and slept on. He had thus frankly admitted to himself his motive for dreaming.

Here is another dream, the stimulus for which acts during sleep itself: One of my women patients, who had had to undergo an unsuccessful operation on the jaw, was to wear a cooling apparatus on the affected cheek, according to the orders of the physicians. But she was in the habit of throwing it off as soon as she had got to sleep. One day I was asked to reprove her for doing so; for she had again thrown the apparatus on the floor. The patient defended herself as follows: "This time I really couldn't help it; it was the result of a dream which I had in the night. In the dream, I was in a box at the opera and was taking a lively interest in the performance. But Mr. Karl Meyer was lying in the sanatorium and complaining pitifully on account of pains in his jaw. I said to myself, 'Since I haven't the pains, I don't need the apparatus either,' that's why I threw it away." This dream of the poor sufferer is similar to the idea in the expression which comes to our lips when we are in a disagreeable situation: "I know something that's a great deal more fun." The dream presents this great deal more fun. Mr. Karl Meyer, to whom the dreamer attributed her pains, was the most indifferent young man of her acquaintance whom she could recall.

It is no more difficult to discover the fulfilment of wishes in several dreams which I have collected from healthy persons. A friend who knew my theory of dreams and had imparted it to his wife, said to me one day: "My wife asked me to tell you that she dreamt yesterday that she was having her menses. You will know what that means." Of course I know: if the young wife dreams that she is having her menses, the menses have stopped. I can understand that she would have liked to enjoy her freedom for a time longer before the discomforts of motherhood began. It was a clever way of giving notice of her first pregnancy. Another friend writes that his wife had recently dreamt that she noticed milk stains on the bosom of her waist. This is also an indication of pregnancy, but this time not of the first one; the young mother wishes to have more nourishment for the second child than she had for the first.

A young woman, who for weeks had been cut off from company because she was nursing a child that was suffering from an infectious disease, dreams, after its safe termination, of a company of people in which A. Daudet, Bourget, M. Provost, and others are present, all of whom are very pleasant to her and entertain her admirably. The different authors in the dream also have the features which their pictures give them. M. Prevost, with whose picture she is not familiar, looks like—the disinfecting man who on the previous day had cleaned the sick rooms and had entered them as the first visitor after a long period. Apparently the dream might be perfectly translated thus: "It is about time now for something more entertaining than this eternal nursing."

Perhaps this selection will suffice to prove that often and under the most complex conditions dreams are found which can be understood only as fulfilments of wishes, and which present their contents without concealment. In most cases these are short and simple dreams, which stand in pleasant contrast to the confused and teeming dream compositions which have mainly attracted the attention of the authors. But it will pay to spend some time upon these simple dreams. The most simple dreams of all, I suppose, are to be expected in the case of children, whose psychic activities are certainly less complicated than those of adults. The psychology of children, in my opinion, is to be called upon for services similar to those which a study of the anatomy and development of the lower animals renders to the investigation of the structure of the highest classes of animals. Until now only a few conscious efforts have been made to take advantage of the psychology of children for such a purpose.

The dreams of little children are simple fulfilments of wishes, and as compared, therefore, with the dreams of adults, are not at all interesting. They present no problem to be solved, but are naturally invaluable as affording proof that the dream in its essence signifies the fulfilment of a wish. I have been able to collect several examples of such dreams from the material furnished by my own children.

For two dreams, one of my daughters, at that time eight and a half years old, the other of a boy five and a quarter years of age, I am indebted to an excursion to the beautiful Hallstatt in the summer of 1896. I must make the preliminary statement that during this summer we were living on a hill near Aussee, from which, when the weather was good, we enjoyed a splendid view of the Dachstein from the roof of our house. The Simony Hut could easily be recognised with a telescope. The little ones often tried to see it through the telescope—I do not know with what success. Before the excursion I had told the children that Hallstatt lay at the foot of the Dachstein. They looked forward to the day with great joy. From Hallstatt we entered the valley of Eschern, which highly pleased the children with its varying aspects. One of them, however, the boy of five, gradually became discontented. As often as a mountain came in view, he would ask: "Is that the Dachstein? "whereupon I would have to answer: "No, only a foot-hill." After this question had been repeated several times, he became altogether silent; and he was quite unwilling to come along on the flight of steps to the waterfall. I thought he was tired out. But the next morning, he approached me radiant with joy, and said: "Last night I dreamt that we were at Simony Hut." I understood him now; he had expected, as I was speaking of the Dachstein, that on the excursion to Hallstatt, he would ascend the mountain and would come face to face with the hut, about which there had been so much discussion at the telescope. When he learned that he was expected to be regaled with foot-hills and a waterfall, he was disappointed and became discontented. The dream compensated him for this. I tried to learn some details of the dream; they were scanty. "Steps must be climbed for six hours," as he had heard.

On this excursion wishes, destined to be satisfied only in dreams, had arisen also in the mind of the girl of eight and a half years. We had taken with us to Hallstatt the twelve-year-old boy of our neighbour— an accomplished cavalier, who, it seems to me, already enjoyed the full sympathy of the little woman. The next morning, then, she related the following dream: "Just think, I dreamt that Emil was one of us, that he said papa and mamma to you, and slept at our house in the big room like our boys. Then mamma came into the room and threw a large handful of chocolate bars under our beds." The brothers of the girl, who evidently had not inherited a familiarity with dream interpretation, declared just like the authors: "That dream is nonsense." The girl defended at least a part of the dream, and it is worth while, from the point of view of the theory of neuroses, to know which part: "That about Emil belonging to us is nonsense, but that about the bars of chocolate is not." It was just this latter part that was obscure to me. For this mamma furnished me the explanation. On the way home from the railway station the children had stopped in front of a slot machine, and had desired exactly such chocolate bars wrapped in paper with a metallic lustre, as the machine, according to their experience, had for sale. But the mother had rightly thought that the day had brought enough wish-fulfilment, and had left this wish to be satisfied in dreams. This little scene had escaped me. I at once understood that portion of the dream which had been condemned by my daughter. I had myself heard the well-behaved guest enjoining the children to wait until papa or mamma had come up. For the little one the dream made a lasting adoption based on this temporary relation of the boy to us. Her tender nature was as yet unacquainted with any form of being together except those mentioned in the dream, which are taken from her brothers. Why the chocolate bars were thrown under the bed could not, of course, be explained without questioning the child.

From a friend I have learnt of a dream very similar to that of my boy. It concerned an eight-year-old girl. The father had undertaken a walk to Dornbach with the children, intending to visit the Rohrerhütte, but turned back because it had grown too late, and promised the children to make up for their disappointment some other time. On the way back, they passed a sign which showed the way to the Hameau. The children now asked to be taken to that place also, but had to be content, for the same reason, with a postponement to another day. The next morning, the eight-year-old girl came to the father, satisfied, saying: "Papa, I dreamt last night that you were with us at the Rohrerhütte and on the Hameau." Her impatience had thus in the dream anticipated the fulfilment of the promise made by her father.

Another dream, which the picturesque beauty of the Aussee inspired in my daughter, at that time three and a quarter years old, is equally straightforward. The little one had crossed the lake for the first time, and the trip had passed too quickly for her. She did not want to leave the boat at the landing, and cried bitterly. The next morning she told us: "Last night I was sailing on the lake." Let us hope that the duration of this dream ride was more satisfactory to her.

My eldest boy, at that time eight years of age, was already dreaming of the realisation of his fancies. He had been riding in a chariot with Achilles, with Diomed as charioteer. He had, of course, on the previous day shown a lively interest in the *Myths of Greece,* which had been given to his elder sister.

If it be granted that the talking of children in sleep likewise belongs to the category of dreaming, I may report the following as one of the most recent dreams in my collection. My youngest girl, at that time nineteen months old, had vomited one morning, and had therefore been kept without food throughout the day. During the night which followed upon this day of hunger, she was heard to call excitedly in her sleep: "Anna Feud, strawberry, huckleberry, omelette, pap!" She used her name in this way in order to express her idea of property; the menu must have included about everything which would seem to her a desirable meal; the fact that berries appeared in it twice was a demonstration against the domestic sanitary regulations, and was based on the circumstance, by no means overlooked by her, that the nurse ascribed her indisposition to an over-plentiful consumption of strawberries; she thus in the dream took revenge for this opinion which was distasteful to her.[2]

If we call childhood happy because it does not yet know sexual desire, we must not forget how abundant a source of disappointment and self-denial, and thus of dream stimulation, the other of the great life-impulses may become for it.[3] Here is a second example showing this. My nephew of twenty-two months had been given the task of congratulating me upon my birthday, and of handing me, as a present, a little basket of cherries, which at that time of the year were not yet in season. It seemed difficult for him, for he repeated again and again: "Cherries in it," and could not be induced to let the little basket go out of his hands. But he knew how to secure his compensation. He had, until now, been in the habit of telling his mother every morning that he had dreamt of the "white soldier," an officer of the guard in a white cloak, whom he had once admired on the street. On the day after the birthday, he awakened joyfully with the information which could have had its origin only in a dream: "He(r)man eat up all the cherries!"[4]

What animals dream of I do not know. A proverb for which I am indebted to one of my readers claims to know, for it raises the question: "What does the goose dream of?" the answer being: "Of maize!" The whole theory that the dream is the fulfilment of a wish is contained in these sentences.[5]

We now perceive that we should have reached our theory of the hidden meaning of the dream by the shortest road if we had merely consulted colloquial usage. The wisdom of proverbs, it is true, sometimes speaks contemptuously enough of the dream—it apparently tries to justify science in expressing the opinion that "Dreams are mere bubbles;" but still for colloquial usage the dream is the gracious fulfiller of wishes. "I should never have fancied that in the wildest dream," exclaims one who finds his expectations surpassed in reality.

Notes:

1. The facts about dreams of thirst were known also to Weygandt, who expresses himself about them (p. 11) as follows: "It is just the sensation of thirst which is most accurately registered of all; it always causes a representation of thirst quenching. The manner in which the dream

pictures the act of thirst quenching is manifold, and is especially apt to be formed according to a recent reminiscence. Here also a universal phenomenon is that disappointment in the slight efficacy of the supposed refreshments sets in immediately after the idea that thirst has been quenched." But he overlooks the fact that the reaction of the dream to the stimulus is universal. If other persona who are troubled by thirst at night awake without dreaming beforehand, this does not constitute an objection to my experiment, but characterises those others as persons who sleep poorly.

2. The dream afterwards accomplished the same purpose in the case of the grandmother, who is older than the child by about seventy years, as it did in the case of the granddaughter. After she had been forced to go hungry for several days on account of the restlessness of her floating kidney, she dreamed, apparently with a transference into the happy time of her flowering maidenhood, that she had been "asked out," invited as a guest for both the important meals, and each time had been served with the most delicious morsels.

3. A more searching investigation into the psychic life of the child teaches us, to be sure, that sexual motive powers in infantile forma, which have been too long overlooked, play a sufficiently great part in the psychic activity of the child. This raises some doubt as to the happiness of the child, as imagined later by the adults. *Cf.* the author's "Three Contributions to the Sexual Theory," translated by A. A. Brill, *Journal of Nervous and Mental Diseases Publishing Company.*

4. It should not be left unmentioned that children sometimes show complex and more obscure dreams, while, on the other hand, adults will often under certain conditions show dreams of an infantile character. How rich in unsuspected material the dreams of children of from four to five years might be is shown by examples in my "Analyse der Phobie eines fünfjahrigen Knaben" (*Jahrbuch,* ed. by Bleuler & Freud, 1909), and in Jung's "Ueber Konflikte der kindlichen Seele" (ebda. ii. vol., 1910). On the other hand, it seems that dreams of an infantile type reappear especially often in adults if they are transferred to unusual conditions of life. Thus Otto Nordenskjold, in his book *Antarctic* (1904), writes as follows about the crew who passed the winter with him. "Very characteristic for the trend of our inmost thoughts were our dreams, which were never more vivid

and numerous than at present. Even those of our comrades with whom dreaming had formerly been an exception had long stories to tell in the morning when we exchanged our experiences in the world of phantasies. They all referred to that outer world which was now so far from us, but they often fitted into our present relations. An especially characteristic dream was the one in which one of our comrades believed himself back on the bench at school, where the task was assigned him of skinning miniature seals which were especially made for the purposes of instruction. Eating and drinking formed the central point around which most of our dreams were grouped. One of us, who was fond of going to big dinner parties at night, was exceedingly glad if he could report in the morning 'that he had had a dinner consisting of three courses.' Another dreamed of tobacco—of whole mountains of tobacco; still another dreamed of a ship approaching on the open sea under full sail. Still another dream deserves to be mentioned. The letter carrier brought the mail, and gave a long explanation of why he had had to wait so long for it; he had delivered it at the wrong place, and only after great effort had been able to get it back. To be sure, we occupied ourselves in sleep with still more impossible things, but the lack of phantasy in almost all the dreams which I myself dreamed or heard others relate was quite striking. It would surely have been of great psychological interest if all the dreams could have been noted. But one can readily understand how we longed for sleep. It alone could afford us everything that we all most ardently desired."

5. A Hungarian proverb referred to by Ferenczi states more explicitly that "the pig dreams of acorns, the goose of maize."

4

Distortion in Dreams

If I make the assertion that wish fulfilment is the meaning of every dream, that, accordingly, there can be no dreams except wish dreams, I am sure at the outset to meet with the most emphatic contradiction. Objections will be made to this effect: "The fact that there are dreams which must be understood as fulfilments of wishes is not new, but, on the contrary, has long since been recognised by the authors. *Cf.* Radestock (pp. 137–138), Volkelt (pp. 110–111), Tissié (p. 70), M. Simon (p. 42) on the hunger dreams of the imprisoned Baron Trenck), and the passage in Griesinger (p. 11). The assumption that there can be nothing but dreams of wish fulfilment, however, is another of those unjustified generalisations by which you have been pleased to distinguish yourself of late. Indeed dreams which exhibit the most painful content, but not a trace of wish fulfilment, occur plentifully enough. The pessimistic philosopher, Edward von Hartman, perhaps stands furthest from the theory of wish fulfilment. He expresses himself in his *Philosophy of the Unconscious,* Part II. (stereotyped edition, p. 34), to the following effect:—

"'As regards the dream, all the troubles of waking life are transferred by it to the sleeping state; only the one thing, which can in some measure reconcile a cultured person to life-scientific and artistic enjoyment is not transferred. . . . ' But even less discontented observers have laid emphasis on the fact that in dreams pain and disgust are more frequent than pleasure; so Scholz (p. 39), Volkelt (p. 80), and others. Indeed two ladies, Sarah Weed and Florence Hallam, have found from the elaboration of their dreams a mathematical expression for the preponderance of displeasure in dreams. They designate 58 per cent. of the dreams as disagreeable, and only 28.6 per cent. as positively pleasant. Besides those dreams which continue the painful sensations of life during sleep, there are also dreams of fear, in which this most terrible of all disagreeable sensations tortures us until we awake, and it is with just these dreams of fear that children are so often persecuted (*Cf.* Debacker concerning the Pavor Nocturnus), though it is in the case of children that you have found dreams of wishing undisguised."

Indeed it is the anxiety dreams which seem to prevent a generalisation of the thesis that the dream is a wish-fulfilment, which we have established by means of the examples in the last section; they seem even to brand this thesis as an absurdity.

It is not difficult, however, to escape these apparently conclusive objections. Please observe that our doctrine does not rest upon an acceptance of the manifest dream content, but has reference to the thought content which is found to lie behind the dream by the process of interpretation. Let us contrast the *manifest* and the *latent dream content.* It is true that there are dreams whose content is of the most painful nature. But has anyone ever tried to interpret these dreams, to disclose their latent thought content? If not, the two objections are no longer valid against us; there always remains the possibility that even painful and fearful dreams may be discovered to be wish fulfilments upon interpretation.[1]

In scientific work it is often advantageous, when the solution of one problem presents difficulties, to take up a second problem, just as it is easier to crack two nuts together instead of separately. Accordingly we are confronted not merely with the problem: How can painful and fearful dreams be the fulfilments of wishes? but we may also, from our discussion so far, raise the question: Why do not the dreams which show an indifferent content, but turn out to be wish-fulfilments, show this meaning undisguised? Take the fully reported dream of Irma's injection; it is in no way painful in its nature, and can be recognised, upon interpretation, as a striking wish-fulfilment. Why, in the first place, is an interpretation necessary? Why does not the dream say directly what it means? As a matter of fact, even the dream of Irma's injection does not at first impress us as representing a wish of the dreamer as fulfilled. The reader will not have received this impression, and even I myself did not know it until I had undertaken the analysis. If we call this peculiarity of the dream of needing an explanation *the fact of the distortion* of dreams, then a second question arises: What is the origin of this disfigurement of dreams?

If one's first impressions on this subject were consulted, one might happen upon several possible solutions; for example, that there is an inability during sleep to find adequate expression for the dream thoughts. The analysis of certain dreams, however, compels us to give the disfigurement of dreams another explanation. I shall show this by employing a second dream of my own, which again involves numerous indiscretions, but which compensates for this personal sacrifice by affording a thorough elucidation of the problem.

Preliminary Statement.—In the spring of 1897 I learnt that two professors of our university had proposed me for appointment as Professor extraord. (assistant professor). This news reached me unexpectedly and pleased me considerably as an expression of appreciation on the part of two eminent men which could not be explained by personal interest. But, I immediately thought, I must not permit myself to attach any expectation to this event. The university government had during the last few years left proposals of this kind unconsidered, and several colleagues, who were ahead of me in years, and who were at least my equals in merit, had been waiting in vain during this time for their appointment. I had no reason to suppose I should fare better. I resolved then to comfort myself. I am not, so far as I know, ambitious, and I engage in medical practice with satisfying results even without the recommendation of a title. Moreover, it was not a question whether I considered the grapes sweet or sour, for they undoubtedly hung much too high for me.

One evening I was visited by a friend of mine, one of those colleagues whose fate I had taken as a warning for myself. As he had long been a candidate for promotion to the position of professor, which in our society raises the physician to a demigod among his patients, and as he was less resigned than I, he was in the habit of making representations from time to time, at the offices of the university government, for the purpose of advancing his interests. He came to me from a visit of that kind. He said that this time he had driven the exalted gentleman into a corner, and had asked him directly whether considerations of creed were not really responsible for the deferment of his appointment. The answer had been that to be sure—in the present state of public opinion—His Excellency was not in a position, &c. "Now I at least know what I am at," said my friend in closing his narrative, which told me nothing new, but which was calculated to confirm me in my resignation. For the same considerations of creed applied to my own case.

On the morning after this visit, I had the following dream, which was notable on account of its form. It consisted of two thoughts and two images, so that a thought and an image alternated. But I here record only the first half of the dream, because the other half has nothing to do with the purpose which the citation of the dream should serve.

I. *Friend R. is my uncle—I feel great affection for him.*

II. *I see before me his face somewhat altered.*

It seems to be elongated; a yellow beard, which surrounds it, is emphasised with peculiar distinctness.

Then follow the other two portions, again a thought and an image, which I omit.

The interpretation of this dream was accomplished in the following manner:

As the dream occurred to me in the course of the forenoon, I laughed outright and said: "The dream is nonsense." But I could not get it out of my mind, and the whole day it pursued me, until, at last, in the evening I reproached myself with the words: "If in the course of dream interpretation one of your patients had nothing better to say than 'That is nonsense,' you would reprove him, and would suspect that behind the dream there was hidden some disagreeable affair, the exposure of which he wanted to spare himself. Apply the same thing in your own case; your opinion that the dream is nonsense probably signifies merely an inner resistance to its interpretation. Do not let yourself be deterred." I then proceeded to the interpretation.

"R. is my uncle." What does that mean. I have had only one uncle, my uncle Joseph.[2] His story, to be sure, was a sad one. He had yielded to the temptation, more than thirty years before, of engaging in dealings which the law punishes severely, and which on that occasion also it had visited with punishment. My father, who thereupon became grey from grief in a few days, always used to say that Uncle Joseph was never a wicked man, but that he was indeed a simpleton; so he expressed himself. If, then, friend R. is my uncle Joseph, that is equivalent to saying: "R. is a simpleton." Hardly credible and very unpleasant! But there is that face which I see in the dream, with its long features and its yellow beard. My uncle actually had such a face—long and surrounded by a handsome blond beard. My friend R. was quite dark, but when dark-haired persons begin to grow grey, they pay for the glory of their youthful years. Their black beard undergoes an unpleasant change of color, each hair separately; first it becomes reddish brown, then yellowish brown, and then at last definitely grey. The beard of my friend R. is now in this stage, as is my own moreover, a fact which I notice with regret. The face which I see in the dream is at once that of my friend R. and that of my uncle. It is like a composite photograph of Galton, who, in order to emphasise family resemblances, had several faces photographed on the same plate. No doubt is thus possible, I am really of the opinion that my friend R. is a simpleton—like my uncle Joseph.

I have still no idea for what purpose I have constructed this relationship, to which I must unconditionally object. But it is not a very far-reaching one, for my uncle was a criminal, my friend R. is innocent—perhaps with the exception of having been punished for knocking down an apprentice with his bicycle. Could I mean this offence? That would be making ridiculous comparisons. Here I recollect another conversation which I had with another colleague, N., and indeed upon the same subject. I met N. on the street. He likewise has been nominated for a professorship, and having heard of my being honoured, congratulated me upon it. I declined emphatically, saying, "You are the last man to make a joke like this, because you have experienced what the nomination is worth in your own case." Thereupon he said, though probably not in earnest, "You cannot be sure about that. Against me there is a very particular objection. Don't you know that a woman once entered a legal complaint against me? I need not assure you that an inquiry was made; it was a mean attempt at blackmail, and it was all I could do to save the plaintiff herself from punishment. But perhaps the affair will be pressed against me at the office in order that I may not be appointed. You, however, are above reproach." Here I have come upon a criminal, and at the same time upon the interpretation and trend of the dream. My uncle Joseph represents for me both colleagues who have not been appointed to the professorship, the one as a simpleton, the other as a criminal. I also know now for what purpose I need this representation. If considerations of creed are a determining factor in the postponement of the appointment of my friends, then my own appointment is also put in question: but if I can refer the rejection of the two friends to other causes, which do not apply to my case, my hope remains undisturbed. This is the procedure of my dream; it makes the one, R., a simpleton, the other, N., a criminal; since, however, I am neither the one nor the other, our community of interest is destroyed, I have a right to enjoy the expectation of being appointed a professor, and have escaped the painful application to my own case of the information which the high official has given to R.

I must occupy myself still further with the interpretation of this dream. For my feelings it is not yet sufficiently cleared up. I am still disquieted by the ease with which I degrade two respected colleagues for the purpose of clearing the way to the professorship for myself. My dissatisfaction with my procedure has indeed diminished since I have learnt to evaluate statements made in dreams. I would argue against anyone who urged that I really consider R. a simpleton, and that I do not credit N.'s account of the blackmail affair. I do not believe either that Irma has been made seriously ill by an injection given her by Otto with a preparation of propyl. Here, as before, it is only the *wish that the case may be as the dream expresses it.* The statement in which my wish is realised sounds less absurd in the second dream than in the first; it is made here with a more skilful utilisation of facts as points of attachment, something like a well-constructed slander, where "there is something in it." For my friend R. had at that time the vote of a professor from the department against him, and my friend N. had himself unsuspectingly furnished me with the material for slander. Nevertheless, I repeat, the dream seems to me to require further elucidation.

I remember now that the dream contains still another portion which so far our interpretation has not taken into account. After it occurs to me that my friend R. is my uncle, I feel great affection for him. To whom does this feeling belong? For my uncle Joseph, of course, I have never had any feelings of affection. For years my friend R. has been beloved and dear to me; but if I were to go to him and express my feelings for him in terms which came anywhere near corresponding to the degree of affection in the dream, he would doubtless be surprised. My affection for him seems untrue and exaggerated, something like my opinion of his psychic qualities, which I express by fusing his personality with that of

120

my uncle; but it is exaggerated in an opposite sense. But now a new state of affairs becomes evident to me. The affection in the dream does not belong to the hidden content, to the thoughts behind the dream; it stands in opposition to this content; it is calculated to hide the information which interpretation may bring. Probably this is its very purpose. I recall with what resistance I applied myself to the work of interpretation, how long I tried to postpone it, and how I declared the dream to be sheer nonsense. I know from my psychoanalytical treatments how such condemnation is to be interpreted. It has no value as affording information, but only as the registration of an affect. If my little daughter does not like an apple which is offered her, she asserts that the apple has a bitter taste, without even having tasted it. If my patients act like the little girl, I know that it is a question of a notion which they want to *suppress.* The same applies to my dream. I do not want to interpret it because it contains something to which I object. After the interpretation of the dream has been completed, I find out what it was I objected to; it was the assertion that R. is a simpleton. I may refer the affection which I feel for R. not to the hidden dream thoughts, but rather to this unwillingness of mine. If my dream as compared with its hidden content is disfigured at this point, and is disfigured, moreover, into something opposite, then the apparent affection in the dream serves the purpose of disfigurement; or, in other words, the disfigurement is here shown to be intended: it is a means of dissimulation. My dream thoughts contain an unfavourable reference to R.; in order that I may not become aware of it, its opposite, a feeling of affection for him, makes its way into the dream.

The fact here recognised might be of universal applicability. As the examples in Section III. have shown, there are dreams which are undisguised wish-fulfilments. Wherever a wish-fulfilment is unrecognisable and concealed, there must be present a feeling of repulsion towards this wish, and in consequence of this repulsion the wish is unable to gain expression except in a disfigured state. I shall try to find a case in social life which is parallel to this occurrence in the inner psychic life. Where in social life can a similar disfigurement of a psychic act be found? Only where two persons are in question, one of whom possesses a certain power, while the other must have a certain consideration for this power. This second person will then disfigure his psychic actions, or, as we may say, he will dissimulate. The politeness which I practise every day is largely dissimulation of this kind. If I interpret my dreams for the benefit of the reader I am forced to make such distortions. The poet also complains about such disfigurement: "You may not tell the best that you know to the youngsters."

The political writer who has unpleasant truths to tell to the government finds himself in the same position. If he tells them without reserve, the government will suppress them—subsequently in case of a verbal expression of opinion, preventatively, if they are to be published in print. The writer must fear *censure;* he therefore modifies and disfigures the expression of his opinion. He finds himself compelled, according to the sensitiveness of this censure, either to restrain himself from certain particular forms of attack or to speak in allusion instead of direct designations. Or he must disguise his objectionable statement in a garb that seems harmless. He may, for instance, tell of an occurrence between two mandarins in the Orient, while he has the officials of his own country in view. The stricter the domination of the censor, the more extensive becomes the disguise, and often the more humorous the means employed to put the reader back on the track of the real meaning.

The correspondence between the phenomena of the censor and those of dream distortion, which may be traced in detail, justifies us in assuming similar conditions for both. We should then assume in each human being, as the primary cause of dream formation, two psychic forces (streams, systems), of which one constitutes the wish expressed by the dream, while the other acts as a censor upon this dream wish, and by means of this censoring forces a distortion of its expression. The only question is as to the basis of the authority of this second instance[3] by virtue of which it may exercise its censorship. If we remember that the hidden dream thoughts are not conscious before analysis, but that the apparent dream content is remembered as conscious, we easily reach the assumption that admittance to consciousness is the privilege of the second instance. Nothing can reach consciousness from the first system which has not first passed the second instance, and the second instance lets nothing pass without exercising its rights and forcing such alterations upon the candidate for admission to consciousness as are pleasant to itself. We are here forming a very definite conception of the "essence" of consciousness; for us the state of becoming conscious is a particular psychic act, different from and independent of becoming fixed or of being conceived, and consciousness appears to us as an organ of sense, which perceives a content presented from another source. It may be shown that psychopathology cannot possibly dispense with these fundamental assumptions. We may reserve a more thorough examination of these for a later time.

If I keep in mind the idea of the two psychic instances and their relations to consciousness, I find in the sphere of politics a very exact analogy for the extraordinary affection which I feel for my friend R., who suffers such degradation in the course of the dream interpretation. I turn my attention to a political state in which a ruler, jealous of his rights, and a live public opinion are in conflict with each other. The people are indignant against an official whom they hate, and demand his dismissal; and in order not to show that he is compelled to respect the public wish, the autocrat will expressly confer upon the official some great honour, for which there would otherwise have been no occasion. Thus the second instance referred to, which controls access to consciousness, honours my friend R. with a profusion of extraordinary tenderness, because the wish activities of the first system, in accordance with a particular interest which they happen to be pursuing, are inclined to put him down as a simpleton.[4]

Perhaps we shall now begin to suspect that dream interpretation is capable of giving us hints about the structure of our psychic apparatus which we have thus far expected in vain from philosophy. We shall not, however, follow this track, but return to our original problem as soon as we have cleared up the subject of dream-disfigurement. The question has arisen how dreams with disagreeable content can be analysed as the fulfilments of wishes. We see now that this is possible in case dream-disfigurement has taken place, in case the disagreeable content serves only as a disguise for what is wished. Keeping in mind our assumptions in regard to the two psychic instances, we may now proceed to say: disagreeable dreams, as a matter of fact, contain something which is disagreeable to the second instance, but which at the same time fulfils a wish of the first instance. They are wish dreams in the sense that every dream originates in the first instance, while the second instance acts towards the dream only in a repelling, not in a creative manner. If we limit ourselves to a consideration of what the second instance contributes to the dream, we can never understand the dream. If we do so, all the riddles which the authors have found in the dream remain unsolved.

That the dream actually has a secret meaning, which turns out to be the fulfilment of a wish, must be proved afresh for every case by means of an analysis. I therefore select several dreams which have painful contents and attempt an analysis of them. They are partly dreams of hysterical subjects, which require long preliminary statements, and now and then also an examination of the psychic processes which occur in hysteria. I cannot, however, avoid this added difficulty in the exposition.

When I give a psychoneurotic patient analytical treatment, dreams are always, as I have said, the subject of our discussion. It must, therefore, give him all the psychological explanations through whose aid I myself have come to an understanding of his symptoms, and here I undergo an unsparing criticism, which is perhaps not less keen than that I must expect from my colleagues. Contradiction of the thesis that all dreams are the fulfilments of wishes is raised by my patients with perfect regularity. Here are several examples of the dream material which is offered me to refute this position.

"You always tell me that the dream is a wish fulfilled," begins a clever lady patient. "Now I shall tell you a dream in which the content is quite the opposite, in which a wish of mine is *not* fulfilled. How do you reconcile that with your theory? The dream is as follows:—

"I want to give a supper, but having nothing at hand except some smoked salmon, I think of going marketing, but I remember that it is Sunday afternoon, when all the shops are closed. I next try to telephone to some caterers, but the telephone is out of order. Thus I must resign my wish to give a supper."

I answer, of course, that only the analysis can decide the meaning of this dream, although I admit that at first sight it seems sensible and coherent, and looks like the opposite of a wish-fulfilment. "But what occurrence has given rise to this dream?" I ask. "You know that the stimulus for a dream always lies among the experiences of the preceding day."

Analysis

The husband of the patient, an upright and conscientious wholesale butcher, had told her the day before that he is growing too fat, and that he must, therefore, begin treatment for obesity. He was going to get up early, take exercise, keep to a strict diet, and above all accept no more invitations to suppers. She proceeds laughingly to relate how her husband at an inn table had made the acquaintance of an artist, who insisted upon painting his portrait because he, the painter, had never found such an expressive head. But her husband had answered in his rough way, that he was very thankful for the honour, but that he was quite convinced that a portion of the backside of a pretty young girl would please the artist better than his whole face.[5] She said that she was at the time very much in love with her husband, and teased him a good deal. She had also asked him not to send her any caviare. What does that mean?

As a matter of fact, she had wanted for a long time to eat a caviare sandwich every forenoon, but had grudged herself the expense. Of course, she would at once get the caviare from her husband, as soon as she asked him for it. But she had begged him, on the contrary, not to send her the caviare, in order that she might tease him about it longer.

This explanation seems far-fetched to me. Unadmitted motives are in the habit of hiding behind such unsatisfactory explanations. We are reminded of subjects hypnotised by Bernheim, who carried out a posthypnotic order, and who, upon being asked for their motives, instead of answering: "I do not know why I did that," had to invent a reason that was obviously inadequate. Something similar is probably the case with the caviare of my patient. I see that she is compelled to create an unfulfilled wish in life. Her dream also shows the reproduction of the wish as accomplished. But why does she need an unfulfilled wish?

The ideas so far produced are insufficient for the interpretation of the dream. I beg for more. After a short pause, which corresponds to the overcoming of a resistance, she reports further that the day before she had made a visit to a friend, of whom she is really jealous, because her husband is always praising this woman so much. Fortunately, this friend is very lean and thin, and her husband likes well-rounded figures. Now of what did this lean friend speak? Naturally of her wish to become somewhat stouter. She also asked my patient: "When are you going to invite us again? You always have such a good table."

Now the meaning of the dream is clear. I may say to the patient: "It is just as though you had thought at the time of the request: 'Of course, I'll invite you, so you can eat yourself fat at my house and become still more pleasing to my husband. I would rather give no more suppers.' The dream then tells you that you cannot give a supper, thereby fulfilling your wish not to contribute anything to the rounding out of your friend's figure. The resolution of your husband to refuse invitations to supper for the sake of getting thin teaches you that one grows fat on the things served in company." Now only some conversation is necessary to confirm the solution. The smoked salmon in the dream has not yet been traced. "How did the salmon mentioned in the dream occur to you?" "Smoked salmon is the favourite dish of this friend," she answered. I happen to know the lady, and may corroborate this by saying that she grudges herself the salmon just as much as my patient grudges herself the caviare.

The dream admits of still another and more exact interpretation, which is necessitated only by a subordinate circumstance. The two interpretations do not contradict one another, but rather cover each other and furnish a neat example of the usual ambiguity of dreams as well as of all other psychopathological formations. We have seen that at the same time that she dreams of the denial of the wish, the patient is in reality occupied in securing an unfulfilled wish (the caviare sandwiches). Her friend, too, had expressed a wish, namely, to get fatter, and it would not surprise us if our lady had dreamt that the wish of the friend was not being fulfilled. For it is her own wish that a wish of her friend's—for increase in weight—should not be fulfilled. Instead of this, however, she dreams that one of her own wishes is not fulfilled. The dream becomes capable of a new interpretation, if in the dream she does not intend herself, but her friend, if she has put herself in the place of her friend, or, as we may say, has identified herself with her friend.

I think she has actually done this, and as a sign of this identification she has created an unfulfilled wish in reality. But what is the meaning of this hysterical identification? To clear this up a thorough exposition is necessary. Identification is a highly important factor in the mechanism of hysterical symptoms; by this means patients are enabled in their symptoms to represent not merely their own experiences, but the experiences of a great number of other persons, and can suffer, as it were, for a whole mass of people, and fill all the parts of a drama by means of their own personalities alone. It will here be objected that this is well-known hysterical imitation, the ability of hysteric subjects to copy all the symptoms which impress them when they occur in others, as though their pity were stimulated to the point of reproduction. But this only indicates the way in which the psychic process is discharged in hysterical imitation; the way in which a psychic act proceeds and the act itself are two different things. The latter is slightly more complicated than one is apt to imagine the imitation of hysterical subjects to be: it corresponds to an unconscious concluded process, as an example will show. The physician who has a female patient with a particular kind of twitching, lodged in the company of other patients in the same room of the hospital, is not surprised when some morning he learns that this peculiar hysterical attack has found imitations. He simply says to himself: The others have seen her and have done likewise: that is psychic infection. Yes, but psychic infection proceeds in somewhat the following manner: As a rule, patients know more about one another than the physician knows about each of them, and they are concerned about each other when the visit of the doctor is over. Some of them have an attack to-day: soon it is known among the rest that a letter from home, a return of love-sickness or the like, is the cause of it. Their sympathy is aroused, and the following syllogism, which does not reach consciousness, is completed in them: "If it is possible to have this kind of an attack from such causes, I too may have this kind of an attack, for I have the same reasons." If this were a cycle capable of becoming conscious, it would perhaps express itself in *fear* of getting the same attack; but it takes place in another psychic sphere, and, therefore, ends in the realisation of the dreaded symptom. Identification is therefore not a simple imitation, but a sympathy based upon the same etiological claim; it expresses an "as though," and refers to some common quality which has remained in the unconscious.

Identification is most often used in hysteria to express sexual community. An hysterical woman identifies herself most readily—although not exclusively—with persons with whom she has had sexual relations, or who have sexual intercourse with the same persons as herself. Language takes such a conception into consideration: two lovers are "one." In the hysterical phantasy, as well as in the dream, it is sufficient for the identification if one thinks of sexual relations, whether or not they become real. The patient, then, only follows the rules of the hysterical thought processes when she gives expression to her jealousy of her friend (which, moreover, she herself admits to be unjustified, in that she puts herself in her place and identifies herself with her by creating a symptom—the denied wish). I might further clarify the process specifically as follows: She puts herself in the place of her friend in the dream, because her friend has taken her own place in relation to her husband, and because she would like to take her friend's place in the esteem of her husband.[6]

The contradiction to my theory of dreams in the case of another female patient, the most witty among all my dreamers, was solved in a simpler manner, although according to the scheme that the non-fulfilment of one wish signifies the fulfilment of another. I had one day explained to her that the dream is a wish-fulfilment. The next day she brought me a dream to the effect that she was travelling with her mother-in-law to their common summer resort. Now I knew that she had struggled violently against spending the summer in the neighbourhood of her mother-in-law. I also knew that she had luckily avoided her mother-in-law by renting an estate in a far-distant country resort. Now the dream reversed this wished-for solution; was not this in the flattest contradiction to my theory of wish-fulfilment in the dream? Certainly, it was only necessary to draw the inferences from this dream in order to get at its interpretation. According to this dream, I was in the wrong. *It was thus her wish that I should be in the wrong, and this wish the dream showed her as fulfilled.* But the wish that I should be in the wrong, which was fulfilled in the theme of the country home, referred to a more serious matter. At that time I had made up my mind, from the material furnished by her analysis, that something of significance for her illness must have occurred at a certain time in her Me. She had denied it because it was not present in her memory. We soon came to see that I was in the right. Her wish that I should be in the wrong, which is transformed into the dream, thus corresponded to the justifiable wish that those things, which at the time had only been suspected, had never occurred at all.

Without an analysis, and merely by means of an assumption, I took the liberty of interpreting a little occurrence in the case of a friend, who had been my colleague through the eight classes of the Gymnasium. He once heard a lecture of mine delivered to a small assemblage, on the novel subject of the dream as the fulfilment of a wish. He went home, dreamt *that he had lost all his suits*—he was a lawyer—and then complained to me about it. I took refuge in the evasion: "One can't win all one's suits," but I thought to myself: "If for eight years I sat as Primus on the first bench, while he moved around somewhere in the middle of the class, may he not naturally have had a wish from his boyhood days that I, too, might for once completely disgrace myself?"

In the same way another dream of a more gloomy character was offered me by a female patient as a contradiction to my theory of the wish-dream. The patient, a young girl, began as follows: "You remember that my sister has now only one boy, Charles: she lost the elder one, Otto, while I was still at her house. Otto was my favourite; it was I who really brought him up. I like the other little fellow, too, but of course not nearly as much as the dead one. Now I dreamt last night that *I saw Charles lying dead before me. He was lying in his little coffin, his hands folded: there were candles all about, and, in short, it was just like the time of little Otto's death, which shocked me so profoundly.* Now tell me, what does this mean? You know me: am I really bad enough to wish my sister to lose the only child she has left? Or does the dream mean that I wish Charles to be dead rather than Otto, whom I like so much better?"

I assured her that this interpretation was impossible. After some reflection I was able to give her the interpretation of the dream, which I subsequently made her confirm.

Having become an orphan at an early age, the girl had been brought up in the house of a much older sister, and had met among the friends and visitors who came to the house, a man who made a lasting impression upon her heart. It looked for a time as though these barely expressed relations were to end in marriage, but this happy culmination was frustrated by the sister, whose motives have never found a complete explanation. After the break, the man who was loved by our patient avoided the house: she herself became independent some time after little Otto's death, to whom her affection had now turned. But she did not succeed in freeing herself from the inclination for her sister's friend in which she had become involved. Her pride commanded her to avoid him; but it was impossible for her to transfer her love to the other suitors who presented themselves in order. Whenever the man whom she loved, who was a member of the literary profession, announced a lecture anywhere, she was sure to be found in the audience; she also seized every other opportunity to see him from a distance unobserved by him. I remembered that on the day before she had told me that the Professor was going to a certain concert, and that she was also going there, in order to enjoy the sight of him. This was on the day of the dream; and the concert was to take place on the day on which she told me the dream. I could now easily see the correct interpretation, and I asked her whether she could think of any event which had happened after the death of little Otto. She answered immediately: "Certainly; at that time the Professor returned after a long absence, and I saw him once more beside the coffin of little Otto." It was exactly as I had expected. I interpreted the dream in the following manner: If now the other boy were to die, the same thing would be repeated. You would spend the day with your sister, the Professor would surely come in order to offer condolence, and you would see him again under the same circumstances as at that time. The dream signifies nothing but this wish of yours to see him again, against which you are fighting inwardly. I know that you are carrying the ticket for to-day's concert in your bag. Your dream is a dream of impatience; it has anticipated the meeting which is to take place to-day by several hours."

In order to disguise her wish she had obviously selected a situation in which wishes of that sort are commonly suppressed—a situation which is so filled with sorrow that love is not thought of. And yet, it is very easily probable that even in the actual situation at the bier of the second, more dearly loved boy, which the dream copied faithfully, she had not been able to suppress her feelings of affection for the visitor whom she had missed for so long a time.

A different explanation was found in the case of a similar dream of another female patient, who was distinguished in her earlier years by her quick wit and her cheerful demeanour, and who still showed these qualities at least in the notions which occurred to her in the course of treatment. In connection with a longer dream, it seemed to this lady that she saw her fifteen-year-old daughter lying dead before her in a box. She was strongly inclined to convert this dream-image into an objection to the theory of wish-fulfilment, but herself suspected that the detail of the box must lead to a different conception of the dream.[7] In the course of the analysis it occurred to her that on the evening before, the conversation of the company had turned upon the English word "box," and upon the numerous translations of it into German, such as box, theatre box, chest, box on the ear, &c. From other components of the same dream it is now possible to add that the lady had guessed the relationship between the English word "box" and the German *Büchse,* and had then been haunted by the memory that *Büchse* (as well as "box") is used in vulgar speech to designate the female genital organ. It was therefore possible, making a certain allowance for her notions on the subject of topographical anatomy, to assume that the child in the box signified a child in the womb of the mother. At this stage of the explanation she no longer denied that the picture of the dream really corresponded to one of her wishes. Like so many other young women, she was by no means happy when she became pregnant, and admitted to me more than once the wish that her child might die before its birth; in a fit of anger following a violent scene with her husband she had even struck her abdomen with her fists in order to hit the child within. The dead child was, therefore, really the fulfilment of a wish, but a wish which had been put aside for fifteen years, and it is not surprising that the fulfilment of the wish was no longer recognised after so long an interval. For there had been many changes meanwhile.

The group of dreams to which the two last mentioned belong, having as content the death of beloved relatives, will be considered again under the head of "Typical Dreams." I shall there be able to show by new examples that in spite of their undesirable content, all these dreams must be interpreted as wish-fulfilments. For the following dream, which again was told me in order to deter me from a hasty generalisation of the theory of wishing in dreams. I am indebted, not to a patient, but to an intelligent jurist of my acquaintance. *"I dream,"* my informant tells me, *"that I am walking in front of my house with a lady on my arm. Here a closed wagon is waiting, a gentleman steps up to me, gives his authority as an agent of the police, and demands that I should follow him. I only ask for time in which to arrange my affairs.* Can you possibly suppose this is a wish of mine to be arrested?" "Of course not," I must admit. "Do you happen to know upon what charge you were arrested?" "Yes; I believe for infanticide." "Infanticide? But you know that only a mother can commit this crime upon her newly born child?" "That is true."[8] "And under what circumstances did you dream; what happened on the evening before?" "I would rather not tell you that; it is a delicate matter." "But I must have it, otherwise we must forgo the interpretation of the dream." "Well, then, I will tell you. I spent the night, not at home, but at the house of a lady who means very much to me. When we awoke in the morning, something again passed between us. Then I went to sleep again, and dreamt what I have told you." "The woman is married?" "Yes." "And you do not wish her to conceive a child?" "No; that might betray us." "Then you do not practise normal coitus?" "I take the precaution to withdraw before ejaculation." "Am I permitted to assume that you did this trick several times during the night, and that in the morning you were not quite sure whether you had succeeded?" "That might be the case." "Then your dream is the fulfilment of a wish. By means of it you secure the assurance that you have not begotten a child, or, what amounts to the same thing, that you have killed a child. I can easily demonstrate the connecting links. Do you remember, a few days ago we were talking about the distress of matrimony (Ehenot), and about the inconsistency of permitting the practice of coitus as long as no impregnation takes place, while every delinquency after the ovum and the semen meet and a foetus is formed is punished as a crime? In connection with this, we also recalled the mediaeval controversy about the moment of time at which the soul is really lodged in the foetus, since the concept of murder becomes admissible only from that point on. Doubtless you also know the gruesome poem by Lenau, which puts infanticide and the prevention of children on the same plane." "Strangely enough, I had happened to think of Lenau during the afternoon." "Another echo of your dream. And now I shall demonstrate to you another subordinate wish-fulfilment in your dream. You walk in front of your house with the lady on your arm. So you take her home, instead of spending the night at her house, as you do in actuality. The fact that the wish-fulfilment, which is the essence of the dream, disguises itself in such an unpleasant form, has perhaps more than one reason. From my essay on the etiology of anxiety neuroses, you will see that I note interrupted coitus as one of the factors which cause the development of neurotic fear. It would be consistent with this that if after repeated cohabitation of the kind mentioned you should be left in an uncomfortable mood, which now becomes an element in the composition of your dream. You also make use of this unpleasant state of mind to conceal the wish-fulfilment. Furthermore, the mention of infanticide has not yet been explained. Why does this crime, which is peculiar to females, occur to you?" "I shall confess to you that I was involved in such an affair years ago. Through my fault a girl tried to protect herself from the consequences of a *liaison* with me by securing an abortion. I had nothing to do with carrying out the plan, but I was naturally for a long time worried lest the affair might be discovered." "I understand; this recollection furnished a second reason why the supposition that you had done your trick badly must have been painful to you."

A young physician, who had heard this dream of my colleague when it was told, must have felt implicated by it, for he hastened to imitate it in a dream of his own, applying its mode of thinking to another subject. The day before he had handed in a declaration of his income, which was perfectly honest, because he had little to declare. He dreamt that an acquaintance of his came from a meeting of the tax commission and informed him that all the other declarations of income had passed uncontested, but that his own had awakened general suspicion, and that he would be punished with a heavy fine. The dream is a poorly-concealed fulfilment of the wish to be known as a physician with a large income. It likewise recalls the story of the young girl who was advised against accepting her suitor because he was a man of quick temper who would surely treat her to blows after they were married. The answer of the girl was: "I wish he *would* strike me!" Her wish to be married is so strong that she takes into the bargain the discomfort which is said to be connected with matrimony, and which is predicted for her, and even raises it to a wish.

If I group the very frequently occurring dreams of this sort, which seem flatly to contradict my theory, in that they contain the denial of a wish or some occurrence decidedly unwished for, under the head of "counter wish-dreams," I observe that they may all be referred to two principles, of which one has not yet been mentioned, although it plays a large part in the dreams of human beings. One of the motives inspiring these dreams is the wish that I should appear in the wrong. These dreams regularly occur in the course of my treatment if the patient shows a resistance against me, and I can count with a large degree of certainty upon causing such a dream after I have once explained to the patient my theory that the dream is a wish-fulfilment.[9] I may even expect this to be the case in a dream merely in order to fulfil the wish that I may appear in the wrong. The last dream which I shall tell from those occurring in the course of treatment again shows this very thing. A young girl who has struggled hard to continue my treatment, against the will of her relatives and the authorities whom she has consulted, dreams as follows: *She is forbidden at home to come to me any more. She then reminds me of the promise I made her to treat her for nothing if necessary, and I say to her: "I can show no consideration in money matters."*

It is not at all easy in this case to demonstrate the fulfilment of a wish, but in all cases of this kind there is a second problem, the solution of which helps also to solve the first. Where does she get the words which she puts into my mouth? Of course I have never told her anything like that, but one of her brothers, the very one who has the greatest influence over her, has been kind enough to make this remark about me. It is then the purpose of the dream that this brother should remain in the right; and she does not try to justify this brother merely in the dream; it is her purpose in life and the motive for her being ill.

The other motive for counter wish-dreams is so clear that there is danger of overlooking it, as for some time happened in my own case. In the sexual make-up of many people there is a masochistic component, which has arisen through the conversion of the aggressive, sadistic component into its opposite. Such people are called "ideal" masochists, if they seek pleasure not in the bodily pain which may be inflicted upon them, but in humiliation and in chastisement of the soul. It is obvious that such persons can have counter wish-dreams and disagreeable dreams, which, however, for them are nothing but wish-fulfilments, affording satisfaction for their masochistic inclinations. Here is such a dream. A young man, who has in earlier years tormented his elder brother, towards whom he was homosexually inclined, but who has undergone a complete change of character, has the following dream, which consists of three parts: (1) *He is "insulted" by his brother.* (2) *Two adults are caressing each other with homosexual intentions.* (3) *His brother has sold the enterprise whose management the young man reserved for his own future.* He awakens from the last-mentioned dream with the most unpleasant feelings, and yet it is a masochistic wish-dream, which might be translated: It would serve me quite right if my brother were to make that sale against my interest, as a punishment for all the torments which he has suffered at my hands.

I hope that the above discussion and examples will suffice—until further objection can be raised—to make it seem credible that even dreams with a painful content are to be analysed as the fulfilments of wishes. Nor will it seem a matter of chance that in the course of interpretation one always happens upon subjects of which one does not like to speak or think. The disagreeable sensation which such dreams arouse is simply identical with the antipathy which endeavours—usually with success—to restrain us from the treatment or discussion of such subjects, and which must be overcome by all of us, if, in spite of its unpleasantness, we find it necessary to take the matter in hand. But this disagreeable sensation, which occurs also in dreams, does not preclude the existence of a wish; everyone has wishes which he would not like to tell to others, which he does not want to admit even to himself. We are, on other grounds, justified in connecting the disagreeable character of all these dreams with the fact of dream disfigurement, and in concluding that these dreams are distorted, and that the wish-fulfilment in them is disguised until recognition is impossible for no other reason than that a repugnance, a will to suppress, exists in relation to the subject-matter of the dream or in relation to the wish which the dream creates. Dream disfigurement, then, turns out in reality to be an act of the censor. We shall take into consideration everything which the analysis of disagreeable dreams has brought to light if we reword our formula as follows: *The dream is the (disguised) fulfilment of a (suppressed, repressed) wish.*[10]

Now there still remain as a particular species of dreams with painful content, dreams of anxiety, the inclusion of which under dreams of wishing will find least acceptance with the uninitiated. But I can settle the problem of anxiety dreams in very short order; for what they may reveal is not a new aspect of the dream problem; it is a question in their case of understanding neurotic anxiety in general. The fear which we experience in the dream is only seemingly explained by the dream content. If we subject the content of the dream to analysis, we become aware that the dream fear is no more justified by the dream content than the fear in a phobia is justified by the idea upon which the phobia depends. For example, it is true that it is possible to fall out of a window, and that some care must be exercised when one is near a window, but it is inexplicable why the anxiety in the corresponding phobia is so great, and why it follows its victims to an extent so much greater than is warranted by its origin. The same explanation, then, which applies to the phobia applies also to the dream of anxiety. In both cases the anxiety is only superficially attached to the idea which accompanies it and comes from another source.

On account of the intimate relation of dream fear to neurotic fear, discussion of the former obliges me to refer to the latter. In a little essay on "The Anxiety Neurosis,"[11] I maintained that neurotic fear has its origin in the sexual life, and corresponds to a libido which has been turned away from its object and has not succeeded in being applied. From this formula, which has since proved its validity more and more clearly, we may deduce the conclusion that the content of anxiety dreams is of a sexual nature, the libido belonging to which content has been transformed into fear. Later on I shall have opportunity to support this assertion by the analysis of several dreams of neurotics. I shall have occasion to revert to the determinations in anxiety dreams and their compatibility with the theory of wish-fulfilment when I again attempt to approach the theory of dreams.

Notes:

1. It is quite incredible with what stubbornness readers and critics exclude this consideration, and leave unheeded the fundamental differentiation between the manifest and the latent dream content.

2. It is remarkable how my memory narrows here for the purposes of analysis—while I am awake. I have known five of my uncles, and have loved and honoured one of them. But at the moment when I overcame my resistance to the interpretation of the dream I said to myself, "I have only one uncle, the one who is intended in the dream."

3. The word is here used in the original Latin sense *instantia,* meaning energy, continuance or persistence in doing. (Translator.)

4. Such hypocritical dreams are not unusual occurrences with me or with

others. While I am working up a certain scientific problem, I am visited for many nights in rapid succession by a somewhat confusing dream which has as its content reconciliation with a friend long ago dropped. After three or four attempts, I finally succeeded in grasping the meaning of this dream. It was in the nature of an encouragement to give up the little consideration still left for the person in question, to drop him completely, but it disguised itself shamefacedly in the opposite feeling. I have reported a "hypocritical oedipus dream" of a person, in which the hostile feelings and the wishes of death of the dream thoughts were replaced by manifest tenderness. ("Typisches Beispiel eines verkappten Oedipustraumes," *Zentralblatt für Psychoanalyse,* Bd. 1, Heft 1–11, 1910.) Another class of hypocritical dreams will be reported in another place.

5. To sit for the painter. Goethe: "And if he has no backside, how can the nobleman sit?"

6. I myself regret the introduction of such passages from the psychopathology of hysteria, which, because of their fragmentary representation and of being torn from all connection with the subject, cannot have a very enlightening influence. If these passages are capable of throwing light upon the intimate relations between the dream and the psychoneuroses, they have served the purpose for which I have taken them up.

7. Something like the smoked salmon in the dream of the deferred supper.

8. It often happens that a dream is told incompletely, and that a recollection of the omitted portions appears only in the course of the analysis. These portions subsequently fitted in, regularly furnish the key to the interpretation. *Cf.* below, about forgetting in dreams.

9. Similar "counter wish-dreams" have been repeatedly reported to me within the last few years by my pupils who thus reacted to their first encounter with the "wish theory of the dream."

10. We may mention here the simplification and modification of this fundamental formula, propounded by Otto Rank: "On the basis and with the help of repressed infantile sexual material, the dream regularly represents as fulfilled actual, and as a rule also erotic, wishes, in a disguised and symbolic form." ("Ein Traum, der sich selbst deutet," *Jahrbuch,* v., Bleuler-Freud, II. B., p. 519, 1910.)

11. See *Selected Papers on Hysteria and other Psychoneuroses,* p. 133, translated by A. A. Brill, *Journal of Nervous and Mental Diseases,* Monograph Series.

5

The Material
and Sources of Dreams

After coming to realise from the analysis of the dream of Irma's injection that the dream is the fulfilment of a wish, our interest was next directed to ascertaining whether we had thus discovered a universal characteristic of the dream, and for the time being we put aside every other question which may have been aroused in the course of that interpretation. Now that we have reached the goal upon one of these paths, we may turn back and select a new starting-point for our excursions among the problems of the dream, even though we may lose sight for a time of the theme of wish-fulfilment, which has been as yet by no means exhaustively treated.

Now that we are able, by applying our process of interpretation, to discover a *latent* dream content which far surpasses the *manifest* dream content in point of significance, we are impelled to take up the individual dream problems afresh, in order to see whether the riddles and contradictions which seemed, when we had only the manifest content, beyond our reach may not be solved for us satisfactorily.

The statements of the authors concerning the relation of the dream to waking life, as well as concerning the source of the dream material, have been given at length in the introductory chapter. We may recall that there are three peculiarities of recollection in the dreams, which have been often remarked but never explained:

1. That the dream distinctly prefers impressions of the few days preceding (Robert, Strümpell, Hildebrandt, also Weed-Hallam).
2. That it makes its selection according to principles other than those of our waking memory, in that it recalls not what is essential and important, but what is subordinate and disregarded.
3. That it has at its disposal the earliest impressions of our childhood, and brings to light details from this period of life which again seem trivial to us, and which in waking life were considered long ago forgotten.[1]

These peculiarities in the selection of the dream material have of course been observed by the authors in connection with the manifest dream content.

(A) Recent and Indifferent Impressions in the Dream

If I now consult my own experience concerning the source of the elements which appear in the dream, I must at once express the opinion that some reference to the experiences of *the day which has most recently passed* is to be found in every dream. Whatever dream I take up, whether my own or another's, this experience is always re-affirmed. Knowing this fact, I can usually begin the work of interpretation by trying to learn the experience of the previous day which has stimulated the dream; for many cases, indeed, this is the quickest way. In the case of the two dreams which I have subjected to close analysis in the preceding chapter (of Irma's injection, and of my uncle with the yellow beard) the reference to the previous day is so obvious that it needs no further elucidation. But in order to show that this reference may be regularly demonstrated, I shall examine a portion of my own dream chronicle. I shall report the dreams only so far as is necessary for the discovery of the dream stimulus in question.

1. I make a visit at a house where I am admitted only with difficulty, &c., and meanwhile I keep a woman waiting for me.

    Source.—A conversation in the evening with a female relative

138

to the effect that she would have to *wait* for some aid which she demanded until, &c.

2. I have written a *monograph* about a certain (obscure) species of plant.

    Source.—I have seen in the show-window of a book store a *monograph* upon the genus cyclamen.

3. I see two women on the street, *mother and daughter,* the latter of whom is my patient.

    Source.—A female patient who is under treatment has told me what difficulties her *mother* puts in the way of her continuing the treatment.

4. At the book store of S. and R. I subscribe to a periodical which costs 20 *florins* annually.

    Source.—During the day my wife has reminded me that I still owe her 20 *florins* of her weekly allowance.

5. I receive a *communication,* in which I am treated as a member, from the Social Democratic Committee.

    Source.—I have received *communications* simultaneously from the Liberal Committee on Elections and from the president of the Humanitarian Society, of which I am really a member.

6. A man on a *steep rock in the middle of the ocean,* after the manner of Boecklin.

    Source.—*Dreyfus on Devil's Island;* at the same time news from my relatives in *England,* &c.

The question might be raised, whether the dream is invariably connected with the events of the previous day, or whether the reference may be extended to impressions from a longer space of time in the immediate past. Probably this matter cannot claim primary importance, but I should like to decide in favour of the exclusive priority of the day before the dream (the dream-day). As often as I thought I had found a case where an impression of two or three days before had been the source of the dream, I could convince myself, after careful investigation, that this impression had been remembered the day before, that a demonstrable reproduction had been interpolated between the day of the event and the time of the dream, and, furthermore, I was able to point out the recent occasion upon which the recollection of the old impression might have occurred. On the other hand, I was unable to convince myself that a regular interval (H. Swoboda calls the first one of this kind eighteen hours) of biological significance occurs between the stimulating impression of the day and its repetition in the dream.[2]

I am, therefore, of the opinion that the stimulus for every dream is to be found among those experiences "upon which one has not yet slept" for a night.

Thus the impressions of the immediate past (with the exception of the day before the night of the dream) stand in no different relation to the dream content from those of times which are as far removed in the past as you please. The dream may select its material from all times of life, provided only, that a chain of thought starting from one of the experiences of the day of the dream (one of the "recent" impressions) reaches back to these earlier ones.

But why this preference for recent impressions? We shall reach some conjectures on this point if we subject one of the dreams already mentioned to a more exact analysis. I select the dream about the monograph.

Content of the dream

I have written a monograph upon a certain plant. The book lies before me, I am just turning over a folded coloured plate. A dried specimen of the plant is bound with every copy, as though from a herbarium.

Analysis

In the forenoon I saw in the show-window of a book store a book entitled, *The Genus Cyclamen,* apparently a monograph on this plant.

The cyclamen is the favourite flower of my wife. I reproach myself for so seldom thinking to bring her flowers, as she wishes. In connection with the theme "bringing flowers," I am reminded of a story which I recently told in a circle of friends to prove my assertion that forgetting is very often the purpose of the unconscious, and that in any case it warrants a conclusion as to the secret disposition of the person who forgets. A young woman who is accustomed to receive a bunch of flowers from her husband on her birthday, misses this token of affection on a festive occasion of this sort, and thereupon bursts into tears. The husband comes up, and is unable to account for her tears until she tells him, "To-day is my birthday." He strikes his forehead and cries, "Why, I had completely forgotten it," and wants to go out to get her some flowers. But she is not to be consoled, for she sees in the forgetfulness of her husband a proof that she does not play the same part in his thoughts as formerly. This Mrs. L. met my wife two days before, and told her that she was feeling well, and asked about me. She was under my treatment years ago.

Supplementary facts: I once actually wrote something like a monograph on a plant, namely, an essay on the coca plant, which drew the attention of K. Roller to the anaesthetic properties of cocaine. I had hinted at this use of the alkaloid in my publication, but I was not sufficiently thorough to pursue the matter further. This suggests that on the forenoon of the day after the dream (for the interpretation of which I did not find time until the evening) I had thought of cocaine in a kind of day phantasy. In case I should ever be afflicted with glaucoma, I was going to go to Berlin, and there have myself operated upon, incognito, at the house of my Berlin friend, by a physician whom he would recommend to me. The surgeon, who would not know upon whom he was operating, would boast as usual how easy these operations had become since the introduction of cocaine; I would not betray by a single sign that I had had a share in making this discovery. With this phantasy were connected thoughts of how difficult it really is for a doctor to claim the medical services of a colleague for his own person. I should be able to pay the Berlin eye specialist, who did not know me, like anyone else. Only after recalling this day-dream do I realise that the recollection of a definite experience is concealed behind it. Shortly after Koller's discovery my father had, in fact, become ill with glaucoma; he was operated upon by my friend, the eye specialist, Dr. Koenigstein. Dr. Koller attended to the cocaine anaesthetisation, and thereupon made the remark that all three of the persons who had shared in the introduction of cocaine had been brought together on one case.

I now proceed to think of the time when I was last reminded of this affair about the cocaine. This was a few days before, when I received a *Festschrift,* with whose publication grateful scholars had commemorated the anniversary of their teacher and laboratory director. Among the honours ascribed to persons connected with the laboratory, I found a notice to the effect that the discovery of the anaesthetic properties of cocaine had been made there by K. Koller. Now I suddenly become aware that the dream is connected with an experience of the previous evening. I had just accompanied Dr. Koenigstein to his home, and had spoken to him about a matter which strongly arouses my interest whenever it is mentioned. While I was talking with him in the vestibule, Professor Gärtner and his young wife came up. I could not refrain from congratulating them both upon their healthy appearance. Now Professor Gärtner is one of the authors of the *Festschrift* of which I have just spoken, and may well have recalled it to me. Likewise Mrs. L., whose birthday disappointment I have referred to, had been mentioned, in another connection, to be sure, in the conversation with Dr. Koenigstein.

I shall now try to explain the other determinations of the dream content. *A dried specimen* of the plant accompanies the monograph as though it were a *herbarium.* A recollection of the *gymnasium* (school) is connected with the herbarium. The director of our *gymnasium* once called the scholars of the higher classes together in order to have them inspect and clean the herbarium. Small worms had been found—bookworms. The director did not seem to have much confidence in my help, for he left only a few leaves for me. I know to this day that there were crucifers on them. My interest in botany was never very great. At my preliminary examination in botany, I was required to identify a crucifer, and did not recognise it. I would have fared badly if my theoretical knowledge had not helped me out. Crucifers suggest composites. The artichoke is really a composite, and the one which I might call my favourite flower. My wife, who is more thoughtful than I, often brings this favourite flower of mine home from the market.

I see the monograph which I have written lying before me. This, too, is not without its reference. The friend whom I pictured wrote to me yesterday from Berlin: "I think a great deal about your dream book. *I see it lying before me finished, and am turning over its leaves."* How I envied him this prophetic power! If I could only see it lying already finished before me!

The folded Coloured Plate.—While I was a student of medicine, I suffered much from a fondness for studying in *monographs* exclusively. In spite of my limited means, I subscribed to a number of the medical archives, in which the coloured plates gave me much delight. I was proud of this inclination for thoroughness. So, when I began to publish on my own account, I had to draw the plates for my own treatises, and I remember one of them turned out so badly that a kindly-disposed colleague ridiculed me for it. This suggests, I don't know exactly how, a very early memory from my youth. My father once thought it would be a joke to hand over a book with *coloured plates (Description of a Journey in Persia)* to me and my eldest sister for destruction. This was hardly to be justified from an educational point of view. I was at the time five years old, and my sister three, and the picture of our blissfully tearing this book to pieces (like an artichoke, I must add, leaf by leaf) is almost the only one from this time of life which has remained fresh in my memory. When I afterwards became a student, I developed a distinct fondness for collecting and possessing books (an analogy to the inclination for studying from monographs, a hobby which occurs in the dream thoughts with reference to cyclamen and artichoke). I became a book-worm (*cf.* herbarium). I have always referred this first passion of my life—since I am engaging in retrospect—to this childhood impression, or rather I have recognised in this childish scene a "concealing recollection" for my subsequent love of books.[3] Of course I also learned at an early age that our passions are often our sorrows. When I was seventeen years old I had a very respectable bill at the book store, and no means with which to pay it, and my father would hardly accept the excuse that my inclination had not been fixed on something worse. But the mention of this later youthful experience immediately brings me back to my conversation that evening with my friend Dr. Koenigstein. For the talk on the evening of the dream-day brought up the same old reproach that I am too fond of my hobbies.

For reasons which do not belong here, I shall not continue the interpretation of this dream, but shall simply indicate the path which leads to it. In the course of the interpretation, I was reminded of my conversation with Dr. Koenigstein, and indeed of more than one portion of it. If I consider the subjects touched upon in this conversation, the meaning of the dream becomes clear to me. All the thought associations which have been started, about the hobbies of my wife and of myself, about the cocaine, about the difficulty of securing medical treatment from one's colleagues, my preference for monographic studies, and my neglect of certain subjects such as botany—all this continues and connects with some branch of this widely ramified conversation. The dream again takes on the character of a justification, of a pleading for my rights, like the first analysed dream of Irma's injection; it even continues the theme which that dream started, and discusses it with the new subject matter which has accrued in the interval between the two dreams. Even the apparently indifferent manner of expression of the dream receives new importance. The meaning is now: "I am indeed the man who has written that valuable and successful treatise (on cocaine)," just as at that time I asserted for my justification: "I am a thorough and industrious student;" in both cases, then: "I can afford to do that." But I may dispense with the further interpretation of the dream, because my only purpose in reporting it was to examine the relation of the dream content to the experience of the previous day which arouses it. As long as I know only the manifest content of this dream, but one relation to a day impression becomes obvious; after I have made the interpretation, a second source of the dream becomes evident in another experience of the same day. The first of these impressions to which the dream refers is an indifferent one, a subordinate circumstance. I see a book in a shop window whose title holds me for a moment, and whose contents could hardly interest me. The second experience has great psychic value; I have talked earnestly with my friend, the eye specialist, for about an hour, I have made allusions in this conversation which must have touched both of us closely, and which awakened memories revealing the most diverse feelings of my inner self. Furthermore, this conversation was broken off unfinished because some friends joined us. What, now, is the relation of these two impressions of the day to each other and to the dream which followed during the next night?

I find in the manifest content merely an allusion to the indifferent impression, and may thus reaffirm that the dream preferably takes up into its content non-essential experiences. In the dream interpretation, on the contrary, everything converges upon an important event which is justified in demanding attention. If I judge the dream in the only correct way, according to the latent content which is brought to light in the analysis, I have unawares come upon a new and important fact. I see the notion that the dream deals only with the worthless fragments of daily experience shattered; I am compelled also to contradict the assertion that our waking psychic life is not continued in the dream, and that the dream instead wastes psychic activity upon a trifling subject matter. The opposite is true; what has occupied our minds during the day also dominates our dream thoughts, and we take pains to dream only of such matters as have given us food for thought during the day.

Perhaps the most obvious explanation for the fact that I dream about some indifferent impression of the day, while the impression which is justifiably stirring furnishes the occasion for dreaming, is that this again is a phenomenon of the dream-disfigurement, which we have above traced to a psychic power acting as a censor. The recollection of the monograph on the genus cyclamen is employed as though it were an allusion to the conversation with my friend, very much as mention of the friend in the dream of the deferred supper is represented by the allusion "smoked salmon." The only question is, by what intermediate steps does the impression of the monograph come to assume the relation of an allusion to the conversation with the eye specialist, since such a relation is not immediately evident. In the example of the deferred supper, the relation is set forth at the outset; "smoked salmon," as the favourite dish of the friend, belongs at once to the series of associations which the person of the friend would call up in the lady who is dreaming. In our new example we have two separated impressions, which seem at first glance to have nothing in common except that they occur on the same day. The monograph catches my attention in the forenoon; I take part in the conversation in the evening. The answer supplied by the analysis is as follows: Such relations between the two impressions do not at first exist, but are established subsequently between the presentation content of the one impression and the presentation content of the other. I have recently emphasised the components in this relation in the course of recording the analysis. With the notion of the monograph on cyclamen I should probably associate the idea that cyclamen is my wife's favourite flower only under some outside influence, and this is perhaps the further recollection of the bunch of flowers missed by Mrs. L. I do not believe that these underlying thoughts would have been sufficient to call forth a dream.

> "There needs no ghost, my lord, come from the grave
> To tell us this,"

as we read in *Hamlet*. But behold! I am reminded in the analysis that the name of the man who interrupted our conversation was Gärtner (Gardener), and that I found his wife in blooming health;[4] I even remember now that one of my female patients, who bears the pretty name of Flora, was for a time the main subject of our conversation. It must have happened that I completed the connection between the two events of the day, the indifferent and the exciting one, by means of these links from the series of associations belonging to the idea of botany. Other relations are then established, that of cocaine, which can with perfect correctness form a go-between connecting the person of Dr. Koenigstein with the botanical monograph which I have written, and strengthen the fusion of the two series of associations into one, so that now a portion of the first experience may be used as an allusion to the second.

I am prepared to find this explanation attacked as arbitrary or artificial. What would have happened if Professor Gärtner and his blooming wife had not come up, and if the patient who was talked about had been called, not Flora, but Anna? The answer is easy, however. If these thought-relations had not been present, others would probably have been selected. It is so easy to establish relations of this sort, as the joking questions and conundrums with which we amuse ourselves daily suffice to show. The range of wit is unlimited. To go a step further: if it had been impossible to establish interrelations of sufficient abundance between the two impressions of the day, the dream would simply have resulted differently; another of the indifferent impressions of the day, such as come to us in multitudes and are forgotten, would have taken the place of the monograph in the dream, would have secured a connection with the content of the talk, and would have represented it in the dream. Since it was the impression of the monograph and no other that had this fate, this impression was probably the most suitable for the establishment of the connection. One need not be astonished, like Lessing's Hanschen Schlau, because "it is the rich people of the world who possess the most money."

Still the psychological process by which, according to our conception, the indifferent experience is substituted for the psychologically important one, seems odd to us and open to question. In a later chapter we shall undertake the task of making this seemingly incorrect operation more intelligible. We are here concerned only with consequences of this procedure, whose assumption we have been forced to make by the regularly recurring experiences of dream analysis. But the process seems to be that, in the course of those intermediate steps, a displacement—let us say of the psychic accent—has taken place, until ideas that are at first weakly charged with intensity, by taking over the charge from ideas which have a stronger initial intensity, reach a degree of strength, which enables them to force their way into consciousness. Such displacements do not at all surprise us when it is a question of the bestowal of affects or of the motor actions in general. The fact that the woman who has remained single transfers her affection to animals, that the bachelor becomes a passionate collector, that the soldier defends a scrap of coloured cloth, his flag, with his life-blood, that in a love affair a momentary clasping of hands brings bliss, or that in *Othello* a lost handkerchief causes a burst of rage—all these are examples of psychic displacement which seem unquestionable to us. But if, in the same manner and according to the same fundamental principles, a decision is made as to what is to reach our consciousness and what is to be withheld from it, that is to say, what we are to think—this produces an impression of morbidity, and we call it an error of thought if it occurs in waking life. We may here anticipate the result of a discussion which will be undertaken later—namely, to the effect that the psychic process which we have recognised as dream displacement proves to be not a process morbidly disturbed, but a process differing from the normal merely in being of a more primitive nature.

We thus find in the fact that the dream content takes up remnants of trivial experiences a manifestation of dream disfigurement (by means of displacement), and we may recall that we have recognised this dream disfigurement as the work of a censor which controls the passage between two psychic instances. We accordingly expect that dream analysis will regularly reveal to us the genuine, significant source of the dream in the life of the day, the recollection of which has transferred its accent to some indifferent recollection. This conception brings us into complete opposition to Robert's theory, which thus becomes valueless for us. The fact which Robert was trying to explain simply doesn't exist; its assumption is based upon a misunderstanding, upon the failure to substitute the real meaning of the dream for its apparent content. Further objection may be made to Robert's doctrine: If it were really the duty of the dream, by means of a special psychic activity, to rid our memory of the "slag" of the recollections of the day, our sleep would have to be more troubled and employed in a more strained effort than we may suppose it to be from our waking life. For the number of indifferent impressions received during the day, against which we should have to protect our memory, is obviously infinitely large; the night would not be long enough to accomplish the task. It is very much more probable that the forgetting of indifferent impressions takes place without any active interference on the part of our psychic powers.

Still something cautions us against taking leave of Robert's idea without further consideration. We have left unexplained the fact that one of the indifferent day-impressions—one from the previous day indeed—regularly furnished a contribution to the dream-content. Relations between this impression and the real source of the dream do not always exist from the beginning; as we have seen, they are established only subsequently, in the course of the dream-work, as though in order to serve the purpose of the intended displacement. There must, therefore, be some necessity to form connections in this particular direction, of the recent, although indifferent impression; the latter must have special fitness for this purpose because of some property. Otherwise it would be just as easy for the dream thoughts to transfer their accent to some inessential member of their own series of associations.

The following experiences will lead us to an explanation. If a day has brought two or more experiences which are fitted to stimulate a dream, then the dream fuses the mention of both into a single whole; it obeys an *impulse to fashion a whole out of them;* for instance: One summer afternoon I entered a railroad compartment, in which I met two friends who were unknown to each other. One of them was an influential colleague, the other a member of a distinguished family, whose physician I was; I made the two gentlemen acquainted with each other; but during the long ride I was the go-between in the conversation, so that I had to treat a subject of conversation now with the one, now with the other. I asked my colleague to recommend a common friend who had just begun his medical practice. He answered that he was convinced of the young man's thoroughness, but that his plain appearance would make his entrance into households of rank difficult. I answered: "That is just why he needs recommendation." Soon afterwards I asked the other fellow-traveller about the health of his aunt—the mother of one of my patients—who was at the time prostrated by a serious illness. During the night after this journey I dreamt that the young friend, for whom I had asked assistance, was in a splendid salon, and was making a funeral oration to a select company with the air of a man of the world—the oration being upon the old lady (now dead for the purposes of the dream) who was the aunt of the second fellow-traveller. (I confess frankly that I had not been on good terms with this lady.) My dream had thus found connections between the two impressions of the day, and by means of them composed a unified situation.

In view of many similar experiences, I am driven to conclude that a kind of compulsion exists for the dream function, forcing it to bring together in the dream all the available sources of dream stimulation into a unified whole.[5] In a subsequent chapter (on the dream function) we shall become acquainted with this impulse for putting together as a part of condensation another primary psychic process.

I shall now discuss the question whether the source from which the dream originates, and to which our analysis leads, must always be a recent (and significant) event, or whether a subjective experience, that is to say, the recollection of a psychologically valuable experience—a chain of thought—can take the part of a dream stimulus. The answer, which results most unequivocally from numerous analyses, is to the following effect. The stimulus for the dream may be a subjective occurrence, which has been made recent, as it were, by the mental activity during the day. It will probably not be out of place here to give a synopsis of various conditions which may be recognised as sources of dreams.

The source of a dream may be:

(a) A recent and psychologically significant experience which is directly represented in the dream.[6]

(b) Several recent, significant experiences, which are united by the dream into a whole.[7]

(c) One or more recent and significant experiences, which are represented in the dream by the mention of a contemporary but indifferent experience.[8]

(d) A subjective significant experience (a recollection, train of thought), which is *regularly* represented in the dream by the mention of a recent but indifferent impression.[9]

As may be seen, in dream interpretation the condition is firmly adhered to throughout that each component of the dream repeats a recent impression of the day. The element which is destined to representation in the dream may either belong to the presentations surrounding the actual dream stimulus itself—and, furthermore, either as an essential or an inessential element of the same—or it may originate in the neighbourhood of an indifferent impression, which, through associations more or less rich, has been brought into relation with the thoughts surrounding the dream stimulus. The apparent multiplicity of the conditions here is produced by *the alternative according to whether displacement has or has not taken place,* and we may note that this alternative serves to explain the contrasts of the dream just as readily as the ascending series from partially awake to fully awake brain cells in the medical theory of the dream (*cf.* p. 74).

Concerning this series, it is further notable that the element which is psychologically valuable, but not recent (a train of thought, a recollection) may be replaced, for the purposes of dream formation, by a recent, but psychologically indifferent, element, if only these two conditions be observed: 1. That the dream shall contain a reference to something which has been recently experienced; 2. That the dream stimulus shall remain a psychologically valuable train of thought. In a single case (*a*) both conditions are fulfilled by the same impression. If it be added that the same indifferent impressions which are used for the dream, as long as they are recent, lose this availability as soon as they become a day (or at most several days) older, the assumption must be made that the very freshness of an impression gives it a certain psychological value for dream formation, which is somewhat equivalent to the value of emotionally accentuated memories or trains of thought. We shall be able to see the basis of this value of *recent* impressions for dream formation only with the help of certain psychological considerations which will appear later.[10]

Incidentally our attention is called to the fact that important changes in the material comprised by our ideas and our memory may be brought about unconsciously and at night. The injunction that one should sleep for a night upon any affair before making a final decision about it is obviously fully justified. But we see that at this point we have proceeded from the psychology of dreaming to that of sleep, a step for which there will often be occasion.

Now there arises an objection threatening to invalidate the conclusions we have just reached. If indifferent impressions can get into the dream only in case they are recent, how does it happen that we find also in the dream content elements from earlier periods in our lives, which at the time when they were recent possessed, as Strümpell expresses it, no psychic value, which, therefore, ought to have been forgotten long ago, and which, therefore, are neither fresh nor psychologically significant?

This objection can be fully met if we rely upon the results furnished by psychoanalysis of neurotics. The solution is as follows: The process of displacement which substitutes indifferent material for that having psychic significance (for dreaming as well as for thinking) has already taken place in those earlier periods of life, and has since become fixed in the memory. Those elements which were originally indifferent are in fact no longer so, since they have acquired the value of psychologically

150

significant material. That which has actually remained indifferent can never be reproduced in the dream.

It will be correct to suppose from the foregoing discussion that I maintain that there are no indifferent dream stimuli, and that, accordingly, there are no harmless dreams. This I believe to be the case, thoroughly and exclusively, allowance being made for the dreams of children and perhaps for short dream reactions to nocturnal sensations. Whatever one may dream, it is either manifestly recognisable as psychically significant or it is disfigured, and can be judged correctly only after a complete interpretation, when, as before, it may be recognised as possessing psychic significance. The dream never concerns itself with trifles; we do not allow ourselves to be disturbed in our sleep by matters of slight importance. Dreams which are apparently harmless turn out to be sinister if one takes pains to interpret them; if I may be permitted the expression, they all have "the mark of the beast." As this is another point on which I may expect opposition, and as I am glad of an opportunity to show dream-disfigurement at work, I shall here subject a number of dreams from my collection to analysis.

I

An intelligent and refined young lady, who, however, in conduct, belongs to the class we call reserved, to the "still waters," relates the following dream:—

Her husband asks: "Should not the piano be tuned?" She answers: "It won't pay; the hammers would have to be newly buffed too." This repeats an actual event of the previous day. Her husband had asked such a question, and she had answered something similar. But what is the significance of her dreaming it? She tells of the piano, indeed, that it is a *disgusting old box* which has a bad tone; it is one of the things which her husband had before they were married,[11] &c., but the key to the true solution lies in the phrase: *It won't pay.* This originated in a visit made the day before to a lady friend. Here she was asked to take off her coat, but she declined, saying, *"It won't pay.* I must go in a moment." At this point, I recall that during yesterday's analysis she suddenly took hold of her coat, a button of which had opened. It is, therefore, as if she had said, "Please don't look in this direction; it won't pay." Thus *"box"* develops into *"chest,"* or breast-box *("bust"),* and the interpretation of the dream leads directly to a time in her bodily development when she was dissatisfied with her shape. It also leads to earlier periods, if we take into consideration *"disgusting"* and *"bad tone,"* and remember how often in allusions and in dreams the two small hemispheres of the feminine body take the place—as a substitute and as an antithesis—of the large ones.

II

I may interrupt this dream to insert a brief harmless dream of a young man. He dreamt that *he was putting on his winter overcoat again, which was terrible.* The occasion for this dream is apparently the cold weather, which has recently set in again. On more careful examination we note that the two short portions of the dream do not fit together well, for what is there "terrible" about wearing a heavy or thick coat in the cold? Unfortunately for the harmlessness of this dream, the first idea educed in analysis is the recollection that on the previous day a lady had secretly admitted to him that her last child owed its existence to the bursting of a condom. He now reconstructs his thoughts in accordance with this suggestion: A thin condom is dangerous, a thick one is bad. The condom is an "overcoat" (*Ueberzieher*), for it is put over something; *Ueberzieher* is also the name given in German to a thin overcoat. An experience like the one related by the lady would indeed be "terrible" for an unmarried man.—We may now return to our other harmless dreamer.

III

She puts a candle into a candlestick; but the candle is broken, so that it does not stand straight. The girls at school say she is clumsy; the young lady replies that it is not her fault.

Here, too, there is an actual occasion for the dream; the day before she had actually put a candle into a candlestick; but this one was not broken. A transparent symbolism has been employed here. The candle is an object which excites the feminine genitals; its being broken, so that it does not stand straight, signifies impotence on the man's part ("it is not her fault"). But does this young woman, carefully brought up, and a stranger to all obscenity, know of this application of the candle? She happens to be able to tell how she came by this information. While riding in a boat on the Rhine, another boat passes containing students who are singing or rather yelling, with great delight: "When the Queen of Sweden with closed shutters and the candles of Apollo. . . ."

She does not hear or understand the last word. Her husband is asked to give her the required explanation. These verses are then replaced in the dream content by the harmless recollection of a command which she once executed clumsily at a girls' boarding school, this occurring by means of the common features *closed shutters.* The connection between the theme of onanism and that of impotence is clear enough. "Apollo" in the latent dream content connects this dream with an earlier one in which the virgin Pallas figured. All this is obviously not harmless.

IV

Lest it may seem too easy a matter to draw conclusions from dreams concerning the dreamer's real circumstances, I add another dream coming from the same person which likewise appears harmless. *"I dreamt of doing something,"* she relates, *"which I actually did during the day, that is to say, I filled a little trunk so full of books that I had difficulty in closing it. My dream was just like the actual occurrence."* Here the person relating the dream herself attaches chief importance to the correspondence between the dream and reality. All such criticisms upon the dream and remarks about it, although they have secured a place in waking thought, regularly belong to the latent dream content, as later examples will further demonstrate We are told, then, that what the dream relates has actually taken place during the day. It would take us too far afield to tell how we reach the idea of using the English language to help us in the interpretation of this dream. Suffice it to say that it is again a question of a little box (*cf.* p. 147, the dream of the dead child in the box) which has been filled so full that nothing more can go into it. Nothing in the least sinister this time.

In all these "harmless" dreams the sexual factor as a motive for the exercise of the censor receives striking prominence. But this is a matter of primary importance, which we must postpone.

(B) Infantile Experiences as the Source of Dreams

As the third of the peculiarities of the dream content, we have cited from all the authors (except Robert) the fact that impressions from the earliest times of our lives, which seem not to be at the disposal of the waking memory, may appear in the dream. It is, of course, difficult to judge how often or how seldom this occurs, because the respective elements of the dream are not recognised according to their origin after waking. The proof that we are dealing with childhood impressions must thus be reached objectively, and the conditions necessary for this happen to coincide only in rare instances. The story is told by A. Maury, as being particularly conclusive, of a man who decided to visit his birthplace after twenty years' absence. During the night before his departure, he dreams that he is in an altogether strange district, and that he there meets a strange man with whom he has a conversation. Having afterward returned to his home, he was able to convince himself that this strange district really existed in the neighbourhood of his home town, and the strange man in the dream turned out to be a friend of his dead father who lived there. Doubtless, a conclusive proof that he had seen both the man and the district in his childhood. The dream, moreover, is to be interpreted as a dream of impatience, like that of the girl who carries her ticket for the concert of the evening in her pocket (p. 124), of the child whose father had promised him an excursion to the Hameau, and the like. The motives explaining why just this impression of childhood is reproduced for the dreamer cannot, of course, be discovered without an analysis.

One of the attendants at my lectures, who boasted that his dreams were very rarely subject to disfigurement, told me that he had sometime before in a dream seen *his former tutor in bed with his nurse,* who had been in the household until he was eleven years old. The location of this scene does not occur to him in the dream. As he was much interested, he told the dream to his elder brother, who laughingly confirmed its reality. The brother said he remembered the affair very well, for he was at the time six years old. The lovers were in the habit of making him, the elder boy, drunk with beer, whenever circumstances were favourable for nocturnal relations. The smaller child, at that time three years old—our dreamer—who slept in the same room as the nurse, was not considered an obstacle.

In still another case it may be definitely ascertained, without the aid of dream interpretation, that the dream contains elements from childhood; that is, if it be a so-called *perennial* dream, which being first dreamt in childhood, later appears again and again after adult age has been reached. I may add a few examples of this sort to those already familiar, although I have never made the acquaintance of such a perennial dream in my own case. A physician in the thirties tells me that a yellow lion, about which he can give the most detailed information, has often appeared in his dream-life from the earliest period of his childhood to the present day. This lion, known to him from his dreams, was one day discovered *in natura* as a long-forgotten object made of porcelain, and on that occasion the young man learned from his mother that this object had been his favourite toy in early childhood, a fact which he himself could no longer remember.

If we now turn from the manifest dream content to the dream thoughts which are revealed only upon analysis, the co-operation of childhood experiences may be found to exist even in dreams whose content would not have led us to suspect anything of the sort. I owe a particularly delightful and instructive example of such a dream to my honoured colleague of the "yellow lion." After reading Nansen's account of his polar expedition, he dreamt that he was giving the bold explorer electrical treatment in an ice field for an ischaemia of which the latter complained! In the analysis of this dream, he remembered a story of his childhood, without which the dream remains entirely unintelligible. When he was a child, three or four years old, he was listening attentively to a conversation of older people about trips of exploration, and presently asked papa whether exploration was a severe illness. He had apparently confused "trips" with "rips," and the ridicule of his brothers and sisters prevented his ever forgetting the humiliating experience.

The case is quite similar when, in the analysis of the dream of the monograph on the genus cyclamen, I happen upon the recollection, retained from childhood, that my father allowed me to destroy a book embellished with coloured plates when I was a little boy five years old. It will perhaps be doubted whether this recollection actually took part in the composition of the dream content, and it will be intimated that the process of analysis has subsequently established the connection. But the abundance and intricacy of the ties of association vouch for the truth of my explanation: cyclamen—favourite flower—favourite dish—artichoke; to pick to pieces like an artichoke, leaf by leaf (a phrase which at that time rang in our ears a propos of the dividing up of the Chinese Empire)—herbarium—bookworm, whose favourite dish is books. I may state further that the final meaning of the dream, which I have not given here, has the most intimate connection with the content of the childhood scene.

In another series of dreams we learn from analysis that the wish itself, which has given rise to the dream, and whose fulfilment the dream turns out to be, has originated in childhood—until one is astonished to find that the child with all its impulses lives on in the dream.

I shall now continue the interpretation of a dream which has already proved instructive—I refer to the dream in which friend R. is my uncle (p. 133). We have carried its interpretation far enough for the wish-motive, of being appointed professor, to assert itself tangibly; and we have explained the affection displayed in the dream for friend R. as a fiction of opposition and spite against the aspersion of the two colleagues, who appear in the dream thoughts. The dream was my own; I may, therefore, continue the analysis by stating that my feelings were not quite satisfied by the solution reached. I know that my opinion of these colleagues who are so badly treated in the dream thoughts would have been expressed in quite different terms in waking life; the potency of the wish not to share their fate in the matter of appointment seemed to me too slight to account for the discrepancy between my estimate in the dream and that of waking. If my desire to be addressed by a new title proves so strong it gives proof of a morbid ambition, which I did not know to exist in me, and which I believe is far from my thoughts. I do not know how others, who think they know me, would judge me, for perhaps I have really been ambitious; but if this be true, my ambition has long since transferred itself to other objects than the title and rank of assistant-professor.

Whence, then, the ambition which the dream has ascribed to me? Here I remember a story which I heard often in my childhood, that at my birth an old peasant's wife had prophesied to my happy mother (I was her first-born) that she had given to the world a great man. Such prophecies must occur very frequently; there are so many mothers happy in expectation, and so many old peasant wives whose influence on earth has waned, and who have therefore turned their eyes towards the future. The prophetess was not likely to suffer for it either. Might my hunger for greatness have originated from this source? But here I recollect an impression from the later years of my childhood, which would serve still better as an explanation. It was of an evening at an inn on the Prater,[12] where my parents were accustomed to take me when I was eleven or twelve years old. We noticed a man who went from table to table and improvised verses upon any subject that was given to him. I was sent to bring the poet to our table and he showed himself thankful for the message. Before asking for his subject he threw off a few rhymes about me, and declared it probable, if he could trust his inspiration, that I would one day become a "minister." I can still distinctly remember the impression made by this second prophecy.

It was at the time of the election for the municipal ministry; my father had recently brought home pictures of those elected to the ministry—Herbst, Giskra, Unger, Berger, and others—and we had illuminated them in honour of these gentlemen. There were even some Jews among them; every industrious Jewish schoolboy therefore had the making of a minister in him. Even the fact that until shortly before my enrolment in the University I wanted to study jurisprudence, and changed my plans only at the last moment, must be connected with the impressions of that tune. A minister's career is under no circumstances open to a medical man. And now for my dream! I begin to see that it transplants me from the sombre present to the hopeful time of the municipal election, and fulfils my wish of that time to the fullest extent. In treating my two estimable and learned colleagues so badly, because they are Jews, the one as a simpleton and the other as a criminal—in doing this I act as though I were the minister of education, I put myself in his place. What thorough revenge I take upon his Excellency! He refuses to appoint me professor extraordinarius, and in return I put myself in his place in the dream.

Another case establishes the fact that although the wish which actuates the dream is a present one, it nevertheless draws great intensification from childhood memories. I refer to a series of dreams which are based upon the longing to go to Rome. I suppose I shall still have to satisfy this longing by means of dreams for a long time to come, because, at the time of year which is at my disposal for travelling, a stay at Rome is to be avoided on account of considerations of health.[13] Thus I once dreamt of seeing the Tiber and the bridge of St. Angelo from the window of a railroad compartment; then the train starts, and it occurs to me that I have never entered the city at all. The view which I saw in the dream was modelled after an engraving which I had noticed in passing the day before in the parlour of one of my patients. On another occasion some one is leading me upon a hill and showing me Rome half enveloped in mist, and so far in the distance that I am astonished at the distinctness of the view. The content of this dream is too rich to be fully reported here. The motive, "to see the promised land from afar," is easily recognisable in it. The city is Lübeck, which I first saw in the mist; the original of the hill is the Gleichenberg. In a third dream, I am at last in Rome, as the dream tells me. To my disappointment, the scenery which I see is anything but urban. *A little river with black water, on one side of which are black rocks, on the other large white flowers. I notice a certain Mr. Zucker* (with whom I am superficially acquainted), *and make up my mind to ask him to show me the way into the city.* It is apparent that I am trying in vain to see a city in the dream which I have never seen in waking life. If I resolve the landscape into its elements, the white flowers indicate Ravenna, which is known to me, and which, for a time at least, deprived Rome of its leading place as capital of Italy. In the swamps around Ravenna we had seen the most beautiful water-lilies in the middle of black pools of water; the dream makes them grow on meadows, like the narcissi of our own Aussee, because at Ravenna it was such tedious work to fetch them out of the water. The black rock, so close to the water, vividly recalls the valley of the Tepl at Karlsbad. "Karlsbad" now enables me to account for the peculiar circumstance that I ask Mr. Zucker the way. In the material of which the dream is composed appear also two of those amusing Jewish s, which conceal so much profound and often bitter worldly wisdom, and which we are so fond of quoting in our conversation and letters. One is the story of the "constitution," and tells how a poor Jew sneaks into the express train for Karlsbad without a ticket, how he is caught and is treated more and more unkindly at each call for tickets by the conductor, and how he tells a friend, whom he meets at one of the stations during his miserable journey, and who asks him where he is travelling: "To Karlsbad, if my constitution will stand it." Associated with this in memory is another story about a Jew who is ignorant of French, and who has express instructions to ask in Paris for the way to the Rue Richelieu. Paris was for many years the object of my own longing, and I took the great satisfaction with which I first set foot on the pavement in Paris as a warrant that I should also attain the fulfilment of other wishes. Asking for the way is again a direct allusion to Rome, for of course all roads lead to Rome. Moreover, the name Zucker (English, sugar) again points to *Karlsbad,* whither we send all persons afflicted with the *constitutional* disease, diabetes (*Zuckerkrankheit,* sugar-disease). The occasion for this dream was the proposal of my Berlin friend that we should meet in Prague at Easter. A further allusion to sugar and diabetes was to be found in the matters which I had to talk over with him.

A fourth dream, occurring shortly after the last one mentioned, brings me back to Rome. I see a street-corner before me and am astonished to see so many German placards posted there. On the day before I had written my friend with prophetic vision that Prague would probably not be a comfortable resort for German travellers. The dream, therefore, simultaneously expressed the wish to meet him at Rome instead of at the Bohemian city, and a desire, which probably originated during my student days, that the German language might be accorded more tolerance in Prague. Besides I must have understood the Czech language in the first three years of my childhood, because I was born in a small village of Moravia, inhabited by Slavs. A Czech nursery rhyme, which I heard in my seventeenth year, became, without effort on my part, so imprinted upon my memory that I can repeat it to this day, although I have no idea of its meaning. There is then no lack in these dreams also of manifold relations to impressions from the first years of my life.

It was during my last journey to Italy, which, among other places, took me past Lake Trasimenus, that I at last found what re-enforcement my longing for the Eternal City had received from the impressions of my youth; this was after I had seen the Tiber, and had turned back with painful emotions when I was within eighty kilometers of Rome. I was just broaching the plan of travelling to Naples via Rome the next year, when this sentence, which I must have read in one of our classical authors, occurred to me: "It is a question which of the two paced up and down in his room the more impatiently after he had made the plan to go to Rome— Assistant-Headmaster Winckelman or the great general Hannibal." I myself had walked in Hannibal's footsteps; like him I was destined never to see Rome, and he too had gone to Campania after the whole world had expected him in Rome. Hannibal, with whom I had reached this point of similarity, had been my favourite hero during my years at the Gymnasium; like so many boys of my age, I bestowed my sympathies during the Punic war, not on the Romans, but on the Carthaginians. Then, when I came finally to understand the consequences of belonging to an alien race, and was forced by the anti-semitic sentiment among my class-mates to assume a definite attitude, the figure of the Semitic commander assumed still greater proportions in my eyes. Hannibal and Rome symbolised for me as a youth the antithesis between the tenaciousness of the Jews and the organisation of the Catholic Church. The significance for our emotional life which the anti-semitic movement has since assumed helped to fix the thoughts and impressions of that earlier time. Thus the wish to get to Rome has become the cover and symbol in my dream-life for several warmly cherished wishes, for the realisation of which one might work with the perseverance and single-mindedness of the Punic general, and whose fulfilment sometimes seems as little favoured by fortune as the wish of Hannibal's life to enter Rome.

And now for the first time I happen upon the youthful experience which, even to-day, still manifests its power in all these emotions and dreams. I may have been ten or twelve years old when my father began to take me with him on his walks, and to reveal to me his views about the things of this world in his conversation. In this way he once told me, in order to show into how much better times I had been born than he, the following: "While I was a young man, I was walking one Saturday on a street in the village where you were born; I was handsomely dressed and wore a new fur cap. Along comes a Christian, who knocks my cap into the mud with one blow and shouts: "Jew, get off the sidewalk." "And what did you do?" "I went into the street and picked up the cap," was the calm answer. That did not seem heroic on the part of the big strong man, who was leading me, a little fellow, by the hand. I contrasted this situation, which did not please me, with another more in harmony with my feelings—the scene in which Hannibal's father, Hamilcar Barka made his boy swear at the domestic altar to take vengeance on the Romans. Since that time Hannibal has had a place in my phantasies.

I think I can follow my enthusiasm for the Carthaginian general still further back into my childhood, so that possibly we have here the transference of an already formed emotional relation to a new vehicle. One of the first books which fell into my childish hands, after I learned to read, was Thiers' *Konsulat und Kaiserreich* (Consulship and Empire); I remember I pasted on the flat backs of my wooden soldiers little labels with the names of the Imperial marshals, and that at that time Masséna (as a Jew Menasse) was already my avowed favourite. Napoleon himself follows Hannibal in crossing the Alps. And perhaps the development of this martial ideal can be traced still further back into my childhood, to the wish which the now friendly, now hostile, intercourse during my first three years with a boy a year older than myself must have actuated in the weaker of the two playmates.

The deeper one goes in the analysis of dreams, the more often one is put on the track of childish experiences which play the part of dream sources in the latent dream content.

We have learned (p. 24) that the dream very rarely reproduces experiences in such a manner that they constitute the sole manifest dream content, unabridged and unchanged. Still some authentic examples showing this process have been reported, and I can add some new ones which again refer to infantile scenes. In the case of one of my patients, a dream once gave a barely disfigured reproduction of a sexual occurrence, which was immediately recognised as an accurate recollection. The memory of it indeed had never been lost in waking life, but it had been greatly obscured, and its revivification was a result of the preceding work of analysis. The dreamer had at the age of twelve visited a bed-ridden school-mate, who had exposed himself by a movement in bed, probably only by chance. At the sight of the genitals, he was seized by a kind of compulsion, exposed himself and took hold of the member belonging to the other boy, who, however, looked at him with surprise and indignation, whereupon he became embarrassed and let go. A dream repeated this scene twenty-three years later, with all the details of the emotions occurring in it, changing it, however, in this respect, that the dreamer took the passive part instead of the active one, while the person of the school-mate was replaced by one belonging to the present.

As a rule, of course, a childhood scene is represented in the manifest dream content only by an allusion, and must be extricated from the dream by means of interpretation. The citation of examples of this kind cannot have a very convincing effect, because every guarantee that they are experiences of childhood is lacking; if they belong to an earlier time of life, they are no longer recognised by our memory. Justification for the conclusion that such childish experiences generally exist in dreams is based upon a great number of factors which become apparent in psychoanalytical work, and which seem reliable enough when regarded as a whole. But when, for the purposes of dream interpretation, such references of dreams to childish experiences are torn from their context, they will perhaps not make much impression, especially since I never give all the material upon which the interpretation depends. However, I shall not let this prevent me from giving some examples.

I

The following dream is from another female patient: *She is in a large room, in which there are all kinds of machines, perhaps, as she imagines, an orthopaedic institute. She hears that I have no time, and that she must take the treatment along with five others. But she resists, and is unwilling to lie down on the bed—or whatever it is—which is intended for her. She stands in a corner and waits for me to say "It is not true." The others, meanwhile, laugh at her, saying it is all foolishness on her part. At the same time it is as if she were called upon to make many small squares.*

The first part of the content of this dream is an allusion to the treatment and a transference on me. The second contains an allusion to a childhood scene; the two portions are connected by the mention of the bed. The orthopaedic institute refers to one of my talks in which I compared the treatment as to its duration and nature with an orthopaedic treatment. At the beginning of the treatment I had to tell her that *for the present* I had little time for her, but that later on I would devote a whole hour to her daily. This aroused in her the old sensitiveness, which is the chief characteristic of children who are to be hysterical. Their desire for love is insatiable. My patient was the youngest of six brothers and sisters (hence, *"with five others"*), and as such the favourite of her father, but in spite of that she seems to have found that her beloved father devoted too little time and attention to her. The detail of her waiting for me to say "It is not true," has the following explanation: A tailor's apprentice had brought her a dress, and she had given him the money for it. Then she asked her husband whether she would have to pay the money again if the boy were to lose it. To tease her, her husband answered "Yes" (the teasing in the dream), and she asked again and again, and *waited for him to say "It is not true."* The thought of the latent dream-content may now be construed as follows: Will she have to pay me the double amount if I devote twice the time to her? a thought which is stingy or filthy. (The uncleanliness of childhood is often replaced in the dream by greediness for money; the word filthy here supplies the bridge.) If all that about waiting until I should say, &c., serves as a dream circumlocution for the word "filthy," the standing-in-a-corner and not lying down-on-the-bed are in keeping; for these two features are component parts of a scene of childhood, in which she had soiled her bed, and for punishment was put into a corner, with the warning that papa would not love her any more, and her brothers and sisters laughed at her, &c. The little squares refer to her young niece, who has shown her the arithmetical trick of writing figures in nine squares, I believe it is, in such a way that upon being added together in any direction they make fifteen.

II

Here is the dream of a man: *He sees two boys tussling with each other, and they are cooper's boys, as he concludes from the implements which are lying about; one of the boys has thrown the other down, the prostrate one wears ear-rings with blue stones. He hurries after the wrongdoer with lifted cane, in order to chastise him. The latter takes refuge with a woman who is standing against a wooden fence, as though it were his mother. She is the wife of a day labourer, and she turns her back to the man who is dreaming. At last she faces about and stares at him with a horrible look, so that he runs away in fright; in her eyes the red flesh of the lower lid seems to stand out.*

The dream has made abundant use of trivial occurrences of the previous day. The day before he actually saw two boys on the street, one of whom threw the other one down. When he hurried up to them in order to settle the quarrel, both of them took flight. Coopers' boys: this is explained only by a subsequent dream, in the analysis of which he used the expression, *"To knock the bottom out of the barrel."* Ear-rings with blue stones, according to his observation, are chiefly worn by prostitutes. Furthermore, a familiar doggerel rhyme about two boys comes up: "The other boy, his name was Mary" (that is, he was a girl). The woman standing up: after the scene with the two boys, he took a walk on the bank of the Danube, and took advantage of being alone to urinate *against a wooden fence.* A little later during his walk, a decently dressed elderly lady smiled at him very pleasantly, and wanted to hand him her card with her address.

Since in the dream the woman stood as he had while urinating, it is a question of a woman urinating, and this explains the "horrible look," and the prominence of the red flesh, which can only refer to the genitals which gap in squatting. He had seen genitals in his childhood, and they had appeared in later recollection as "proud flesh" and as "wound." The dream unites two occasions upon which, as a young boy, the dreamer had had opportunity to see the genitals of little girls, in throwing one down, and while another was urinating; and, as is shown by another association, he had kept in memory a punishment or threat of his father's, called forth by the sexual curiosity which the boy manifested on these occasions.

III

A great mass of childish memories, which have been hastily united in a phantasy, is to be found behind the following dream of a young lady.

She goes out in trepidation, in order to do some shopping. On the Graben[15] she sinks to her knees as though broken down. Many people collect around her, especially the hackney-coach drivers; but no one helps her to get up. She makes many unavailing attempts; finally she must have succeeded, for she is put into a hackney-coach which is to take her home. A large, heavily laden basket (something like a market-basket) is thrown after her through the window.

This is the same woman who is always harassed in her dreams as she was harassed when a child. The first situation of the dream is apparently taken from seeing a horse that had fallen, just as "broken down" points to horse-racing. She was a rider in her early years, still earlier she was probably also a horse. Her first childish memory of the seventeen-year-old son of the porter, who, being seized on the street by an epileptic fit, was brought home in a coach, is connected with the idea of falling down. Of this, of course, she has only heard, but the idea of epileptic fits and of falling down has obtained great power over her phantasies, and has later influenced the form of her own hysterical attacks. When a person of the female sex dreams of falling, this almost regularly has a sexual significance; she becomes a "fallen woman," and for the purpose of the dream under consideration this interpretation is probably the least doubtful, for she falls on the Graben, the place in Vienna which is known as the concourse of prostitutes. The market-basket admits of more than one interpretation; in the sense of refusal (German, *Korb* = basket—snub, refusal), she remembers the many snubs which she first gave her suitors, and which she later, as she thinks, received herself. Here belongs also the detail that *no one will help her up,* which she herself interprets as being disdained. Furthermore, the market-basket recalls phantasies that have already appeared in the course of analysis, in which she imagines she has married far beneath her station, and now goes marketing herself. But lastly the market-basket might be interpreted as the mark of a servant. This suggests further childhood memories—of a cook who was sent away because she stole; she, too, sank to her knees and begged for mercy. The dreamer was at that time twelve years old. Then there is a recollection of a chamber-maid, who was dismissed because she had an affair with the coachman of the household, who, incidently, married her afterwards. This recollection, therefore, gives us a clue to the coachman in the dream (who do not, in contrast with what is actually the case, take the part of the fallen woman). But there still remains to be explained the throwing of the basket, and the throwing of it through the window. This takes her to the transference of baggage on the railroad, to the *Fensterln,*[16] in the country, and to minor impressions received at a country resort, of a gentleman throwing some blue plums to a lady through her window, and of the

164

dreamer's little sister being frightened because a cretin who was passing looked in at the window. And now from behind this there emerges an obscure recollection, from her tenth year, of a nurse who made love at the country resort with a servant of the household, of which the child had opportunity to see something, and who was "fired" (thrown out) (in the dream the opposite: "thrown into"), a story which we had also approached by several other paths. The baggage, moreover, or the trunk of a servant, is disparagingly referred to in Vienna as "seven plums." "Pack up your seven plums and get out."

My collection, of course, contains an abundant supply of such patients' dreams, whose analysis leads to childish impressions that are remembered obscurely or not at all, and that often date back to the first three years of life. But it is a mistake to draw conclusions from them which are to apply to the dream in general; we are in every case dealing with neurotic, particularly with hysterical persons; and the part played by childhood scenes in these dreams might be conditioned by the nature of the neurosis, and not by that of the dream. However, I am struck quite as often in the course of interpreting my own dreams, which I do not do on account of obvious symptoms of disease, by the fact that I unsuspectingly come upon a scene of childhood in the latent dream content, and that a whole series of dreams suddenly falls into line with conclusions drawn from childish experiences. I have already given examples of this, and shall give still more upon various occasions. Perhaps I cannot close the whole chapter more fittingly than by citing several of my own dreams, in which recent happenings and long-forgotten experiences of childhood appear together as sources of dreams.

I

After I have been travelling and have gone to bed hungry and tired, the great necessities of life begin to assert their claims in sleep, and I dream as follows: *I go into a kitchen to order some pastry. Here three women are standing, one of whom is the hostess, and is turning something in her hand as though she were making dumplings. She answers that I must wait until she has finished* (not distinctly as a speech). *I become impatient and go away insulted. I put on an overcoat; but the first one which I try is too long. I take it off, and am somewhat astonished to find that it has fur trimming. A second one has sewn into it a long strip of cloth with Turkish drawings. A stranger with a long face and a short pointed beard comes up and prevents me from putting it on, declaring that it belongs to him. I now show him that it is embroidered all over in Turkish fashion. He asks, "What business are the Turkish (drawings, strips of cloth . . .) of yours? But we then become quite friendly with each other.*

In the analysis of this dream there occurs to me quite unexpectedly the novel which I read, that is to say, which I began with the end of the first volume, when I was perhaps thirteen years old. I have never known the name of the novel or of its author, but the conclusion remains vividly in my memory. The hero succumbs to insanity, and continually calls the names of the three women that have signified the greatest good and ill fortune for him during life. Pélagie is one of these names. I still do not know what to make of this name in the analysis. À propos of the three women there now come to the surface the three Parcae who spin the fate of man, and I know that one of the three women, the hostess in the dream, is the mother who gives life, and who, moreover, as in my case, gives the first nourishment to the living creature. Love and hunger meet at the mother's breast. A young man—so runs an anecdote—who became a great admirer of womanly beauty, once when the conversation turned upon a beautiful wet nurse who had nourished him as a child, expressed himself to the effect that he was sorry that he had not taken better advantage of his opportunity at the time. I am in the habit of using the anecdote to illustrate the factor of subsequence in the mechanism of psychoneuroses. . . . One of the Parcae, then, is rubbing the palms of her hands together as though she were making dumplings. A strange occupation for one of the Fates, which is urgently in need of an explanation! This is now found in another and earlier childhood memory. When I was six years old, and was receiving my first instructions from my mother, I was asked to believe that we are made of earth, and that therefore we must return to earth. But this did not suit me, and I doubted her teaching. Thereupon my mother rubbed the palms of her hands together—just as in making dumplings, except that there was no dough between them— and showed me the blackish scales of epidermis which were thus rubbed off as a proof that it is earth of which we are made. My astonishment at this demonstration *ad oculos* was without limit, and I acquiesced in the idea which I was later to hear expressed in words: "Thou owest nature a death."[17] Thus the women are really Parcae whom I visit in the kitchen, as I have done so often in my childhood years when I was hungry, and when

166

my mother used to order me to wait until lunch was ready. And now for the dumplings! At least one of my teachers at the University, the very one to whom I am indebted for my histological knowledge (epidermis), might be reminded by the name Knoedl (German, *Knoedel* = dumplings) of a person whom he had to prosecute for committing a plagiarism of his writings. To commit plagiarism, to appropriate anything one can get, even though it belongs to another, obviously leads to the second part of the dream, in which I am treated like a certain overcoat thief, who for a tune plied his trade in the auditoria. I wrote down the expression plagiarism—without any reason—because it presented itself to me, and now I perceive that it must belong to the latent dream-content, because it will serve as a bridge between different parts of the manifest dream-content. The chain of associations—Pélagie—plagiarism—plagiostomi[18] (sharks)—fish bladder—connects the old novel with the affair of Knoedl and with the overcoats (German, *Überzieher* = thing drawn over—overcoat or condom), which obviously refer to an object belonging to the technique of sexual life.[19] This, it is true, is a very forced and irrational connection, but it is nevertheless one which I could not establish in waking life if it had not been already established by the activity of the dream. Indeed, as though nothing were sacred for this impulse to force connections, the beloved name, Bruecke (bridge of words, see above), now serves to remind me of the institution in which I spent my happiest hours as a student, quite without any cares ("So you will ever find more pleasure at the breasts of knowledge without measure"), in the most complete contrast to the urgent desires which vex me while I dream. And finally there comes to the surface the recollection of another dear teacher, whose name again sounds like something to eat (Fleischl—German, *Flei*sch = meat—like Knoedl), and of a pathetic scene, in which the scales of epidermis play a part (mother—hostess), and insanity (the novel), and a remedy from the Latin kitchen which numbs the sensation of hunger, to wit, cocaine.

In this manner I could follow the intricate trains of thought still further, and could fully explain the part of the dream which is missing in the analysis; but I must refrain, because the personal sacrifices which it would require are too great. I shall merely take up one of the threads, which will serve to lead us directly to the dream thoughts that lie at the bottom of the confusion. The stranger, with the long face and pointed beard, who wants to prevent me from putting on the overcoat, has the features of a tradesman at Spalato, of whom my wife made ample purchases of Turkish cloths. His name was Popovic, a suspicious name, which, by the way, has given the humorist Stettenheim a chance to make a significant remark: "He told me his name, and blushingly shook my hand."[20] Moreover, there is the same abuse of names as above with Pélagie, Knoedl, Bruecke, Fleischl. That such playing with names is childish nonsense can be asserted without fear of contradiction; if I indulge in it, this indulgence amounts to an act of retribution, for my own name has numberless times fallen a victim to such weak-minded attempts at humour. Goethe once remarked how sensitive a man is about his name with which, as with his skin, he feels that he has grown up, whereupon Herder composed the following on his name:

"Thou who art born of *gods,* of *Goths,* or of *Kot* (mud)—
Thy *god* like images, too, are dust."

I perceive that this digression about the abuse of names was only intended to prepare for this complaint. But let us stop here. . . . The purchase at Spalato reminds me of another one at Cattaro, where I was too cautious, and missed an opportunity for making some desirable acquisitions. (Missing an opportunity at the breast of the nurse, see above.) Another dream thought, occasioned in the dreamer by the sensation of hunger, is as follows: *One should let nothing which one can have escape, even if a little wrong is done; no opportunity should be missed, life is so short, death inevitable.* Owing to the fact that this also has a sexual significance, and that desire is unwilling to stop at a wrong, this philosophy of *carpe diem* must fear the censor and must hide behind a dream. This now makes articulate counter-thoughts of all kinds, recollections of a time when spiritual food alone was sufficient for the dreamer; it suggests repressions of every kind, and even threats of disgusting sexual punishments.

II

A second dream requires a longer preliminary statement:

I have taken a car to the West Station in order to begin a vacation journey to the Aussee, and I reach the station in time for the train to Ischl, which leaves earlier. Here I see Count Thun, who is again going to see the Emperor at Ischl. In spite of the rain, he has come in an open carriage, has passed out at once through the door for local trains, and has motioned back the gate-keeper, who does not know him and who wants to take his ticket, with a little wave of his hand. After the train to Ischl has left, I am told to leave the platform and go back into the hot waiting-room; but with difficulty I secure permission to remain. I pass the time in watching the people who make use of bribes to secure a compartment; I make up my mind to insist on my rights—that is, to demand the same privilege. Meanwhile I sing something to myself, which I afterwards recognise to be the aria from Figaro's Wedding:

> "If my lord Count wishes to try a dance,
>   Try a dance,
>   Let him but say so,
>   I'll play him a tune."

(Possibly another person would not have recognised the song.)

During the whole afternoon I have been in an insolent, combative mood; I have spoken roughly to the waiter and the cabman, I hope without hurting their feelings; now all kinds of bold and revolutionary thoughts come into my head, of a kind suited to the words of Figaro and the comedy of Beaumarchais, which I had seen at the Comédie Française. The speech about great men who had taken the trouble to be born; the aristocratic prerogative, which Count Almaviva wants to apply in the case of Susan; the jokes which our malicious journalists of the Opposition make upon the name of Count Thun (German, *thun* = doing) by calling him Count Do-Nothing. I really do not envy him; he has now a difficult mission with the Emperor, and I am the real Count Do-Nothing, for I am taking a vacation. With this, all kinds of cheerful plans for the vacation. A gentleman now arrives who is known to me as a representative of the Government at the medical examinations, and who has won the flattering nickname of "Governmental bed-fellow" by his activities in this capacity. By insisting on his official station he secures half of a first-class compartment, and I hear one guard say to the other: "Where are we going to put the gentleman with the first-class half-compartment?" A pretty favouritism; I am paying for a whole first-class compartment. Now I get a whole compartment for myself, but not in a through coach, so that there is no toilet at my disposal during the night. My complaints to the guard are without result; I get even by proposing that at least there be a hole made in the floor of this compartment for the possible needs of the travellers. I really awake at a quarter of three in the morning with a desire to urinate, having had the following dream:

Crowd of people, meeting of students. . . . *A certain Count (Thun or Taafe) is making a speech. Upon being asked to say something about the Germans, he declares with contemptuous mien that their favourite flower is Colt's-foot, and then puts something like a torn leaf, really the crumpled skeleton of a leaf, into his buttonhole. I make a start, I make a start then,*[21] *but I am surprised at this idea of mine.* Then more indistinctly: *It seems as though it were the vestibide (Aula), the exits are jammed, as though it were necessary to flee. I make my way through a suite of handsomely furnished rooms, apparently governmental chambers, with furniture of a colour which is between brown and violet, and at last I come to a passage where a housekeeper, an elderly, fat woman(Frauenzimmer), is seated. I try to avoid talking to her, but apparently she thinks I have a right to pass because she asks whether she shall accompany me with the lamp. I signify to her to tell her that she is to remain standing on the stairs, and in this I appear to myself very clever, for avoiding being watched at last. I am downstairs now, and I find a narrow, steep way along which I go.*

Again indistinctly . . . *It is as if my second task were to get away out of the city, as my earlier was to get out of the house. I am riding in a one-horse carriage, and tell the driver to take me to a railway station. "I cannot ride with you on the tracks," I say, after he has made the objection that I have tired him out. Here it seems as though I had already driven with him along a course which is ordinarily traversed on the railroad. The stations are crowded; I consider whether I shall go to Krems or to Znaim, but I think that the court will be there, and I decide in favour of Graz or something of the sort. Now I am seated in the coach, which is something like a street-car, and I have in my buttonhole a long braided thing, on which are violet-brown violets of stiff material, which attracts the attention of many people. Here the scene breaks off.*

I am again in front of the railroad station, but I am with an elderly gentleman. I invent a scheme for remaining unrecognised, but I also see this plan already carried out. Thinking and experiencing are here, as it were, the same thing. He pretends to be blind, at least in one eye, and I hold a male urinal in front of him (which we have had to buy in the city or did buy), I am thus a sick attendant, and have to give him the urinal because he is blind. If the conductor sees us in this position, he must pass us by without drawing attention. At the same time the attitude of the person mentioned is visually observed. Then I awake with a desire to urinate.

The whole dream seems a sort of phantasy, which takes the dreamer back to the revolutionary year 1848, the memory of which had been renewed by the anniversary year 1898, as well as by a little excursion to Wachau, where I had become acquainted with Emmersdorf, a town which I wrongly supposed to be the resting-place of the student leader Fischof, to whom several features of the dream content might refer. The thought associations then lead me to England, to the house of my brother, who was accustomed jokingly to tell his wife of "Fifty years ago," according to the title of a poem by Lord Tennyson, whereupon the children were in the habit of correcting: "Fifteen years ago." This phantasy, however, which subtilely attaches itself to the thoughts which the sight of the Count Thun has given rise to, is only like the façade of Italian churches which is superimposed without being organically connected with the building behind it; unlike these façades, however, the phantasy is filled with gaps and confused, and the parts from within break through at many places. The first situation of the dream is concocted from several scenes, into which I am able to separate it. The arrogant attitude of the Count in the dream is copied from a scene at the Gymnasium which took place in my fifteenth year. We had contrived a conspiracy against an unpopular and ignorant teacher, the leading spirit in which was a schoolmate who seems to have taken Henry VIII. of England as his model. It fell to me to carry out the *coup-d'état,* and a discussion of the importance of the Danube (German *Donau*) for Austria (Wachau!) was the occasion upon which matters came to open indignation. A fellow-conspirator was the only aristocratic schoolmate whom we had—he was called the "giraffe" on account of his conspicuous longitudinal development—and he stood just like the Count in the dream, while he was being reprimanded by the tyrant of the school, the Professor of the German language. The explanation of the favourite flower and the putting into the buttonhole of something which again must have been a flower (which recalls the orchids, which I had brought to a lady friend on the same day, and besides that the rose of Jericho) prominently recalls the scene in Shakespeare's historical plays which opens the civil wars of the Pled and the White Roses; the mention of Henry VIII. has opened the way to this reminiscence. It is not very far now from roses to red and white carnations. Meanwhile two little rhymes, the one German, the other Spanish, insinuate themselves into the analysis: "Roses, tulips, carnations, all flowers fade," and "Isabelita, no llores que se marchitan las flores." The Spanish is taken from *Figaro.*Here in Vienna white carnations have become the insignia of the Anti-Semites, the red ones of the Social Democrats. Behind this is the recollection of an anti-Semitic challenge during a railway trip in beautiful Saxony (Anglo-Saxon). The third scene contributing to the formation of the first situation in the dream takes place in my early student life. There was a discussion in the German students' club about the relation of philosophy to the general sciences. A green youth, full of the materialistic doctrine, I thrust myself forward and defended a very one-sided view. Thereupon a sagacious older school-fellow, who has since shown his capacity for leading men and organising the masses, and who, moreover, bears a name belonging to the animal kingdom, arose and called us down thoroughly; he too, he said, had herded swine in his youth, and had come back repentant to the house of his father. I started up (as in the dream), became very uncivil, and answered that since I knew he had herded swine, I was not surprised at the tone of his discourse. (In the dream I am surprised at my national German sentiment.) There was great commotion; and the demand came from all sides that I take back what I had said, but I remained steadfast. The man who had been insulted was too sensible to take the advice, which was given him, to send a challenge, and let the matter drop.

The remaining elements of this scene of the dream are of more remote origin. What is the meaning of the Count's proclaiming the colt's foot? Here I must consult my train of associations. Colt's-foot (German: *Huflattich*)—lattice—lettuce—salad-dog (the dog that grudges others what he cannot eat himself). Here plenty of opprobrious epithets may be discerned: Gir-affe (German *Affe* = monkey, ape), pig, sow, dog; I might even find means to arrive at donkey, on a detour by way of a name, and thus again at contempt for an academic teacher. Furthermore I translate colt's-foot (*Huflattich*)—I do not know how correctly—by "pisse-en-lit." I got this idea from Zola's *Germinal*, in which children are ordered to bring salad of this kind. The dog—*chien*—has a name sounding like the major function (*chier*, as *pisser* stands for the minor one). Now we shall soon have before us the indecent in all three of its categories; for in the same *Germinal*, which has a lot to do with the future revolution there is described a very peculiar contest, depending upon the production of gaseous excretions, called flatus.[22] And now I must remark how the way to this flatus has been for a long while preparing, beginning with the flowers, and proceeding to the Spanish rhyme of Isabclita to Ferdinand and Isabella, and, by way of Henry VIII., to English history at the time of the expedition of the Armada against England, after the victorious termination of which the English struck a medal with the inscription: *"Afflavit* et dissipati sunt," for the storm had scattered the Spanish fleet. I had thought of taking this phrase for the title of a chapter on "Therapeutics"—to be meant half jokingly—if I should ever have occasion to give a detailed account of my conception and treatment of hysteria.

I cannot give such a detailed solution of the second scene of the dream, out of regard for the censor. For at this point I put myself in the place of a certain eminent gentleman of that revolutionary period, who also had an adventure with an eagle, who is said to have suffered from incontinence of the bowels, and the like; and I believe I *should not be justified at this point in passing* the censor, although it was an aulic councillor (*aula, consilarius aulicus*) who told me the greater part of these stories. The allusion to the suite of rooms in the dream relates to the private car of his Excellency, into which I had opportunity to look for a moment; but it signifies, as so often in dreams, a woman (Frauenzimmer; German *Zimmer*—room is appended to *Frauen*—woman, in order to imply a slight amount of contempt).[23] In the person of the housekeeper I give scant recognition to an intelligent elderly lady for the entertainment and the many good stories which I have enjoyed at her house. . . . The feature of the lamp goes back to Grillparzer, who notes a charming experience of a similar nature, which he afterwards made use of in "Hero and Leander" (the billows of the ocean and of love—the Armada and the storm).[24]

I must also forgo detailed analysis of the two remaining portions of the dream; I shall select only those elements which lead to two childhood scenes, for the sake of which alone I have taken up the dream. The reader will guess that it is sexual matter which forces me to this suppression; but he need not be content with this explanation. Many things which must be treated as secrets in the presence of others are not treated such with one's self, and here it is not a question of considerations inducing me to hide the solution, but of motives of the inner censor concealing the real content of the dream from myself. I may say, then, that the analysis shows these three portions of the dream to be impertinent boasting, the exuberance of an absurd grandiose idea which has long since been suppressed in my waking life, which, however, dares show itself in the manifest dream content by one or two projections (*I seem clever to myself*), and which makes the arrogant mood of the evening before the dream perfectly intelligible. It is boasting, indeed, in all departments; thus the mention of Graz refers to the phrase: What is the price of Graz? which we are fond of using when we feel over-supplied with money. Whoever will recall Master Rabelais's unexcelled description of the "Life and Deeds of Gargantua and his Son Pantagruel," will be able to supply the boastful content intimated in the first portion of the dream. The following belongs to the two childhood scenes which have been promised. I had bought a new trunk for this journey, whose colour, a brownish violet, appears in the dream several times. (Violet-brown violets made of stiff material, next to a thing which is called "girl-catcher"—the furniture in the governmental chambers). That something new attracts people's attention is a well-known belief of children. Now I have been told the folio wing story of my childhood; I remember hearing the story rather than the occurrence itself. I am told that at the age of two I still occasionally wetted my bed, that I was often reproached on this subject, and that I consoled my father by promising to buy him a beautiful new red bed in N. (the nearest large city). (Hence the detail inserted in the dream that *we bought the urinal in the city or had to buy it;* one must keep one's promises. Attention is further called to the identity of the male urinal and the feminine trunk, box). All the megalomania of the child is contained in this promise. The significance of the dream of difficulty in urinating in the case of the child has been already considered in the interpretation of an earlier dream (*cf.* the dream on p. 185).

Now there was another domestic occurrence, when I was seven or eight years old, which I remember very well. One evening, before going to bed I had disregarded the dictates of discretion not to satisfy my wants in the bedroom of my parents and in their presence, and in his reprimand for this delinquency my father made the remark: "That boy will never amount to anything." It must have terribly mortified my ambition, for allusions to this scene return again and again in my dreams, and are regularly coupled with enumerations of my accomplishments and successes, as though I wanted to say: "You see, I have amounted to something after all." Now this childhood scene furnishes the elements for the last image of the dream, in which of course, the rôles are interchanged for the sake of revenge. The elderly man, obviously my father, for the blindness in one eye signifies his glaucoma[25] on one side is now urinating before me as I once urinated before him. In glaucoma I refer to cocaine, which stood my father in good stead in his operation, as though I had thereby fulfilled my promises. Besides that I make sport of him; since he is blind I must hold the urinal in front of him, and I gloat over allusions to my discoveries in the theory of hysteria, of which I am so proud.[26]

If the two childhood scenes of urinating are otherwise closely connected with the desire for greatness, their rehabilitation on the trip to the Aussee was further favoured by the accidental circumstance that my compartment had no water-closet, and that I had to expect embarrassment on the ride as actually happened in the morning. I awoke with the sensation of a bodily need. I suppose one might be inclined to credit these sensations with being the actual stimulus of the dream; I should, however, prefer a different conception—namely, that it was the dream thoughts which gave rise to the desire to urinate. It is quite unusual for me to be disturbed in sleep by any need, at least at the time of this awakening, a quarter of four in the morning. I may forestall further objection by remarking that I have hardly ever felt a desire to urinate after awakening early on other journeys made under more comfortable circumstances. Moreover, I can leave this point undecided without hurting my argument.

Since I have learned, further, from experience in dream analysis that there always remain important trains of thought proceeding from dreams whose interpretation at first seems complete (because the sources of the dream and the actuation of the wish are easily demonstrable), trains of thought reaching back into earliest childhood, I have been forced to ask myself whether this feature does not constitute an essential condition of dreaming. If I were to generalise this thesis, a connection with what has been recently experienced would form a part of the manifest content of every dream and a connection with what has been most remotely experienced, of its latent content; and I can actually show in the analysis of hysteria that in a true sense these remote experiences have remained recent up to the present time. But this conjecture seems still very difficult to prove; I shall probably have to return to the part played by the earliest childhood experiences, in another connection (Chapter 7.).

Of the three peculiarities of dream memory considered at the beginning, one—the preference for the unimportant in the dream content—has been satisfactorily explained by tracing it back to dream disfigurement. We have been able to establish the existence of the other two—the selection of recent and of infantile material—but we have found it impossible to explain them by the motive of dream. Let us keep in mind these two characteristics, which still remain to be explained or evaluated; a place for them will have to be found elsewhere, either in the psychology of the sleeping state, or in the discussion of the structure of the psychic apparatus which we shall undertake later, after we have learned that the inner nature of the apparatus may be observed through dream interpretation as though through a window.

Just here I may emphasize another result of the last few dream analyses. The dream often appears ambiguous; not only may several wish-fulfilments, as the examples show, be united in it, but one meaning or one wish-fulfilment may also conceal another, until at the bottom one comes upon the fulfilment of a wish from the earliest period of childhood; and here too, it may be questioned whether "often" in this sentence may not more correctly be replaced by "regularly."

(C)   Somatic Sources of Dreams

If the attempt be made to interest the cultured layman in the problems of dreaming, and if, with this end in view, he be asked the question from what source dreams originate according to his opinion, it is generally found that the person thus interrogated thinks himself in assured possession of a part of the solution. He immediately thinks of the influence which a disturbed or impeded digestion ("Dreams come from the stomach"), accidental bodily position, and little occurrences during sleep, exercise upon the formation of dreams, and he seems not to suspect that even after the consideration of all these factors there still remains something unexplained.

We have explained at length in the introductory chapter (p. 24), what a rôle in the formation of dreams the scientific literature credits to the account of somatic exciting sources, so that we need here only recall the results of this investigation. We have seen that three kinds of somatic exciting sources are distinguished, objective sensory stimuli which proceed from external objects, the inner states of excitation of the sensory organs having only a subjective basis, and the bodily stimuli which originate internally; and we have noticed the inclination on the part of the authors to force the psychic sources of the dream into the background or to disregard them altogether in favour of these somatic sources of stimulation (p. 42).

In testing the claims which are made on behalf of these classes of somatic sources of stimulation, we have discovered that the significance of the objective stimuli of the sensory organs—whether accidental stimuli during sleep or those stimuli which cannot be excluded from our dormant psychic life—has been definitely established by numerous observations and is confirmed by experiments (p. 27); we have seen that the part played by subjective sensory stimuli appears to be demonstrated by the return of hypnogogic sensory images in dreams, and that although the referring of these dream images and ideas, in the broadest sense, to internal bodily stimulation is not demonstrable in every detail, it can be supported by the well-known influence which an exciting state of the digestive, urinary, and sexual organs exercise upon the contents of our dreams.

"Nerve stimulus" and "bodily stimulus," then, would be the somatic sources of the dream—that is, the only sources whatever of the dream, according to several authors.

But we have already found a number of doubts, which seem to attack not so much the correctness of the somatic theory of stimulation as its adequacy.

However certain all the representatives of this theory may have felt about the actual facts on which it is based—especially in case of the accidental and external nerve stimuli, which may be recognised in the content of the dream without any trouble—nevertheless none of them has been able to avoid the admission that the abundant ideal content of dreams does not admit of explanation by external nerve-stimuli alone. Miss Mary Whiton Calkins has tested her own dreams and those of another person for a period of six weeks with this idea in mind, and has found only from 13.2 per cent. to 6.7 per cent. in which the element of external sensory perception was demonstrable; only two cases in the collection could be referred to organic sensations. Statistics here confirm what a hasty glance at our own experience might have led us to suspect.

The decision has been made repeatedly to distinguish the "dream of nerve stimulus" from the other forms of the dream as a well-established sub-species. Spitta divided dreams into dreams of nerve stimulus and association dreams. But the solution clearly remained unsatisfactory as long as the link between the somatic sources of dreams and their ideal content could not be demonstrated.

Besides the first objection, of the inadequate frequency of external exciting sources, there arises as a second objection the inadequate explanation of dreams offered by the introduction of this sort of dream sources. The representatives of the theory accordingly must explain two things, in the first place, why the external stimulus in the dream is never recognised according to its real nature, but is regularly mistaken for something else (*cf.* the alarm-clock dreams, p. 30), and secondly, why the reaction of the receiving mind to this misrecognised stimulus should result so indeterminately and changefully. As an answer to these questions, we have heard from Strümpell that the mind, as a result of its being turned away from the outer world during sleep, is not capable of giving correct interpretation to the objective sensory stimulus, but is forced to form illusions on the basis of the indefinite incitements from many directions. As expressed in his own words (p. 108):

"As soon as a sensation, a sensational complex, a feeling, or a psychic process in general, arises in the mind during sleep from an outer or inner nerve-stimulus, and is perceived by the mind, this process calls up sensory images, that is to say, earlier perceptions, either unembellished or with the psychic values belonging to them, from the range of waking experiences, of which the mind has remained in possession. It seems to collect about itself, as it were, a greater or less number of such images, from which the impression which originates from the nerve-stimulus receives its psychic value. It is usually said here, as the idiom does of waking thought, that the mind *interprets* impressions of nerve-stimuli in sleep. The result of this interpretation is the so-called nerve-stimulus dream—that is to say, a dream whose composition is conditioned by the fact that a nerve-stimulus brings about its effect in psychic life according to the laws of reproduction."

The opinion of Wundt agrees in all essentials with this theory. He says that the ideas in the dream are probably the result, for the most part, of sensory stimuli, especially of those of general sensation, and are therefore mostly phantastic illusions—probably memory presentations which are only partly pure, and which have been raised to hallucinations. Strümpell has found an excellent simile (p. 84). It is as "if the ten fingers of a person ignorant of music should stray over the keyboard of an instrument"—to illustrate the relation between dream content and dream stimuli, which follows from this theory. The implication is that the dream does not appear as a psychic phenomenon, originating from psychic motives, but as the result of a physiological stimulus, which is expressed in psychic symptomology, because the apparatus which is affected by the stimulus is not capable of any other expression. Upon a similar assumption is based, for example, the explanation of compulsive ideas which Meynert tried to give by means of the famous simile of the dial on which individual figures are prominent because they are in more marked relief.

However popular this theory of somatic dream stimuli may have become, and however seductive it may seem, it is nevertheless easy to show the weak point in it. Every somatic dream stimulus which provokes the psychic apparatus to interpretation through the formation of illusions, is capable of giving rise to an incalculable number of such attempts at interpretation; it can thus attain representation in the dream content by means of an extraordinary number of different ideas. But the theory of Strümpell and Wundt is incapable of instancing any motive which has control over the relation between the external stimulus and the dream idea which has been selected to interpret it, and therefore of explaining the "peculiar choice" which the stimuli "often enough make in the course of their reproductive activity" (Lipps, *Grundtatsachen des Seelerdebens*, p. 170). Other objections may be directed against the fundamental assumption of the whole theory of illusions—the assumption that during sleep the mind is not in a condition to recognise the real nature of the objective sensory stimuli. The old physiologist Burdach proves to us that the mind is quite capable even during sleep of interpreting correctly the sensory impressions which reach it, and of reacting in accordance with the correct interpretation. He establishes this by showing that it is possible to exempt certain impressions which seem important to the individuals, from the neglect of sleeping (nurse and child), and that one is more surely awakened by one's own name than by an indifferent auditory impression, all of which presupposes, of course, that the mind distinguishes among sensations, even in sleep (Chapter 1, p. 52). Burdach infers from these observations that it is not an incapability of interpreting sensory stimuli in the sleeping state which must be assumed, but a lack of interest in them. The same arguments which Burdach used in 1830, later reappear unchanged in the works of Lipps in the year 1883, where they are employed for the purpose of attacking the theory of somatic stimuli. According to this the mind seems to be like the sleeper in the anecdote, who, upon being asked, "Are you asleep?" answers "No," and upon being again addressed with the words, "Then lend me ten florins," takes refuge in the excuse: "I am asleep."

The inadequacy of the theory of somatic dream stimuli may also be demonstrated in another manner. Observations show that I am not urged to dream by external stimulations, even if these stimulations appear in the dream as soon as, and in case that, I dream. In response to the tactile or pressure stimulus which I get while sleeping, various reactions are at my disposal. I can overlook it and discover only upon awakening that my leg has been uncovered or my arm under pressure; pathology shows the most numerous examples where powerfully acting sensory and motor stimuli of different sorts remain without effect during sleep. I can perceive a sensation during sleep through and through sleep, as it were, which happens as a rule with painful stimuli, but without weaving the pain into the texture of the dream; thirdly, I can awaken on account of the stimulus in order to obviate it. Only as a fourth possible reaction, I may be impelled to dream by a nerve stimulus; but the other possibilities are realised at least as often as that of dream formation. This could not be the case if the *motive for dreaming did not lie outside of the somatic sources of dreams.*

Taking proper account of the defect in the explanation of dreams by somatic stimuli which has just been shown, other authors—Scherner, who was joined by the philosopher Volkelt—have tried to determine more exactly the psychic activities which cause the variegated dream images to arise from the somatic stimuli, and have thus transferred the essential nature of dreams back to the province of the mind, and to that of psychic activity. Scherner not only gave a poetically appreciative, glowing and vivid description of the psychic peculiarities which develop in the course of dream formation; he also thought he had guessed the principle according to which the mind proceeds with the stimuli that are at its disposal. The dream activity, according to Scherner—after phantasy has been freed from the shackles imposed upon it during the day, and has been given free rein—strives to represent symbolically the nature of the organ from which the stimulus proceeds. Thus we have a kind of dream-book as a guide for the interpretation of dreams, by means of which bodily sensations, the conditions of the organs and of the stimuli may be inferred from dream images. "Thus the image of a cat expresses an angry discontented mood, the image of a light-coloured bit of smooth pastry the nudity of the body. The human body as a whole is pictured as a house by the phantasy of the dream, and each individual organ of the body as a part of the house. In 'toothache-dreams' a high vaulted vestibule corresponds to the mouth and a stair to the descent of the gullet to the alimentary canal; in the 'headache-dream' the ceiling of a room which is covered with disgusting reptile-like spiders is chosen to denote the upper part of the head" (Volkelt, p. 39). "Several different symbols are used by the dream for the same organ, thus the breathing lungs find their symbol in an oven filled with flames and with a roaring draught, the heart in hollow chests and baskets, and the bladder in round, bag-shaped objects or anything else hollow. It is especially important that at the end of a dream the stimulating organ or its function be represented undisguised and usually on the dreamer's own body. Thus the 'toothache-dream' usually ends by the dreamer drawing a tooth from his own mouth" (p. 35). It cannot be said that this theory has found much favour with the authors. Above all, it seems extravagant; there has been no inclination even to discover the small amount of justification to which it may, in my opinion, lay claim. As may be seen, it leads to a revival of the dream interpretation by means of symbolism, which the ancients used, except that the source from which the interpretation is to be taken is limited to the human body. The lack of a technique of interpretation which is scientifically comprehensible must seriously limit the applicability of Scherner's theory. Arbitrariness in dream interpretation seems in no wise excluded, especially since a stimulus may be expressed by several representations in the content of the dream; thus Scherner's associate, Volkelt, has already found it impossible to confirm the representation of the body as a house. Another objection is that here again dream activity is attributed to the mind as a useless and aimless activity, since according to the theory in question the mind is content with forming phantasies about the stimulus with which it is concerned, without even remotely contemplating anything like a discharge of the stimulus.

But Scherner's theory of the symbolisation of bodily stimuli by the dream receives a heavy blow from another objection. These bodily stimuli are present at all times, and according to general assumption the mind is more accessible to them during sleep than in waking. It is thus incomprehensible why the mind does not dream continually throughout the night, and why it does not dream every night and about all the organs. If one attempts to avoid this objection by making the condition that especial stimuli must proceed from the eye, the ear, the teeth, the intestines in order to arouse dream activity, one is confronted by the difficulty of proving that this increase of stimulation is objective, which is possible only in a small number of cases. If the dream of flying is a symbolisation of the upward and downward motion of the pulmonary lobes, either this dream, as has already been remarked by Strümpell, would be dreamt much oftener, or an accentuation of the function of breathing during the dream would have to be demonstrable. Still another case is possible—the most probable of all—that now and then special motives directing attention to the visceral sensations which are universally present are active, but this case takes us beyond the range of Scherner's theory.

The value of Scherner's and Volkelt's discussions lies in the fact that they call attention to a number of characteristics of the dream content which are in need of explanation, and which seem to promise new knowledge. It is quite true that symbolisations of organs of the body and of their functions are contained in dreams, that water in a dream often signifies a desire to urinate, that the male genital may often be represented by a staff standing erect or by a pillar, &c. In dreams which show a very animated field of vision and brilliant colours, in contrast to the dimness of other dreams, the interpretation may hardly be dismissed that they are "dreams of visual stimulation," any more than it may be disputed that there is a contribution of illusory formations in dreams which contain noise and confusion of voices. A dream like that of Scherner, of two rows of fair handsome boys standing opposite to each other on a bridge, attacking each other and then taking their places again, until finally the dreamer himself sits down on the bridge and pulls a long tooth out of his jaw; or a similar one of Volkelt's, in which two rows of drawers play a part, and which again ends in the extraction of a tooth; dream formations of this sort, which are related in great numbers by the authors, prevent our discarding Scherner's theory as an idle fabrication without seeking to find its kernel of truth. We are now confronted by the task of giving the supposed symbolisation of the dental stimulus an explanation of a different kind.

Throughout our consideration of the theory of the somatic sources of dreams, I have refrained from urging the argument which is inferred from our dream analyses. If we have succeeded in proving, by a procedure which other authors have not applied in their investigation of dreams, that the dream as a psychic action possesses value peculiar to itself, that a wish supplies the motive for its formation, and that the experiences of the previous day furnish the immediate material for its content, any other theory of dreams neglecting such an important method of investigation, and accordingly causing the dream to appear a useless and problematic psychic reaction to somatic stimuli, is dismissible without any particular comment. Otherwise there must be—which is highly improbable—two entirely different kinds of dreams, of which only one has come under our observation, while only the other has been observed by the earlier connoisseurs of the dream. It still remains to provide a place for the facts which are used to support the prevailing theory of somatic dream-stimuli, within our own theory of dreams.

We have already taken the first step in this direction in setting up the thesis that the dream activity is under a compulsion to elaborate all the dream stimuli which are simultaneously present into a unified whole (p. 168). We have seen that when two or more experiences capable of making an impression have been left over from the previous day, the wishes which result from them are united into one dream; similarly, that an impression possessing psychic value and the indifferent experiences of the previous day are united in the dream material, provided there are available connecting ideas between the two. Thus the dream appears to be a reaction to everything which is simultaneously present as actual in the sleeping mind. As far as we have hitherto analysed the dream material, we have discovered it to be a collection of psychic remnants and memory traces, which we were obliged to credit (on account of the preference shown for recent and infantile material) with a character of actuality, though the nature of this was not at the time determinable. Now it will not be difficult to foretell what will happen when new material in the form of sensations is added to these actualities of memory. These stimuli likewise derive importance for the dream because they are actual; they are united with the other psychic actualities in order to make up the material for dream formation. To express it differently, the stimuli which appear during sleep are worked over into the fulfilment of a wish, the other component parts of which are the remnants of daily experience with which we are familiar. This union, however, is not inevitable; we have heard that more than one sort of attitude towards bodily stimuli is possible during sleep. Wherever this union has been brought about, it has simply been possible to find for the dream content that kind of presentation material which will give representation to both classes of dream sources, the somatic as well as the psychic.

The essential nature of the dream is not changed by this addition of somatic material to the psychic sources of the dream; it remains the fulfilment of a wish without reference to the way in which its expression is determined by the actual material.

I shall gladly find room here for a number of peculiarities, which serve to put a different face on the significance of external stimuli for the dream. I imagine that a co-operation of individual, physiological, and accidental factors, conditioned by momentary circumstances, determines how one will act in each particular case of intensive objective stimulation during sleep; the degree of the profoundness of sleep whether habitual or accidental in connection with the intensity of the stimulus, will in one case make it possible to suppress the stimulus, so that it will not disturb sleep; in another case they will force an awakening or will support the attempt to overcome the stimulus by weaving it into the texture of the dream. In correspondence with the multiplicity of these combinations, external objective stimuli will receive expression more frequently in the case of one person than in that of another. In the case of myself, who am an excellent sleeper, and who stubbornly resists any kind of disturbance in sleep, this intermixture of external causes of irritation into my dreams is very rare, while psychic motives apparently cause me to dream very easily. I have indeed noted only a single dream in which an objective, painful source of stimulation is demonstrable, and it will be highly instructive to see what effect the external stimulus had in this very dream.

I am riding on a grey horse, at first timidly and awkwardly, as though I were only leaning against something. I meet a colleague P., who is mounted on a horse and is wearing a heavy woollen suit; he. calls my attention to something (probably to the fact that my riding position is bad). Now I become more and more expert on the horse, which is most intelligent; I sit comfortably, and I notice that I am already quite at home in the saddle. For a saddle I have a kind of padding, which completely fills the space between the neck and the rump of the horse. In this manner I ride with difficulty between two lumber-wagons. After having ridden up the street for some distance, I turn around and want to dismount, at first in front of a little open chapel, which is situated close to the street. Then I actually dismount in front of a chapel which stands near the first; the hotel is in the same street, I could let the horse go there by itself, but I prefer to lead it there. It seems as if I should be ashamed to arrive there on horseback. In front of the hotel is standing a hall-boy who shows me a card of mine which has been found, and who ridicules me on account of it. On the card is written, doubly underlined, "Eat nothing," and then a second sentence (indistinct) something like "Do not work"; at the same time a hazy idea that I am in a strange city, in which I do no work.

It will not be apparent at once that this dream originated under the influence, or rather under the compulsion, of a stimulus of pain. The day before I had suffered from furuncles, which made every movement a torture, and at last a furuncle had grown to the size of an apple at the root of the scrotum, and had caused me the most intolerable pains that accompanied every step; a feverish lassitude, lack of appetite, and the hard work to which I had nevertheless kept myself during the day, had conspired with the pain to make me lose my temper. I was not altogether in a condition to discharge my duties as a physician, but in view of the nature and the location of the malady, one might have expected some performance other than riding, for which I was very especially unfitted. It is this very activity, of riding into which I am plunged by the dream; it is the most energetic denial of the suffering which is capable of being conceived. In the first place, I do not know how to ride, I do not usually dream of it, and I never sat on a horse but once—without a saddle—and then I did not feel comfortable. But in this dream I ride as though I had no furuncle on the perineum, and why? *just because I don't want any.* According to the description my saddle is the poultice which has made it possible for me to go to sleep. Probably I did not feel anything of my pain—as I was thus taken care of—during the first few hours of sleeping. Then the painful sensations announced themselves and tried to wake me up, whereupon the dream came and said soothingly: "Keep on sleeping, you won't wake up anyway! You have no furuncle at all, for you are riding on a horse, and with a furuncle where you have it riding is impossible!" And the dream was successful; the pain was stifled, and I went on sleeping.

But the dream was not satisfied with "suggesting away" the furuncle by means of tenaciously adhering to an idea incompatible with that of the malady, in doing which it behaved like the hallucinatory insanity of the mother who has lost her child, or like the merchant who has been deprived of his fortune by losses.[27] In addition the details of the denied sensation and of the image which is used to displace it are employed by the dream as a means to connect the material ordinarily actually present in the mind with the dream situation, and to give this material representation. I am riding on a *grey* horse—the colour of the horse corresponds exactly to the *pepper-and-salt* costume in which I last met my colleague P. in the country. I have been warned that highly seasoned food is the cause of furunculosis, but in any case it is preferable as an etiological explanation to sugar which ordinarily suggests furunculosis. My friend P. has been pleased to "ride the high horse" with regard to me, ever since he superseded me in the treatment of a female patient, with whom I had performed great feats (in the dream I first sit on the horse side-saddle fashion, like a circus rider), but who really led me wherever she wished, like the horse in the anecdote about the Sunday equestrian. Thus the horse came to be a symbolic representation of a lady patient (in the dream it is most intelligent). "I feel quite at home up here," refers to the position which I occupied in the patient's household until I was replaced by my colleague P. "I thought you were securely seated in the saddle," one of my few well-wishers among the great physicians of this city recently said to me with reference to the same household. And it was a feat to practise psychotherapy for ten hours a day with such pains, but I know that I cannot continue my particularly difficult work for any length of time without complete physical health, and the dream is full of gloomy allusions to the situation which must in that case result (the card such as neurasthenics have and present to doctors): *No work and no food.* With further interpretation I see that the dream activity has succeeded in rinding the way from the wish-situation of riding to very early infantile scenes of quarrelling, which must have taken place between me and my nephew, who is now living in England, and who, moreover, is a year older than I. Besides it has taken up elements from my journeys to Italy; the street in the dream is composed of impressions of Verona and Siena. Still more exhaustive interpretation leads to sexual dream-thoughts, and I recall what significance dream allusions to that beautiful country had in the case of a female patient who had never been in Italy (Itlay—German *gen Italien*— *Genitalien*—genitals). At the same time there are references to the house in which I was physician before my friend P., and to the place where the furuncle is located.

Among the dreams mentioned in the previous chapter there are several which might serve as examples for the elaboration of so-called nerve stimuli. The dream about drinking in full draughts is one of this sort; the somatic excitement in it seems to be the only source of the dream, and the wish resulting from the sensation—thirst—the only motive for dreaming. Something similar is true of the other simple dreams, if the somatic excitement alone is capable of forming a wish. The dream of the sick woman who throws the cooling apparatus from her cheek at night is an instance of a peculiar way of reacting to painful excitements with a wish-fulfilment; it seems as though the patient had temporarily succeeded in making herself analgesic by ascribing her pains to a stranger.

My dream about the three Parcae is obviously a dream of hunger, but it has found means to refer the need for food back to the longing of the child for its mother's breast, and to make the harmless desire a cloak for a more serious one, which is not permitted to express itself so openly. In the dream about Count Thun we have seen how an accidental bodily desire is brought into connection with the strongest, and likewise the most strongly suppressed emotions of the psychic life. And when the First Consul incorporates the sound of an exploding bomb into a dream of battle before it causes him to wake, as in the case reported by Garnier, the purpose for which psychic activity generally concerns itself with sensations occurring during sleep is revealed with extraordinary clearness. A young lawyer, who has been deeply preoccupied with his first great bankruptcy proceeding, and who goes to sleep during the afternoon following, acts just like the great Napoleon. He dreams about a certain G. Reich in *Hussiatyn* (German *husten*—to cough), whom he knows in connection with the bankruptcy proceeding, but *Hussiatyn* forces itself upon his attention still further, with the result that he is obliged to awaken, and hears his wife—who is suffering from bronchial catarrh—coughing violently.

Let us compare the dream of Napoleon I., who, incidentally, was an excellent sleeper, with that of the sleepy student, who was awakened by his landlady with the admonition that he must go to the hospital, who thereupon dreams himself into a bed in the hospital, and then sleeps on, with the following account of his motives: If I am already in the hospital, I shan't have to get up in order to go there. The latter is obviously a dream of convenience; the sleeper frankly admits to himself the motive for his dreaming; but he thereby reveals one of the secrets of dreaming in general. In a certain sense all dreams are dreams of convenience; they serve the purpose of continuing sleep instead of awakening. *The dream is the guardian of sleep, not the disturber of it.* We shall justify this conception with respect to the psychic factors of awakening elsewhere; it is possible, however, at this point to prove its applicability to the influence exerted by objective external excitements. Either the mind does not concern itself at all with the causes of sensations, if it is able to do this in spite of their intensity and of their significance, which is well understood by it; or it employs the dream to deny these stimuli; or thirdly, if it is forced to recognise the stimulus, it seeks to find that interpretation of the stimulus which shall represent the actual sensation as a component part of a situation which is desired and which is compatible with sleep. The actual sensation is woven into the dream *in order to deprive it of its reality.* Napoleon is permitted to go on sleeping; it is only a dream recollection of the thunder of the cannon at Arcole which is trying to disturb him.[28]

The wish to sleep, by which the conscious ego has been suspended and which along with the dream-censor contributes its share to the dream, must thus always be taken into account as a motive for the formation of dreams, and every successful dream is a fulfilment of this wish. The relation of this general, regularly present, and invariable sleep-wish to the other wishes, of which now the one, now the other is fulfilled, will be the subject of a further explanation. In the wish to sleep we have discovered a factor capable of supplying the deficiency in the theory of Strümpell and Wundt, and of explaining the perversity and capriciousness in the interpretation of the outer stimulus. The correct interpretation, of which the sleeping mind is quite capable, would imply an active interest and would require that sleep be terminated; hence, of those interpretations which are possible at all, only those are admitted which are agreeable to the absolute censorship of the somatic wish. It is something like this: It's the nightingale and not the lark. For if it's the lark, love's night is at an end. From among the interpretations of the excitement which are at the moment possible, that one is selected which can secure the best connection with the wish-possibilities that are lying in wait in the mind. Thus everything is definitely determined, and nothing is left to caprice. The misinterpretation is not an illusion, but—if you will—an excuse. Here again, however, there is admitted an action which is a modification of the normal psychic procedure, as in the case where substitution by means of displacement is effected for the purposes of the dream-censor.

If the outer nerve stimuli and inner bodily stimuli are sufficiently intense to compel psychic attention, they represent—that is, in case they result in dreaming and not in awakening—a definite point in the formation of dreams, a nucleus in the dream material, for which an appropriate wish-fulfilment is sought, in a way similar (see above) to the search for connecting ideas between two dream stimuli. To this extent it is true for a number of dreams that the somatic determines what their content is to be.

In this extreme case a wish which is not exactly actual is aroused for the purpose of dream formation. But the dream can do nothing but represent a wish in a situation as fulfilled; it is, as it were, confronted by the task of seeking what wish may be represented and fulfilled by means of the situation which is now actual. Even if this actual material is of a painful or disagreeable character, still it is not useless for the purposes of dream formation. The psychic life has control even over wishes the fulfilment of which brings forth pleasure—a statement which seems contradictory, but which becomes intelligible if one takes into account the presence of two psychic instances and the censor existing between them.

There are in the psychic life, as we have heard, *repressed* wishes which belong to the first system, and to whose fulfilment the second system is opposed. There are wishes of this kind—and we do not mean this in an historic sense, that there have been such wishes and that these have then been destroyed—but the theory of repression, which is essential to the study of psychoneurosis, asserts that such repressed wishes still exist, contemporaneously with an inhibition weighing them down. Language has hit upon the truth when it speaks of the "suppression" of such impulses. The psychic contrivance for bringing such wishes to realisation remains preserved and in a condition to be used. But if it happens that such a suppressed wish is fulfilled, the vanquished inhibition of the second system (which is capable of becoming conscious) is then expressed as a painful feeling. To close this discussion; if sensations of a disagreeable character which originate from somatic sources are presented during sleep, this constellation is taken advantage of by the dream activity to represent the fulfilment—with more or less retention of the censor—of an otherwise suppressed wish.

This condition of affairs makes possible a number of anxiety dreams, while another series of the dream formations which are unfavourable to the wish theory exhibits a different mechanism. For anxiety in dreams may be of a psycho-neurotic nature, or it may originate in psychosexual excitements, in which case the anxiety corresponds to a repressed *libido*. Then this anxiety as well as the whole anxiety dream has the significance of a neurotic symptom, and we are at the dividing-line where the wish-fulfilling tendency of dreams disappears. But in other anxiety-dreams the feeling of anxiety comes from somatic sources (for instance in the case of persons suffering from pulmonary or heart trouble, where there is occasional difficulty in getting breath), and then it is used to aid those energetically suppressed wishes in attaining fulfilment in the form of a dream, the dreaming of which from psychic motives would have resulted in the same release of fear. It is not difficult to unite these two apparently discrepant cases. Of two psychic formations, an emotional inclination and an ideal content, which are ultimately connected, the one, which is presented as actual, supports the other in the dream; now anxiety of somatic origin supports the suppressed presentation content, now the ideal content, which is freed from suppression, and which proceeds with the impetus given by sexual emotion, assists the discharge of anxiety. Of the one case it may be said that an emotion of somatic origin is psychically interpreted; in the other case everything is of psychic origin but the content which has been suppressed is easily replaced by a somatic interpretation which is suited to anxiety. The difficulties which lie in the way of understanding all this have little to do with the dream; they are due to the fact that in discussing these points we are touching upon the problems of the development of anxiety and of repression.

Undoubtedly the aggregate of bodily feelings is to be included among the commanding dream stimuli which originate internally. Not that it is able to furnish the dream content, but it forces the dream thoughts to make a choice from the material destined to serve the purpose of representation in the dream content; it does this by putting within easy reach that part of the material which is suited to its own character, while withholding the other. Moreover this general feeling, which is left over from the day, is probably connected with the psychic remnants which are significant for the dream.

If somatic sources of excitement occurring during sleep—that is, the sensations of sleep—are not of unusual intensity, they play a part in the formation of dreams similar, in my judgment, to that of the impressions of the day which have remained recent but indifferent. I mean that they are drawn into the dream formation, if they are qualified for being united with the presentation content of the psychic dream-source, but in no other case. They are treated as a cheap ever-ready material, which is utilised as often as it is needed, instead of prescribing, as a precious material does, the manner in which it is to be utilised. The case is similar to that where a patron of art brings to an artist a rare stone, a fragment of onyx, in order that a work of art may be made of it. The size of the stone, its colour, and its marking help to decide what bust or what scene shall be represented in it, while in the case where there is a uniform and abundant supply of marble or sandstone the artist follows only the idea which takes shape in his mind. Only in this manner, it seems to me, is the fact explicable that the dream content resulting from bodily excitements that have not been accentuated to a usual degree, does not appear in all dreams and during every night.

Perhaps an example, which takes us back to the interpretation of dreams, will best illustrate my meaning. One day I was trying to understand the meaning of the sensations of being impeded, of not being able to move from the spot, of not being able to get finished, &c., which are dreamt about so often, and which are so closely allied to anxiety. That night I had the following dream: *I am very incompletely dressed, and I go from a dwelling on the ground floor up a flight of stairs to an upper story. In doing this I jump over three steps at a time, and I am glad to find I can mount the steps so quickly. Suddenly I see that a servant girl is coming down the stairs, that is, towards me. I am ashamed and try to hurry away, and now there appears that sensation of being impeded; I am glued to the steps and cannot move from the spot.*

Analysis

The situation of the dream is taken from everyday reality. In a house in Vienna I have two apartments, which are connected only by a flight of stairs outside. My consultation-rooms and my study are on an elevated portion of the ground floor, and one story higher are my living-rooms. When I have finished my work downstairs late at night, I go up the steps into my bedroom. On the evening before the dream I had actually gone this short distance in a somewhat disorderly attire—that is to say, I had taken off my collar, cravat, and cuffs; but in the dream this has changed into a somewhat more advanced degree of undress, which as usual is indefinite. Jumping over the steps is my usual method of mounting stairs; moreover it is the fulfilment of a wish that has been recognised in the dream, for I have reassured myself about the condition of my heart action by the ease of this accomplishment. Moreover the manner in which I climb the stairs is an effective contrast to the sensation of being impeded which occurs in the second half of the dream. It shows me—something which needed no proof—that the dream has no difficulty in representing motor actions as carried out fully and completely; think of flying in dreams!

But the stairs which I go up are not those of my house; at first I do not recognise them; only the person coming toward me reveals to me the location which they are intended to signify. This woman is the maid of the old lady whom I visit twice daily to give hypodermic injections; the stairs, too, are quite similar to those which I must mount there twice daily.

How do this flight of stairs and this woman get into my dream? Being ashamed because one is not fully dressed, is undoubtedly of a sexual character; the servant of whom I dream is older than I, sulky, and not in the least attractive. These questions call up exactly the following occurrences: When I make my morning visit at this house I am usually seized with a desire to clear my throat; the product of the expectoration falls upon the steps. For there is no spittoon on either of these floors, and I take the view that the stairs should not be kept clean at my expense, but by the provision of a spittoon. The housekeeper, likewise an elderly and sulky person, with instincts for cleanliness, takes another view of the matter. She lies in wait for me to see whether I take the liberty referred to, and when she has made sure of it, I hear her growl distinctly. For days thereafter she refuses to show me her customary regard when we meet. On the day before the dream the position of the housekeeper had been strengthened by the servant girl. I had just finished my usual hurried visit to the patient when the servant confronted me in the ante-room and observed: "You might as well have wiped your shoes to-day, doctor, before you came into the room. The red carpet is all dirty again from your feet." This is the whole claim which the flight of stairs and the servant-girl can make for appearing in my dream.

An intimate connection exists between my flying over the stairs and my spitting on the stairs. Pharyngitis and diseases of the heart are both said to be punishments for the vice of smoking, on account of which vice, of course, I do not enjoy a reputation for great neatness with my housekeeper in the one house any more than in the other, both of which the dream fuses into a single image.

I must postpone the further interpretation of this dream until I can give an account of the origin of the typical dream of incomplete dress. I only note as a preliminary result from the dream which has just been cited that the dream sensation of inhibited action is always aroused at a point where a certain connection requires it. A peculiar condition of my motility during sleep cannot be the cause of this dream content, for a moment before I saw myself hurrying over the steps with ease, as though in confirmation of this fact.

(D) Typical Dreams

In general we are not in a position to interpret the dream of another person if he is unwilling to furnish us with the unconscious thoughts which lie behind the dream content, and for this reason the practical applicability of our method of dream interpretation is seriously curtailed.[29] But there are a certain number of dreams—in contrast with the usual freedom displayed by the individual in fashioning his dream world with characteristic peculiarity, and thereby making it unintelligible—which almost every one has dreamed in the same manner, and of which we are accustomed to assume that they have the same significance in the case of every dreamer. A peculiar interest belongs to these typical dreams for the reason that they probably all come from the same sources with every person, that they are thus particularly suited to give us information upon the sources of dreams.

Typical dreams are worthy of the most exhaustive investigation. I shall, however, only give a somewhat detailed consideration to examples of this species, and for this purpose I shall first select the so-called embarrassment dream of nakedness, and the dream of the death of dear relatives.

The dream of being naked or scantily clad in the presence of strangers occurs with the further addition that one is not at all ashamed of it, &c. But the dream of nakedness is worthy of our interest only when shame and embarrassment are felt in it, when one wishes to flee or to hide, and when one feels the strange inhibition that it is impossible to move from the spot and that one is incapable of altering the disagreeable situation. It is only in this connection that the dream is typical; the nucleus of its content may otherwise be brought into all kinds of relations or may be replaced by individual amplifications. It is essentially a question of a disagreeable sensation of the nature of shame, the wish to be able to hide one's nakedness, chiefly by means of locomotion, without being able to accomplish this. I believe that the great majority of my readers will at some time have found themselves in this situation in a dream.

Usually the nature and manner of the experience is indistinct. It is usually reported, "I was in my shirt," but this is rarely a clear image; in most cases the lack of clothing is so indeterminate that it is designated in the report of the dream by a set of alternatives: "I was in my chemise or in my petticoat." As a rule the deficiency in the toilet is not serious enough to justify the feeling of shame attached to it. For a person who has served in the army, nakedness is often replaced by a mode of adjustment that is contrary to regulations. "I am on the street without my sabre and I see officers coming," or "I am without my necktie," or "I am wearing checkered civilian's trousers," &c.

The persons before whom one is ashamed are almost always strangers with faces that have been left undetermined. It never occurs in the typical dream that one is reproved or even noticed on account of the dress which causes the embarrassment to one's self. Quite on the contrary, the people have an air of indifference, or, as I had opportunity to observe in a particularly clear dream, they look stiffly solemn. This is worth thinking about.

The shamed embarrassment of the dreamer and the indifference of the spectators form a contradiction which often occurs in the dream. It would better accord with the feelings of the dreamer if the strangers looked at him in astonishment and laughed at him, or if they grew indignant. I think, however, that the latter unpleasant feature has been obviated by the tendency to wish-fulfilment, while the embarrassment, being retained on some account or other, has been left standing, and thus the two parts fail to agree. We have interesting evidence to show that the dream, whose appearance has been partially disfigured by the tendency to wish-fulfilment, has not been properly understood. For it has become the basis of a fairy tale familiar to us all in Andersen's version,[30] and it has recently received poetic treatment by L. Fulda in the *Talisman*. In Andersen's fairy tale we are told of two impostors who weave a costly garment for the Emperor, which, however, shall be visible only to the good and true. The Emperor goes forth clad in this invisible garment, and, the fabric serving as a sort of touchstone, all the people are frightened into acting as though they did not notice the nakedness of the Emperor.

But such is the situation in our dream. It does not require great boldness to assume that the unintelligible dream content has suggested the invention of a state of undress in which the situation that is being remembered becomes significant. This situation has then been deprived of its original meaning, and placed at the service of other purposes. But we shall see that such misunderstanding of the dream content often occurs on account of the conscious activity of the second psychic system, and is to be recognised as a factor in the ultimate formation of the dream; furthermore, that in the development of the obsessions and phobias similar misunderstandings, likewise within the same psychic personality, play a leading part. The source from which in our dream the material for this transformation is taken can also be explained. The impostor is the dream, the Emperor is the dreamer himself, and the moralising tendency betrays a hazy knowledge of the fact that the latent dream content is occupied with forbidden wishes which have become the victims of repression. The connection in which such dreams appear during my analysis of neurotics leaves no room for doubting that the dream is based upon a recollection from earliest childhood. Only in our childhood was there a time when we were seen by our relatives as well as by strange nurses, servant girls, and visitors, in scanty clothing, and at that time we were not ashamed of our nakedness.[31]

It may be observed in the case of children who are a little older that being undressed has a kind of intoxicating effect upon them, instead of making them ashamed. They laugh, jump about, and strike their bodies; the mother, or whoever is present, forbids them to do this, and says: "Fie, that is shameful—you mustn't do that." Children often show exhibitional cravings; it is hardly possible to go through a village in our part of the country without meeting a two or three-year-old tot who lifts up his or her shirt before the traveller, perhaps in his honour. One of my patients has reserved in his conscious memory a scene from the eighth year of his life in which he had just undressed previous to going to bed, and was about to dance into the room of his little sister in his undershirt when the servant prevented his doing it. In the childhood history of neurotics, denudation in the presence of children of the opposite sex plays a great part; in paranoia the desire to be observed while dressing and undressing may be directly traced to these experiences; among those remaining perverted there is a class which has accentuated the childish impulse to a compulsion—they are the exhibitionists.

This age of childhood in which the sense of shame is lacking seems to our later recollections a Paradise, and Paradise itself is nothing but a composite phantasy from the childhood of the individual. It is for this reason, too, that in Paradise human beings are naked and are not ashamed until the moment arrives when the sense of shame and of fear are aroused; expulsion follows, and sexual life and cultural development begin. Into this Paradise the dream can take us back every night; we have already ventured the conjecture that the impressions from earliest childhood (from the prehistoric period until about the end of the fourth year) in themselves, and independently of everything else, crave reproduction, perhaps without further reference to their content, and that the repetition of them is the fulfilment of a wish. Dreams of nakedness, then, are exhibition dreams.[32]

One's own person, which is seen not as that of a child, but as belonging to the present, and the idea of scanty clothing, which became buried beneath so many later *négligée*. recollections or because of the censor, turns out to be obscure—these two things constitute the nucleus of the exhibition dream. Next come the persons before whom one is ashamed. I know of no example where the actual spectators at those infantile exhibitions reappear in the dream. For the dream is hardly ever a simple recollection. Strangely enough, those persons who are the objects of our sexual interest during childhood are omitted from all the reproductions of the dream, of hysteria, and of the compulsion neurosis; paranoia alone puts the spectators back into their places, and is fanatically convinced of their presence, although they remain invisible. What the dream substitutes for these, the "many strange people," who take no notice of the spectacle which is presented, is exactly the wish-opposite of that single, intimate person for whom the exposure was intended. "Many strange people," moreover, are often found in the dream in any other favourable connection; as a wish-opposite they always signify "a secret."[33] It may be seen how the restoration of the old condition of affairs, as it occurs in paranoia, is subject to this antithesis. One is no longer alone. One is certainly being watched, but the spectators are "many strange, curiously indeterminate people."

Furthermore, repression has a place in the exhibition dream. For the disagreeable sensation of the dream is the reaction of the second psychic instance to the fact that the exhibition scene which has been rejected by it has in spite of this succeeded in securing representation. The only way to avoid this sensation would be not to revive the scene.

Later on we shall again deal with the sensation of being inhibited. It serves the dream excellently in representing the conflict of the will, the negation. According to our unconscious purpose exhibition is to be continued; according to the demands of the censor, it is to be stopped.

The relation of our typical dreams to fairy tales and to other poetic material is neither a sporadic nor an accidental one. Occasionally the keen insight of a poet has analytically recognised the transforming process—of which the poet is usually the tool—and has followed it backwards, that is to say, traced it to the dream. A friend has called my attention to the following passage in G. Keller's *Der Grüne Heinrich:*

"I do not wish, dear Lee, that you should ever come to realise from experience the peculiar piquant truth contained in the situation of Odysseus, when he appears before Nausikaa and her playmates, naked and covered with mud! Would you like to know what it means? Let us consider the incident closely. If you are ever separated from your home, and from everything that is dear to you, and wander about in a strange country, when you have seen and experienced much, when you have cares and sorrows, and are, perhaps, even miserable and forlorn, you will some night inevitably dream that you are approaching your home; you will see it shining and beaming in the most beautiful colours; charming, delicate and lovely figures will come to meet you; and you will suddenly discover that you are going about in rags, naked and covered with dust. A nameless feeling of shame and fear seizes you, you try to cover yourself and to hide, and you awaken bathed in sweat. As long as men exist, this will be the dream of the care-laden, fortune-battered man, and thus Homer has taken his situation from the profoundest depths of the eternal character of humanity."

This profound and eternal character of humanity, upon the touching of which in his listeners the poet usually calculates, is made up of the stirrings of the spirit which are rooted in childhood, in the period which later becomes prehistoric. Suppressed and forbidden wishes of childhood break forth under cover of those wishes of the homeless man which are unobjectionable and capable of becoming conscious, and for that reason the dream which is made objective in the legend of Nausikaa regularly assumes the form of a dream of anxiety.

My own dream, mentioned on p. 218, of hurrying up the stairs, which is soon afterward changed into that of being glued to the steps, is likewise an exhibition dream, because it shows the essential components of such' a dream. It must thus permit of being referred to childish experiences, and the possession of these ought to tell us how far the behaviour of the servant girl towards me—her reproach that I had soiled the carpet—

helped her to secure the position which she occupies in the dream. I am now able to furnish the desired explanation. One learns in psychoanalysis to interpret temporal proximity by objective connection; two thoughts, apparently without connection, which immediately follow one another, belong to a unity which can be inferred; just as an *a* and a *t,* which I write down together, should be pronounced as one syllable, *at.* The same is true of the relation of dreams to one another. The dream just cited, of the stairs, has been taken from a series of dreams, whose other members I am familiar with on account of having interpreted them. The dream which is included in this series must belong to the same connection. Now the other dreams of the series are based upon the recollection of a nurse to whom I was entrusted from some time in the period when I was suckling to the age of two and a half years, and of whom a hazy recollection has remained in my consciousness. According to information which I have recently obtained from my mother, she was old and ugly, but very intelligent and thorough; according to inferences which I may draw from my dreams, she did not always give me the kindest treatment, and said hard words to me when I showed insufficient aptitude for education in cleanliness. Thus by attempting to continue this educational work the servant girl develops a claim to be treated by me, in the dream, as an incarnation of the prehistoric old woman. It is to be assumed that the child bestowed his love upon this governess in spite of her bad treatment of him.[34]

Another series of dreams which might be called typical are those which have the content that a dear relative, parent, brother, or sister, child or the like, has died. Two classes of these dreams must immediately be distinguished—those in which the dreamer remains unaffected by sorrow while dreaming, and those in which he feels profound grief on account of the death, in which he even expresses this grief during sleep by fervid tears.

We may ignore the dreams of the first group; they have no claim to be reckoned as typical. If they are analysed, it is found that they signify something else than what they contain, that they are intended to cover up some other wish. Thus it is with the dream of the aunt who sees the only son of her sister lying on a bier before her (p 146). This does not signify that she wishes the death of her little nephew; it only conceals, as we have learned, a wish to see a beloved person once more after long separation—the same person whom she had seen again after a similar long intermission at the funeral of another nephew. This wish, which is the real content of the dream, gives no cause for sorrow, and for that reason no sorrow is felt in the dream. It may be seen in this case that the emotion which is contained in the dream does not belong to the manifest content of the dream, but to the latent one, and that the emotional content has remained free from the disfigurement which has befallen the presentation content.

It is a different story with the dreams in which the death of a beloved relative is imagined and where sorrowful emotion is felt. These signify, as their content says, the wish that the person in question may die, and as I may here expect that the feelings of all readers and of all persons who have dreamt anything similar will object to my interpretation, I must strive to present my proof on the broadest possible basis.

We have already had one example to show that the wishes represented in the dream as fulfilled are not always actual wishes. They may also be dead, discarded, covered, and repressed wishes, which we must nevertheless credit with a sort of continuous existence on account of their reappearance in the dream. They are not dead like persons who have died in our sense, but they resemble the shades in the *Odyssey* which awaken a certain kind of life as soon as they have drunk blood. In the dream of the dead child in the box (p. 147) we were concerned with a wish that had been actual fifteen years before, and which had been frankly admitted from that time. It is, perhaps, not unimportant from the point of view of dream theory if I add that a recollection from earliest childhood is at the basis even of this dream. While the dreamer was a little child—it cannot be definitely determined at what time—she had heard that during pregnancy of which she was the fruit her mother had fallen into a profound depression of spirits and had passionately wished for the death of her child before birth. Having grown up herself and become pregnant, she now follows the example of her mother.

If some one dreams with expressions of grief that his father or mother, his brother or sister, has died, I shall not use the dream as a proof that he wishes them dead *now*. The theory of the dreams does not require so much; it is satisfied with concluding that the dreamer has wished them , dead— at some one time in childhood. I fear, however, that this limitation will not contribute much to quiet the objectors; they might just as energetically contest the possibility that they have ever had such thoughts as they are sure that they do not cherish such wishes at present. I must, therefore, reconstruct a part of the submerged infantile psychology on the basis of the testimony which the present still furnishes.[35]

Let us at first consider the relation of children to their brothers and sisters. I do not know why we presuppose that it must be a loving one, since examples of brotherly and sisterly enmity among adults force themselves upon every one's experience, and since we so often know that this estrangement originated even during childhood or has always existed. But many grown-up people, who to-day are tenderly attached to their brothers and sisters and stand by them, have lived with them during childhood in almost uninterrupted hostility. The older child has ill-treated the younger, slandered it, and deprived it of its toys; the younger has been consumed by helpless fury against the elder, has envied it and feared it, or its first impulse toward liberty and first feelings of injustice have been directed against the oppressor. The parents say that the children do not agree, and cannot find the reason for it. It is not difficult to see that the character even of a well-behaved child is not what we wish to find in a grown-up person. The child is absolutely egotistical; it feels its wants acutely and strives remorselessly to satisfy them, especially with its competitors, other children, and in the first instance with its brothers and sisters. For doing this we do not call the child wicked—we call it naughty; it is not responsible for its evil deeds either in our judgment or in the eyes of the penal law. And this is justifiably so; for we may expect that within this very period of life which we call childhood, altruistic impulses and morality will come to life in the little egotist, and that, in the words of Meynert, a secondary ego will overlay and restrain the primary one. It is true that morality does not develop simultaneously in all departments, and furthermore, the duration of the unmoral period of childhood is of different length in different individuals. In cases where the development of this morality fails to appear, we are pleased to talk about "degeneration"; they are obviously cases of arrested development. Where the primary character has already been covered up by later development, it may be at least partially uncovered again by an attack of hysteria. The correspondence between the so-called hysterical character and that of a naughty child is strikingly evident. A compulsion neurosis, on the other hand, corresponds to a super-morality, imposed upon the primary character that is again asserting itself, as an increased check.

Many persons, then, who love their brothers and sisters, and who would feel bereaved by their decease, have evil wishes towards them from earlier times in their unconscious wishes, which are capable of being realised in the dream. It is particularly interesting to observe little children up to three years old in their attitude towards their brothers and sisters. So far the child has been the only one; he is now informed that the stork has brought a new child. The younger surveys the arrival, and then expresses his opinion decidedly: "The stork had better take it back again."[36]

I subscribe in all seriousness to the opinion that the child knows enough to calculate the disadvantage it has to expect on account of the new-comer. I know in the case of a lady of my acquaintance who agrees very well with a sister four years younger than herself, that she responded to the news of her younger sister's arrival with the following words: "But I shan't give her my red cap, anyway." If the child comes to this realisation only at a later time, its enmity will be aroused at that point. I know of a case where a girl, not yet three years old, tried to strangle a suckling in the cradle, because its continued presence, she suspected, boded her no good. Children are capable of envy at this time of life in all its intensity and distinctness. Again, perhaps, the little brother or sister has really soon disappeared; the child has again drawn the entire affection of the household to itself, and then a new child is sent by the stork; is it then unnatural for the favourite to wish that the new competitor may have the same fate as the earlier one, in order that he may be treated as well as he was before during the interval? Of course this attitude of the child towards the younger infant is under normal circumstances a simple function of the difference of age. After a certain time the maternal instincts of the girl will be excited towards the helpless new-born child.

Feelings of enmity towards brothers and sisters must occur far more frequently during the age of childhood than is noted by the dull observation of adults.

In case of my own children, who followed one another rapidly, I missed the opportunity to make such observations; I am now retrieving it through my little nephew, whose complete domination was disturbed after fifteen months by the arrival of a female competitor. I hear, it is true, that the young man acts very chivalrously towards his little sister, that he kisses her hand and pets her; but in spite of this I have convinced myself that even before the completion of his second year he is using his new facility in language to criticise this person who seems superfluous to him. Whenever the conversation turns upon her, he chimes in and cries angrily:

"Too (l)ittle, too (l)ittle." During the last few months, since the child has outgrown this unfavourable criticism, owing to its splendid development, he has found another way of justifying his insistence that she does not deserve so much attention. On all suitable occasions he reminds us, "She hasn't any teeth."[37] We have all preserved the recollection of the eldest daughter of another sister of mine—how the child which was at that time six years old sought assurance from one aunt after another for an hour and a half with the question: "Lucy can't understand that yet, can she?" Lucy was the competitor, two and a half years younger.

I have never failed in any of my female patients to find this dream of the death of brothers and sisters denoting exaggerated hostility. I have met with only one exception, which could easily be reinterpreted into a confirmation of the rule. Once in the course of a sitting while I was explaining this condition of affairs to a lady, as it seemed to have a bearing upon the symptoms under consideration, she answered, to my astonishment, that she had never had such dreams. However, she thought of another dream which supposedly had nothing to do with the matter—a dream which she had first dreamed at the age of four, when she was the youngest child, and had since dreamed repeatedly. "A great number of children, all of them the dreamer's brothers and sisters, and male and female cousins, were romping about in a meadow. Suddenly they all got wings, flew up, and were gone." She had no idea of the significance of the dream; but it will not be difficult for us to recognise it as a dream of the death of all the brothers and sisters, in its original form, and little influenced by the censor. I venture to insert the following interpretation: At the death of one out of a large number of children—in this case the children of two brothers were brought up in common as brothers and sisters—is it not probable that our dreamer, at that time not yet four years old, asked a wise, grown-up person: "What becomes of children when they are dead?" The answer probably was: "They get wings and become angels." According to this explanation all the brothers and sisters and cousins in the dream now have wings like angels and—this is the important thing—they fly away. Our little angel-maker remains alone, think of it, the only one after such a multitude! The feature that the children are romping about on a meadow points with little ambiguity to butterflies, as though the child had been led by the same association which induced the ancients to conceive Psyche as having the wings of a butterfly.

Perhaps some one will now object that, although the inimical impulses of children towards their brothers and sisters may well enough be admitted, how does the childish disposition arrive at such a height of wickedness as to wish death to a competitor or stronger playmate, as though all transgressions could be atoned for only by the death-punishment? Whoever talks in this manner forgets that the childish idea of "being dead" has little else but the words in common with our own. The child knows nothing of the horrors of decay, of shivering in the cold grave, of the terror of the infinite Nothing, which the grown-up person, as all the myths concerning the Great Beyond testify, finds it so hard to bear in his conception. Fear of death is strange to the child; therefore it plays with the horrible word and threatens another child: "If you do that again you will die, as Francis died," whereat the poor mother shudders, for perhaps she cannot forget that the great majority of mortals do not succeed in living beyond the years of childhood. It is still possible, even for a child eight years old, on returning from a museum of natural history, to say to its mother: "Mamma, I love you so; if you ever die, I am going to have you stuffed and set you up here in the room so I can always, always see you!" So little does the childish conception of being dead resemble our own.[38]

Being dead means for the child, which has been spared the scenes of suffering previous to dying, the same as "being gone," not disturbing the survivors any more. The child does not distinguish the manner and means by which this absence is brought about, whether by travelling, estrangement, or death. If, during the prehistoric years of a child, a nurse has been sent away and its mother has died a short while after, the two experiences, as is revealed by analysis, overlap in his memory. The fact that the child does not miss very intensely those who are absent has been realised by many a mother to her sorrow, after she has returned home after a summer journey of several weeks, and has been told upon inquiry: "The children have not asked for their mother a single time." But if she really goes to that "undiscovered country from whose bourn no traveller returns," the children seem at first to have forgotten her, and begin only *subsequently* to remember the dead mother.

If, then, the child has motives for wishing the absence of another child, every restraint is lacking which would prevent it from clothing this wish in the form that the child may die, and the psychic reaction to the dream of wishing death proves that, in spite of all the differences in content, the wish in the case of the child is somehow or other the same as it is with adults.

If now the death-wish of the child towards its brothers and sisters has been explained by the childish egotism, which causes the child to regard its brothers and sisters as competitors, how may we account for the same wish towards parents, who bestow love on the child and satisfy its wants, and whose preservation it ought to desire from these very egotistical motives?

In the solution of this difficulty we are aided by the experience that dreams of the death of parents predominantly refer to that member of the parental couple which shares the sex of the dreamer, so that the man mostly dreams of the death of his father, the woman of the death of her mother. I cannot claim that this happens regularly, but the predominating occurrence of this dream in the manner indicated is so evident that it must be explained through some factor that is universally operative. To express the matter boldly, it is as though a sexual preference becomes active at an early period, as though the boy regards his father as a rival in love, and as though the girl takes the same attitude toward her mother—a rival by getting rid of whom he or she cannot but profit.

Before rejecting this idea as monstrous, let the reader consider the actual relations between parents and children. What the requirements of culture and piety demand of this relation must be distinguished from what daily observation shows us to be the fact. More than one cause for hostile feeling is concealed within the relations between parents and children; the conditions necessary for the actuation of wishes which cannot exist in the presence of the censor are most abundantly provided. Let us dwell at first upon the relation between father and son. I believe that the sanctity which we have ascribed to the injunction of the decalogue dulls our perception of reality. Perhaps we hardly dare to notice that the greater part of humanity neglects to obey the fifth commandment. In the lowest as well as in the highest strata of human society, piety towards parents is in the habit of receding before other interests. The obscure reports which have come to us in mythology and legend from the primeval ages of human society give us an unpleasant idea of the power of the father and the ruthlessness with which it was used. Kronos devours his children, as the wild boar devours the brood of the sow; Zeus emasculates his father[39] and takes his place as a ruler. The more despotically the father ruled in the ancient family, the more must the son have taken the position of an enemy, and the greater must have been his impatience, as designated successor, to obtain the mastery himself after his father's death. Even in our own middle-class family the father is accustomed to aid the development of the germ of hatred which naturally belongs to the paternal relation by refusing the son the disposal of his own destiny, or the means necessary for this. A physician often has occasion to notice that the son's grief at the loss of his father cannot suppress his satisfaction at the liberty which he has at last obtained. Every father frantically holds on to whatever of the sadly antiquated *potestas patris* still remains in the society of to-day, and every poet who, like Ibsen, puts the ancient strife between father and son in the foreground of his fiction is sure of his effect. The causes of conflict between mother and daughter arise when the daughter grows up and finds a guardian in her mother, while she desires sexual freedom, and when, on the other hand, the mother has been warned by the budding beauty of her daughter that the time has come for her to renounce sexual claims.

All these conditions are notorious and open to everyone's inspection. But they do not serve to explain dreams of the death of parents found in the case of persons to whom piety towards their parents has long since come to be inviolable. We are furthermore prepared by the preceding discussion to find that the death-wish towards parents is to be explained by reference to earliest childhood.

This conjecture is reaffirmed with a certainty that makes doubt impossible in its application to psychoneurotics through the analyses that have been undertaken with them. It is here found that the sexual wishes of the child—in so far as they deserve this designation in their embryonic state—awaken at a very early period, and that the first inclinations of the girl are directed towards the father, and the first childish cravings of the boy towards the mother. The father thus becomes an annoying competitor for the boy, as the mother does for the girl, and we have already shown in the case of brothers and sisters how little it takes for this feeling to lead the child to the death-wish. Sexual selection, as a rule, early becomes evident in the parents; it is a natural tendency for the father to indulge the little daughter, and for the mother to take the part of the sons, while both work earnestly for the education of the little ones when the magic of sex does not prejudice their judgment. The child is very well aware of any partiality, and resists that member of the parental couple who discourages it. To find love in a grown-up person is for the child not only the satisfaction of a particular craving, but also means that the child's will is to be yielded to in other respects. Thus the child obeys its own sexual impulse, and at the same time re-enforces the feeling which proceeds from the parents, if it makes a selection among the parents that corresponds to theirs.

Most of the signs of these infantile inclinations are usually overlooked; some of them may be observed even after the first years of childhood. An eight-year-old girl of my acquaintance, when her mother is called from the table, takes advantage of the opportunity to proclaim herself her successor. "Now I shall be Mamma; Charles, do you want some more vegetables? Have some, I beg you," &c. A particularly gifted and vivacious girl, not yet four years old, with whom this bit of child psychology is unusually transparent, says outright: "Now mother can go away; then father must marry me and I shall be his wife." Nor does this wish by any means exclude from child life the possibility that the child may love his mother affectionately. If the little boy is allowed to sleep at his mother's side whenever his father goes on a journey, and if after his father's return he must go back to the nursery to a person whom he likes far less, the wish may be easily actuated that his father may always be absent, in order that he may keep his place next to his dear, beautiful mamma; and the father's death is obviously a means for the attainment of this wish; for the child's experience has taught him that "dead" folks, like grandpa, for example, are always absent; they never return.

Although observations upon little children lend themselves, without being forced, to the proposed interpretation, they do not carry the full conviction which psychoanalyses of adult neurotics obtrude upon the physician. The dreams in question are here cited with introductions of such a nature that their interpretation as wish-dreams becomes unavoidable. One day I find a lady sad and weeping. She says: "I do not want to see my relatives any more; they must shudder at me." Thereupon, almost without any transition, she tells that she remembers a dream, whose significance, of course, she does not know. She dreamed it four years before, and it is as follows: *A fox or a lynx is taking a walk on the roof; then something falls down, or she falls down, and after that her mother is carried out of the house dead*—whereat the dreamer cries bitterly. No sooner had I informed her that this dream must signify a wish from her childhood to see her mother dead, and that it is because of this dream that she thinks that her relatives must shudder at her, than she furnished some material for explaining the dream. "Lynx-eye" is an opprobrious epithet which a street boy once bestowed on her when she was a very small child; when she was three years old a brick had fallen on her mother's head so that she bled severely.

I once had opportunity to make a thorough study of a young girl who underwent several psychic states. In the state of frenzied excitement with which the illness started, the patient showed a very strong aversion to her mother; she struck and scolded her as soon as she approached the bed, while at the same time she remained loving and obedient to a much older sister. Then there followed a clear but somewhat apathetic state with very much disturbed sleep. It was in this phase that I began to treat her and to analyse her dreams. An enormous number of these dealt in a more or less abstruse manner with the death of the mother; now she was present at the funeral of an old woman, now she saw her sisters sitting at the table dressed in mourning; the meaning of the dreams could not be doubted. During the further progress of the convalescence hysterical phobias appeared; the most torturing of these was the idea that something happened to her mother. She was always having to hurry home from wherever she happened to be in order to convince herself that her mother was still alive. Now this case, in view of my other experiences, was very instructive; it showed in polyglot translations, as it were, the different ways in which the psychic apparatus reacts to the same exciting idea. In the state of excitement which I conceive as the overpowering of the second psychic instance, the unconscious enmity towards the mother became potent as a motor impulse; then, after calmness set in, following the suppression of the tumult, and after the domination of the censor had been restored, this feeling of enmity had access only to the province of dreams in order to realise the wish that the mother might die; and after the normal condition had been still further strengthened, it created the excessive concern for the mother as a hysterical counter-reaction and manifestation of defence. In the light of these considerations it is no longer inexplicable why hysterical girls are so often extravagantly attached to their mothers.

On another occasion I had opportunity to get a profound insight into the unconscious psychic life of a young man for whom a compulsion-neurosis made life almost unendurable, so that he could not go on the street, because he was harassed by the obsession that he would kill every one he met. He spent his days in arranging evidence for an alibi in case he should be charged with any murder that might have occurred in the city. It is superfluous to remark that this man was as moral as he was highly cultured. The analysis—which, moreover, led to a cure—discovered murderous impulses toward the young man's somewhat over-strict father as the basis of these disagreeable ideas of compulsion—impulses which, to his great surprise, had received conscious expression when he was seven years old, but which, of course, had originated in much earlier years of childhood. After the painful illness and death of the father, the obsessive reproach transferred to strangers in the form of the afore-mentioned phobia, appeared when the young man was thirty-one years old. Anyone capable of wishing to push his own father from a mountain-top into an abyss is certainly not to be trusted to spare the lives of those who are not so closely bound to him; he does well to lock himself into his room.

According to my experience, which is now large, parents play a leading part in the infantile psychology of all later neurotics, and falling in love with one member of the parental couple and hatred of the other help to make up that fateful sum of material furnished by the psychic impulses, which has been formed during the infantile period, and which is of such great importance for the symptoms appearing in the later neurosis. But I do not think that psychoneurotics are here sharply distinguished from normal human beings, in that they are capable of creating something absolutely new and peculiar to themselves. It is far more probable, as is shown also by occasional observation upon normal children, that in their loving or hostile wishes towards their parents psychoneurotics only show in exaggerated form feelings which are present less distinctly and less intensely in the minds of most children. Antiquity has furnished us with legendary material to confirm this fact, and the deep and universal effectiveness of these legends can only be explained by granting a similar universal applicability to the above-mentioned assumption in infantile psychology.

I refer to the legend of King Oedipus and the drama of the same name by Sophocles. Oedipus, the son of Laius, king of Thebes, and of Jocasta, is exposed while a suckling, because an oracle has informed the father that his son, who is still unborn, will be his murderer. He is rescued, and grows up as the king's son at a foreign court, until, being uncertain about his origin, he also consults the oracle, and is advised to avoid his native place, for he is destined to become the murderer of his father and the husband of his mother. On the road leading away from his supposed home he meets King

Laius and strikes him dead in a sudden quarrel. Then he comes to the gates of Thebes, where he solves the riddle of the Sphynx who is barring the way, and he is elected king by the Thebans in gratitude, and is presented with the hand of Jocasta. He reigns in peace and honour for a long time, and begets two sons and two daughters upon his unknown mother, until at last a plague breaks out which causes the Thebans to consult the oracle anew. Here Sophocles' tragedy begins. The messengers bring the advice that the plague will stop as soon as the murderer of Laius is driven from the country. But where is he hidden?

"Where are they to be found? How shall we trace the perpetrators of so old a crime where no conjecture leads to discovery?"[40]

The action of the play now consists merely in a revelation, which is gradually completed and artfully delayed—resembling the work of a psychoanalysis—of the fact that Oedipus himself is the murderer of Laius, and the son of the dead man and of Jocasta. Oedipus, profoundly shocked at the monstrosities which he has unknowingly committed, blinds himself and leaves his native place. The oracle has been fulfilled.

The *Oedipus Tyrannus* is a so-called tragedy of fate; its tragic effect is said to be found in the opposition between the powerful will of the gods and the vain resistance of the human beings who are threatened with destruction; resignation to the will of God and confession of one's own helplessness is the lesson which the deeply-moved spectator is to learn from the tragedy. Consequently modern authors have tried to obtain a similar tragic effect by embodying the same opposition in a story of their own invention. But spectators have sat unmoved while a curse or an oracular sentence has been fulfilled on blameless human beings in spite of all their struggles; later tragedies of fate have all remained without effect.

If the *Oedipus Tyrannus* is capable of moving modern men no less than it moved the contemporary Greeks, the explanation of this fact cannot lie merely in the assumption that the effect of the Greek tragedy is based upon the opposition between fate and human will, but is to be sought in the peculiar nature of the material by which the opposition is shown. There must be a voice within us which is prepared to recognise the compelling power of fate in *Oedipus,* while we justly condemn the situations occurring in *Die Ahnfrau* or in other tragedies of later date as arbitrary inventions. And there must be a factor corresponding to this inner voice in the story of King Oedipus. His fate moves us only for the reason that it might have been ours, for the oracle has put the same curse upon us before our birth as upon him. Perhaps we are all destined to direct our first sexual impulses towards our mothers, and our first hatred and violent wishes towards our fathers; our dreams convince us of it. King Oedipus, who has struck his father Laius dead and has married his mother Jocasta, is nothing but the realised wish of our childhood. But more fortunate than he, we have since succeeded, unless we have become psychoneurotics, in withdrawing our sexual impulses from our mothers and in forgetting our jealousy of our fathers. We recoil from the person for whom this primitive wish has been fulfilled with all the force of the repression which these wishes have suffered within us. By his analysis, showing us the guilt of Oedipus, the poet urges us to recognise our own inner self, in which these impulses, even if suppressed, are still present. The comparison with which the chorus leaves us—

> " . . . Behold! this Oedipus, who unravelled the famous riddle and who was a man of eminent virtue; a man who trusted neither to popularity nor to the fortune of his citizens; see how great a storm of adversity hath at last overtaken him" (Act v. sc. 4).

This warning applies to ourselves and to our pride, to us, who have grown so wise and so powerful in our own estimation since the years of our childhood. Like Oedipus, we live in ignorance of the wishes that offend morality, wishes which nature has forced upon us, and after the revelation of which we want to avert every glance from the scenes of our childhood.

In the very text of Sophocles' tragedy there is an unmistakable reference to the fact that the Oedipus legend originates in an extremely old dream material, which consists of the painful disturbance of the relation towards one's parents by means of the first impulses of sexuality. Jocasta comforts Oedipus—who is not yet enlightened, but who has become worried on account of the oracle—by mentioning to him the dream which is dreamt by so many people, though she attaches no significance to it—

> "For it hath already been the lot of many men in dreams to think themselves partners of their mother's bed. But he passes most easily through life to whom these circumstances are trifles" (Act iv. sc. 3).

The dream of having sexual intercourse with one's mother occurred at that time, as it does to-day, to many people, who tell it with indignation and astonishment. As may be understood, it is the key to the tragedy and the complement to the dream of the death of the father. The story of Oedipus is the reaction of the imagination to these two typical dreams, and just as the dream when occurring to an adult is experienced with feelings of resistance, so the legend must contain terror and self-chastisement. The appearance which it further assumes is the result of an uncomprehending secondary elaboration which tries to make it serve theological purposes (*cf.* the dream material of exhibitionism, p. 220). The attempt to reconcile divine omnipotence with human responsibility must, of course, fail with this material as with every other.[41]

I must not leave the typical dream of the death of dear relatives without somewhat further elucidating the subject of their significance for the theory of the dream in general. These dreams show us a realisation of the very unusual case where the dream thought, which has been created by the repressed wish, completely escapes the censor, and is transferred to the dream without alteration. There must be present peculiar conditions making possible such an outcome. I find circumstances favourable to these dreams in the two following factors: First, there is no wish which we believe further from us; we believe such a wish "would never occur to us in a dream"; the dream censor is therefore not prepared for this monstrosity, just as the legislation of Solon was incapable of establishing a punishment for patricide. Secondly, the repressed and unsuspected wish is in just this case particularly often met by a fragment of the day's experience in the shape of a concern about the life of the beloved person. This concern cannot be registered in the dream by any other means than by taking advantage of the wish that has the same content; but it is possible for the wish to mask itself behind the concern which has been awakened during the day. If one is inclined to think all this a more simple process, and that one merely continues during the night and in dreams what one has been concerned with during the day, the dream of the death of beloved persons is removed from all connection with dream explanation, and an easily reducible problem is uselessly retained.

It is also instructive to trace the relation of these dreams to anxiety dreams. In the dream of the death of dear persons the repressed wish has found a way of avoiding the censor, and the distortion which it causes.

In this case the inevitable concomitant manifestation is that disagreeable sensations are felt in the dream. Thus the dream of fear is brought about only when the censor is entirely or partially overpowered, and, on the other hand, the overpowering of the censor is made easier when fear has already been furnished by somatic sources. Thus it becomes obvious for what purpose the censor performs its office and practises dream distortion; it does this *in order to prevent the development of fear or other forms of disagreeable emotion.*

I have spoken above of the egotism of the infantile mind, and I may now resume this subject in order to suggest that dreams preserve this characteristic—thus showing their connection with infantile life. Every dream is absolutely egotistical; in every dream the beloved ego appears, even though it may be in a disguised form. The wishes that are realised in dreams are regularly the wishes of this ego; it is only a deceptive appearance if interest in another person is thought to have caused the dream. I shall subject to analysis several examples which appear to contradict this assertion.

I

A boy not yet four years old relates the following: *He saw a large dish garnished, and upon it a large piece of roast meat, and the meat was all of a sudden—not cut to pieces—but eaten up. He did not see the person who ate it.*[42]

Who may this strange person be of whose luxurious repast this little fellow dreams? The experiences of the day must give us the explanation of this. For a few days the boy had been living on a diet of milk according to the doctor's prescription; but on the evening of the day before the dream he had been naughty, and as a punishment he had been deprived of his evening meal. He had already undergone one such hunger-cure, and had acted very bravely. He knew that he would get nothing to eat, but he did not dare to indicate by a word that he was hungry. Education was beginning to have its influence upon him; this is expressed even in the dream which shows the beginnings of dream disfigurement. There is no doubt that he himself is the person whose wishes are directed toward this abundant meal, and a meal of roast meat at that. But since he knows that this is forbidden him, he does not dare, as children do in the dream (*cf.* the dream about strawberries of my little Anna, p. 125), to sit down to the meal himself. The person remains anonymous.

II

Once I dream that I see on the show-table of a book store a new number in the Book-lovers' Collection—the collection which I am in the habit of buying (art monographs, monographs on the history of the world, famous art centres, &c.). *The new collection is called Famous Orators (or Orations), and the first number bears the name of Doctor Lecher.*

In the course of analysis it appears improbable that the fame of Dr. Lecher, the long-winded orator of the German Opposition, should occupy my thoughts while I am dreaming. The fact is that, a few days before, I undertook the psychic cure of some new patients, and was now forced to talk for from ten to twelve hours a day. Thus I myself am the long-winded orator.

III

Upon another occasion I dream that a teacher of my acquaintance at the university says: *My son, the Myopic.* Then there follows a dialogue consisting of short speeches and replies. A third portion of the dream follows in which I and my sons appear, and as far as the latent dream content is concerned, father, son, and Professor M. are alike only lay figures to represent me and my eldest son. I shall consider this dream again further on because of another peculiarity.

IV

The following dream gives an example of really base egotistical feelings, which are concealed behind affectionate concern:

My friend Otto looks ill, his face is brown and his eyes bulge.

Otto is my family physician, to whom I owe a debt greater than I can ever hope to repay, since he has guarded the health of my children for years. He has treated them successfully when they were taken sick, and besides that he has given them presents on all occasions which gave him any excuse for doing so. He came for a visit on the day of the dream, and my wife noticed that he looked tired and exhausted. Then comes my dream at night, and attributes to him a few of the symptoms of Basedow's disease. Any one disregarding my rules for dream interpretation would understand this dream to mean that I am concerned about the health of my friend, and that this concern is realised in the dream. It would thus be a contradiction not only of the assertion that the dream is a wish-fulfilment, but also of the assertion that it is accessible only to egotistic impulses. But let the person who interprets the dream in this manner explain to me why I fear that Otto has Basedow's disease, for which diagnosis his appearance does not give the slightest justification? As opposed to this, my analysis furnishes the following material, taken from an occurrence which happened six years ago. A small party of us, including Professor R., were driving in profound darkness through the forest of N., which is several hours distant from our country home. The coachman, who was not quite sober, threw us and the wagon down a bank, and it was only by a lucky accident that we all escaped unhurt. But we were forced to spend the night at the nearest inn, where the news of our accident awakened great sympathy. A gentleman, who showed unmistakable signs of the morbus Basedowii—nothing but a brownish colour of the skin of the face and bulging eyes, no goitre— placed himself entirely at our disposal and asked what he could do for us. Professor R. answered in his decided way: "Nothing but lend me a night-shirt." Whereupon our generous friend replied: "I am sorry but I cannot do that," and went away.

In continuing the analysis, it occurs to me that Basedow is the name not only of a physician, but also of a famous educator. (Now that I am awake I do not feel quite sure of this fact.) My friend Otto is the person whom I have asked to take charge of the physical education of my children—especially during the age of puberty (hence the night-shirt)—in case anything should happen to me. By seeing Otto in the dream with the morbid symptoms of our above-mentioned generous benefactor, I apparently mean to say, "If anything happens to me, just as little is to be expected for my children from him as was to be expected then from Baron L., in spite of his well-meaning offers." The egotistical turn of this dream ought now to be clear.[43]

But where is the wish-fulfilment to be found? It is not in the vengeance secured upon my friend Otto, whose fate it seems to be to receive ill-treatment in my dreams, but in the following circumstances: In representing Otto in the dream as Baron L., I have at the same time identified myself with some one else, that is to say, with Professor R., for I have asked something of Otto, just as R. asked something of Baron L. at the time of the occurrence which has been mentioned. And that is the point. For Professor R. has pursued his way independently outside the schools, somewhat as I have done, and has only in later years received the title which he earned long ago. I am therefore again wishing to be a professor! The very phrase "in later years" is the fulfilment of wish, for it signifies that I shall live long enough to pilot my boy through the age of puberty myself.

I gave only a brief account of the other forms of typical dreams in the first edition of this book, because an insufficient amount of good material was at my disposal. My experience, which has since been increased, now makes it possible for me to divide these dreams into two broad classes—first, those which really have the same meaning every time, and secondly, those which must be subjected to the most widely different interpretations in spite of their identical or similar content. Among the typical dreams of the first sort I shall closely consider the examination dream and the so-called dream of dental irritation.

Every one who has received his degree after having passed the final college examination, complains of the ruthlessness with which he is pursued by the anxiety dream that he will fail, that he must repeat his work, &c. For the holder of the university degree this typical dream is replaced by another, which represents to him that he has to pass the examination for the doctor's degree, and against which he vainly raises the objection in his sleep that he has already been practising for years—that he is already a university instructor or the head of a law firm. These are the ineradicable memories of the punishments which we suffered when we were children for misdeeds which we had committed—memories which were revived in us on that *dies irae, dies illa* of the severe examination at the two critical junctures in our studies. The "examination-phobia" of neurotics is also strengthened by this childish fear. After we have ceased to be schoolboys it is no longer our parents and guardians as at first, or our teachers as later on, who see to our punishment; the inexorable chain of causes and effects in life has taken over our further education. Now we dream of examinations for graduation or for the doctor's degree—and who has not been faint-hearted in these tests, even though he belonged to the righteous?—whenever we fear that an outcome will punish us because we have not done something, or because we have not accomplished something as we should—in short whenever we feel the weight of responsibility.

I owe the actual explanation of examination dreams to a remark made by a well-informed colleague, who once asserted in a scientific discussion that in his experience the examination dream occurs only to persons who have passed the examination, never to those who have gone to pieces on it. The anxiety dream of the examination, which occurs, as is being more and more corroborated, when the dreamer is looking forward to a responsible action on his part the next day and the possibility of disgrace, has therefore probably selected an occasion in the past where the great anxiety has shown itself to have been without justification and has been contradicted by the result. This would be a very striking example of a misconception of the dream content on the part of the waking instance. The objection to the dream, which is conceived as the indignant protest, "But I am already a doctor," &c., would be in reality a consolation which the dreams offer, and which would therefore be to the following effect: "Do not be afraid of the morrow; think of the fear which you had before the final examination, and yet nothing came of it. You are a doctor this minute," &c. The fear, however, which we attribute to the dream, originates in the remnants of daily experience.

The tests of this explanation which I was able to make in my own case and in that of others, although they were not sufficiently numerous, have been altogether successful. I failed, for example, in the examination for the doctor's degree in legal medicine; never once have I been concerned about this matter in my dreams, while I have often enough been examined in botany, zoology, or chemistry, in which subjects I took the examinations with well-founded anxiety, but escaped punishment through the clemency of fortune or of the examiner. In my dreams of college examination, I am regularly examined in history, a subject which I passed brilliantly at the time, but only, I must admit, because my good-natured professor—my one-eyed benefactor in another dream (*cf.* p. 21)—did not overlook the fact that on the list of questions I had crossed out the second of three questions as an indication that he should not insist on it. One of my patients, who withdrew before the final college examinations and made them up later, but who failed in the officer's examination and did not become an officer, tells me that he dreams about the former examination often enough, but never about the latter.

The above-mentioned colleague (Dr. Stekel of Vienna) calls attention to the double meaning of the word "Matura" (*Matura*— examination for college degree: mature, ripe), and claims that he has observed that examination dreams occur very frequently when a sexual test is set for the following day, in which, therefore, the disgrace which is feared might consist in the manifestation of slight potency. A German colleague takes exception to this, as it appears, justly, on the ground that this examination is denominated in Germany the Abiturium and hence lacks this double meaning.

On account of their similar affective impression dreams of missing a train deserve to be placed next to examination dreams. Their explanation also justifies this relationship. They are consolation dreams directed against another feeling of fear perceived in the dream, the fear of dying. "To depart" is one of the most frequent and one of the most easily reached symbols of death. The dream thus says consolingly: "Compose yourself, you are not going to die (to depart)," just as the examination dream calms us by saying "Fear not, nothing will happen to you even this time." The difficulty in understanding both kinds of dreams is due to the fact that the feeling of anxiety is directly connected with the expression of consolation. Stekel treats fully the symbolisms of death in his recently published book *Die Sprache des Traumes*.

The meaning of the "dreams of dental irritation," which I have had to analyse often enough with my patients, escaped me for a long time, because, much to my astonishment, resistances that wore altogether too great obstructed their interpretation.

At last overwhelming evidence convinced me that, in the case of men, nothing else than cravings for masturbation from the time of puberty furnishes the motive power for these dreams. I shall analyse two such dreams, one of which is likewise "a dream of flight." The two dreams are of the same person—a young man with a strong homosexuality, which, however, has been repressed in life.

He is witnessing a performance of Fidelio *from the parquette of the opera house; he is sitting next to L., whose personality is congenial to him, and whose friendship he would like to have. He suddenly flies diagonally clear across the parquette; he then puts his hand in his mouth and draws out two of his teeth.*

He himself describes the flight by saying it was as if he were "thrown" into the air. As it was a performance of *Fidelio* he recalls the poet's words:

"He who a charming wife acquired—"

But even the acquisition of a charming wife is not among the wishes of the dreamer. Two other verses would be more appropriate:

"He who succeeds in the lucky (big) throw,
A friend of a friend to be . . ."

The dream thus contains the "lucky (big) throw," which is not, however, a wish-fulfilment only. It also conceals the painful reflection that in his striving after friendship he has often had the misfortune to be "thrown down," and the fear lest this fate may be repeated in the case of the young man next whom he has enjoyed the performance of *Fidelio*. This is now followed by a confession which quite puts this refined dreamer to shame, to the effect that once, after such a rejection on the part of a friend, out of burning desire he merged into sexual excitement and masturbated twice in succession.

The other dream is as follows: *Two professors of the university who are known to him are treating him in my stead. One of them does something with his penis; he fears an operation. The other one thrusts an iron bar at his mouth so that he loses two teeth. He is bound with four silken cloths.*

The sexual significance of this dream can hardly be doubted. The silken cloths are equivalent to an identification with a homosexual of his acquaintance. The dreamer, who has never achieved coition, but who has never actually sought sexual intercourse with men, conceives sexual intercourse after the model of the masturbation which he was once taught during the time of puberty.

I believe that the frequent modifications of the typical dream of dental irritation—that, for example, of another person drawing the tooth from the dreamer's mouth, are made intelligible by means of the same explanation. It may, however, be difficult to see how "dental irritation" can come to have this significance. I may then call attention to a transference from below to above which occurs very frequently. This transference is at the service of sexual repression, and by means of it all kinds of sensations and intentions occurring in hysteria which ought to be enacted in the genitals can be realised upon less objectionable parts of the body. It is also a case of such transference when the genitals are replaced by the face in the symbolism of unconscious thought. This is assisted by the fact that the buttocks resemble the cheeks, and also by the usage of language which calls the nymphae "lips," as resembling those that enclose the opening of the mouth. The nose is compared to the penis in numerous allusions, and in one place as in the other the presence of hair completes the resemblance. Only one part of the anatomy—the teeth—

218

are beyond all possibility of being compared with anything, and it is just this coincidence of agreement and disagreement which makes the teeth suitable for representation under pressure of sexual repression.

I do not wish to claim that the interpretation of the dream of dental irritation as a dream of masturbation, the justification of which I cannot doubt, has been freed of all obscurity.[44] I carry the explanation as far as I am able, and must leave the rest unsolved. But I must also refer to another connection revealed by an idiomatic expression. In our country there is in use an indelicate designation for the act of masturbation, namely: To pull one out, or to pull one down.[45] I am unable to say whence these colloquialisms originate, and on what symbolisms they are based, but the teeth would well fit in with the first of the two.[46]

Dreams in which one is flying or hovering, falling, swimming, or the like, belong to the second group of typical dreams. What do these dreams signify? A general statement on this point cannot be made. They signify something different in each case, as we shall hear: only the sensational material which they contain always comes from the same source.

It is necessary to conclude, from the material obtained in psychoanalysis, that these dreams repeat impressions from childhood—that is, that they refer to the movement games which have such extraordinary attractions for the child. What uncle has never made a child fly by running across the room with it with arms outstretched, or has never played falling with it by rocking it on his knee and then suddenly stretching out his leg, or by lifting it up high and then pretending to withdraw support. At this the children shout with joy, and demand more untiringly, especially if there is a little fright and dizziness attached to it; in after years they create a repetition of this in the dream, but in the dream they omit the hands which have held them, so that they now freely float and fall. The fondness of all small children for games like rocking and see-sawing is well known; and if they see gymnastic tricks at the circus their recollection of this rocking is refreshed. With some boys the hysterical attack consists simply in the reproduction of such tricks, which they accomplish with great skill. Not infrequently sexual sensations are excited by these movement games, harmless as they are in themselves.[47] To express the idea by a word which is current among us, and which covers all of these matters: It is the wild playing ("Hetzen") of childhood which dreams about flying, falling, vertigo, and the like repeat, and the voluptuous feelings of which have now been turned into fear. But as every mother knows, the wild playing of children has often enough culminated in quarrelling and tears.

I therefore have good reason for rejecting the explanation that the condition of our dermal sensations during sleep, the sensations caused by the movements of the lungs, and the like, give rise to dreams of flying and falling. I see that these very sensations have been reproduced from the memory with which the dream is concerned—that they are, therefore, a part of the dream content and not of the dream sources.

This material, similar in its character and origin consisting of sensations of motion, is now used for the representation of the most manifold dream thoughts. Dreams of flying, for the most part characterised by delight, require the most widely different interpretations—altogether special interpretations in the case of some persons, and even interpretations of a typical nature in that of others. One of my patients was in the habit of dreaming very often that she was suspended above the street at a certain height, without touching the ground. She had grown only to a very small stature, and shunned every kind of contamination which accompanies intercourse with human beings. Her dream of suspension fulfilled both of her wishes, by raising her feet from the ground and by allowing her head to tower in the upper regions. In the case of other female dreamers the dream of flying had the significance of a longing: If I were a little bird; others thus become angels at night because they have missed being called that by day. The intimate connection between flying and the idea of a bird makes it comprehensible that the dream of flying in the case of men usually has a significance of coarse sensuality.[48] We shall also not be surprised to hear that this or that dreamer is always very proud of his ability to fly.

Dr. Paul Federn (Vienna) has propounded the fascinating theory that a great many flying dreams are erection dreams, since the remarkable phenomena of erection which so constantly occupy the human phantasy must strongly impress upon it a notion of the suspension of gravity (*cf.* the winged phalli of the ancients).

Dreams of falling are most frequently characterised by fear. Their interpretation, when they occur in women, is subject to no difficulty because women always accept the symbolic sense of falling, which is a circumlocution for the indulgence of an erotic temptation. We have not yet exhausted the infantile sources of the dream of falling; nearly all children have fallen occasionally, and then been picked up and fondled; if they fell out of bed at night, they were picked up by their nurse and taken into her bed.

People who dream often of swimming, of cleaving the waves, with great enjoyment, &c., have usually been persons who wetted their beds, and they now repeat in the dream a pleasure which they have long since learned to forgo. We shall soon learn from one example or another to what representation the dreams of swimming easily lend themselves.

The interpretation of dreams about fire justifies a prohibition of the nursery which forbids children to burn matches in order that they may not wet the bed at night. They too are based on the reminiscence of *enuresis nocturnus* of childhood. In the *Bruchstück einer Hysterieanalyse,* 1905,[49] I have given the complete analysis and synthesis of such a fire-dream in connection with the infantile history of the dreamer, and have shown to the representation of what emotions this infantile material has been utilised in maturer years.

It would be possible to cite a considerable number of other "typical" dreams, if these are understood to refer to the frequent recurrence of the same manifest dream content in the case of different dreamers, as, for example: dreams of passing through narrow alleys, of walking through a whole suite of rooms; dreams of the nocturnal burglar against whom nervous people direct precautionary measures before going to sleep; dreams of being chased by wild animals (bulls, horses), or of being threatened with knives, daggers, and lances. The last two are characteristic as the manifest dream content of persons suffering from anxiety, &c. An investigation dealing especially with this material would be well worth while. In lieu of this I have two remarks to offer, which, however, do not apply exclusively to typical dreams.

I

The more one is occupied with the solution of dreams, the more willing one must become to acknowledge that the majority of the dreams of adults treat of sexual material and give expression to erotic wishes. Only one who really analyses dreams, that is to say, who pushes forward from their manifest content to the latent dream thoughts, can form an opinion on this subject—never the person who is satisfied with registering the manifest content (as, for example, Näcke in his works on sexual dreams). Let us recognise at once that this fact is not to be wondered at, but that it is in complete harmony with the fundamental assumptions of dream explanation. No other impulse has had to undergo so much suppression from the time of childhood as the sex impulse in its numerous components,[50] from no other impulse have survived so many and such intense unconscious wishes, which now act in the sleeping state in such a manner as to produce dreams. In dream interpretation, this significance of sexual complexes must never be forgotten, nor must they, of course, be exaggerated to the point of being considered exclusive.

Of many dreams it can be ascertained by a careful interpretation that they are even to be taken bisexually, inasmuch as they result in an irrefutable secondary interpretation in which they realise homosexual feelings—that is, feelings that are common to the normal sexual activity of the dreaming person. But that all dreams are to be interpreted bisexually, as maintained by W. Stekel,[51] and Alf. Adler,[52] seems to me to be a generalisation as indemonstrable as it is improbable, which I should not like to support. Above all I should not know how to dispose of the apparent fact that there are many dreams satisfying other than—in the widest sense—erotic needs, as dreams of hunger, thirst, convenience, &c. Likewise the similar assertions "that behind every dream one finds the death sentence" (Stekel), and that every dream shows "a continuation from the feminine to the masculine line" (Adler), seem to me to proceed far beyond what is admissible in the interpretation of dreams.

We have already asserted elsewhere that dreams which are conspicuously innocent invariably embody coarse erotic wishes, and we might confirm this by means of numerous fresh examples. But many dreams which appear indifferent, and which would never be suspected of any particular significance, can be traced back, after analysis, to unmistakably sexual wish-feelings, which are often of an unexpected nature. For example, who would suspect a sexual wish in the following dream until the interpretation had been worked out? The dreamer relates: *Between two stately palaces stands a little house, receding somewhat, whose doors are closed. My wife leads me a little way along the street up to the little home, and pushes in the door, and then I slip quickly and easily into the interior of a courtyard that slants obliquely upwards.*

Anyone who has had experience in the translating of dreams will, of course, immediately perceive that penetrating into narrow spaces, and opening locked doors, belong to the commonest sexual symbolism, and will easily find in this dream a representation of attempted coition from behind (between the two stately buttocks of the female body). The narrow slanting passage is of course the vagina; the assistance attributed to the wife of the dreamer requires the interpretation that in reality it is only consideration for the wife which is responsible for the detention from such an attempt. Moreover, inquiry shows that on the previous day a young girl had entered the household of the dreamer who had pleased him, and who had given him the impression that she would not be altogether opposed to an approach of this sort. The little house between the two palaces is taken from a reminiscence of the Hradschin in Prague, and thus points again to the girl who is a native of that city.

If with my patients I emphasise the frequency of the Oedipus dream—of having sexual intercourse with one's mother—I get the answer: "I cannot remember such a dream." Immediately afterwards, however, there arises the recollection of another disguised and indifferent dream, which has been dreamed repeatedly by the patient, and the analysis shows it to be a dream of this same content—that is, another Oedipus dream. I can assure the reader that veiled dreams of sexual intercourse with the mother are a great deal more frequent than open ones to the same effect.[53]

There are dreams about landscapes and localities in which emphasis is always laid upon the assurance: "I have been there before." In this case the locality is always the genital organ of the mother; it can indeed be asserted with such certainty of no other locality that one "has been there before."

A large number of dreams, often full of fear, which are concerned with passing through narrow spaces or with staying in the water, are based upon fancies about the embryonic life, about the sojourn in the mother's womb, and about the act of birth. The following is the dream of a young man who in his fancy has already while in embryo taken advantage of his opportunity to spy upon an act of coition between his parents.

"He is in a deep shaft, in which there is a window, as in the Semmering Tunnel. At first he sees an empty landscape through this window, and then he composes a picture into it, which is immediately at hand and which fills out the empty space. The picture represents a field which is being thoroughly harrowed by an implement, and the delightful air, the accompanying idea of hard work, and the bluish-black clods of earth make a pleasant impression. He then goes on and sees a primary school opened . . . and he is surprised that so much attention is devoted in it to the sexual feelings of the child, which makes him think of me."

Here is a pretty water-dream of a female patient, which was turned to extraordinary account in the course of treatment.

At her summer resort at the . . . Lake, she hurls herself into the dark water at a place where the pale moon is reflected in the water.

Dreams of this sort are parturition dreams; their interpretation is accomplished by reversing the fact reported in the manifest dream content; thus, instead of "throwing one's self into the water," read "coming out of the water," that is, "being born." The place from which one is born is recognised if one thinks of the bad sense of the French "la lune." The pale moon thus becomes the white "bottom" (Popo), which the child soon recognises as the place from which it came. Now what can be the meaning of the patient's wishing to be born at her summer resort? I asked the dreamer this, and she answered without hesitation: "Hasn't the treatment made me as though I were born again?" Thus the dream becomes an invitation to continue the cure at this summer resort, that is, to visit her there; perhaps it also contains a very bashful allusion to the wish to become a mother herself.[54]

Another dream of parturition, with its interpretation, I take from the work of E. Jones. *"She stood at the seashore watching a small boy, who seemed to be hers, wading into the water. This he did till the water covered him, and she could only see his head bobbing up and down near the surface. The scene then changed to the crowded hall of a hotel. Her husband left her, and she 'entered into conversation with' a stranger."* The second half of the dream was discovered in the analysis to represent a flight from her husband, and the entering into intimate relations with a third person, behind whom was plainly indicated Mr. X.'s brother mentioned in a former dream. The first part of the dream was a fairly evident birth phantasy. In dreams as in mythology, the delivery of a child *from* the uterine waters is commonly presented by distortion as the entry of the child *into* water; among many others, the births of Adonis, Osiris, Moses, and Bacchus are well-known illustrations of this. The bobbing up and down of the head in the water at once recalled to the patient the sensation of quickening she had experienced in her only pregnancy. Thinking of the boy going into the water induced a reverie in which she saw herself taking him out of the water, carrying him into the nursery, washing him and dressing him, and installing him in her household.

The second half of the dream, therefore, represents thoughts concerning the elopement, which belonged to the first half of the underlying latent content; the first half of the dream corresponded with the second half of the latent content, the birth phantasy. Besides this inversion in order, further inversions took place in each half of the dream. In the first half the child *entered* the water, and then his head bobbed; in the underlying dream thoughts first the quickening occurred, and then the child left the water (a double inversion). In the second half her husband left her; in the dream thoughts she left her husband.

Another parturition dream is related by Abraham of a young woman looking forward to her first confinement (p. 22): From a place in the floor of the house a subterranean canal leads directly into the water (parturition path, amniotic liquor). She lifts up a trap in the floor, and there immediately appears a creature dressed in a brownish fur, which almost resembles a seal. This creature changes into the younger brother of the dreamer, to whom she has always stood in maternal relationship.

Dreams of "saving" are connected with parturition dreams. To save, especially to save from the water, is equivalent to giving birth when dreamed by a woman; this sense is, however, modified when the dreamer is a man.[55]

Robbers, burglars at night, and ghosts, of which we are afraid before going to bed, and which occasionally even disturb our sleep, originate in one and the same childish reminiscence. They are the nightly visitors who have awakened the child to set it on the chamber so that it may not wet the bed, or have lifted the cover in order to see clearly how the child is holding its hands while sleeping. I have been able to induce an exact recollection of the nocturnal visitor in the analysis of some of these anxiety dreams. The robbers were always the father, the ghosts more probably corresponded to feminine persons with white night-gowns.

II

When one has become familiar with the abundant use of symbolism for the representation of sexual material in dreams, one naturally raises the question whether there are not many of these symbols which appear once and for all with a firmly established significance like the signs in stenography; and one is tempted to compile a new dream-book according to the cipher method. In this connection it may be remarked that this symbolism does not belong peculiarly to the dream, but rather to unconscious thinking, particularly that of the masses, and it is to be found in greater perfection in the folklore, in the myths, legends, and manners of speech, in the proverbial sayings, and in the current witticisms of a nation than in its dreams.[56]

The dream takes advantage of this symbolism in order to give a disguised representation to its latent thoughts. Among the symbols which are used in this manner there are of course many which regularly, or almost regularly, mean the same thing. Only it is necessary to keep in mind the curious plasticity of psychic material. Now and then a symbol in the dream content may have to be interpreted not symbolically, but according to its real meaning; at another time the dreamer, owing to a peculiar set of recollections, may create for himself the right to use anything whatever as a sexual symbol, though it is not ordinarily used in that way. Nor are the most frequently used sexual symbols unambiguous every time.

After these limitations and reservations I may call attention to the following: Emperor and Empress (King and Queen) in most cases really represent the parents of the dreamer;[57] the dreamer himself or herself is the prince or princess. All elongated objects, sticks, tree-trunks, and umbrellas (on account of the stretching-up which might be compared to an erection! all elongated and sharp weapons, knives, daggers, and pikes, are intended to represent the male member. A frequent, not very intelligible, symbol for the same is a nail-file (on account of the rubbing and scraping?). Little cases, boxes, caskets, closets, and stoves correspond to the female part. The symbolism of lock and key has been very gracefully employed by Uhland in his song about the "Graf en Eberstein," to make a common smutty joke. The dream of walking through a row of rooms is a brothel or harem dream. Staircases, ladders, and flights of stairs, or climbing on these, either upwards or downwards, are symbolic representations of the sexual act.[58] Smooth walls over which one is climbing, façades of houses upon which one is letting oneself down, frequently under great anxiety, 'correspond to the erect human body, and probably repeat in the dream reminiscences of the upward climbing of little children on their parents or foster parents. "Smooth" walls are men. Often in a dream of anxiety one is holding on firmly to some projection from a house. Tables, set tables, and boards are women, perhaps on account of the opposition which does away with the bodily contours. Since "bed and board" (*mensa et thorus*) constitute marriage, the former are often put for the latter in the dream, and as far as practicable the sexual presentation complex is transposed to the eating complex. Of articles of dress the woman's hat may frequently be definitely interpreted as the male genital. In dreams of men one often finds the cravat as a symbol for the penis; this indeed is not only because cravats hang down long, and are characteristic of the man, but also because one can select them at pleasure, a freedom which is prohibited by nature in the original of the symbol. Persons who make use of this symbol in the dream are very extravagant with cravats, and possess regular collections of them.[59] All complicated machines and apparatus in dream are very probably genitals, in the description of which dream symbolism shows itself to be as tireless as the activity of wit. Likewise many landscapes in dreams, especially with bridges or with wooded mountains, can be readily recognised as descriptions of the genitals. Finally where one finds incomprehensible neologisms one may think of combinations made up of components having a sexual significance. Children also in the dream often signify the genitals, as men and women are in the habit of fondly referring to their genital organ as their "little one." As a very recent symbol of the male genital may be mentioned the flying machine, utilisation of which is justified by its relation to flying as well as occasionally by its form. To play with a little child or to beat a little one is often the dream's representation of onanism. A number of other symbols, in part not sufficiently verified, are given by Stekel, who illustrates them with examples. Right and left, according to him, are to be conceived in the dream in an ethical sense. "The right way always signifies the road to righteousness, the left the one to crime. Thus the left may signify homosexuality, incest, and perversion, while the right signifies marriage, relations with a prostitute, &c. The meaning is always determined by the individual moral view-point of the dreamer" (*l.c.,* p. 466). Relatives in the dream generally play the rôle of genitals (p. 473). Not to be able to catch up with a wagon is interpreted by Stekel as regret not to be able to come up to a difference in age (p. 479). Baggage with which one travels is the burden of sin by which one is oppressed (*ibid.*). Also numbers, which frequently occur in the dream, are assigned by Stekel a fixed symbolical meaning, but these interpretations seem neither sufficiently verified nor of general validity, although the interpretation in individual cases can generally be recognised as probable. In a recently published book by W. Stekel, *Die Sprache des Traumes,* which I was unable to utilise, there is a list (p. 72) of the most common sexual symbols, the object of which is to prove that all sexual symbols can be bisexually used. He states: "Is there a symbol which (if in any way permitted by the phantasy) may not be used simultaneously in the masculine and the feminine sense!" To be sure the clause in parentheses takes away much of the absoluteness of this assertion, for this is not at all permitted by the phantasy. I do not, however, think it superfluous to state that in my experience Stekel's general statement has to give way to the recognition of a greater manifoldness. Besides those symbols, which are just as frequent for the male as for the female genitals, there are others which preponderately, or almost exclusively, designate one of the sexes, and there are still others of which only the male or only the female signification is known. To use long, firm objects and weapons as symbols of the female genitals, or hollow objects (chests, boxes, pouches, &c.), as symbols of the male genitals, is indeed not allowed by the fancy.

It is true that the tendency of the dream and the unconscious fancy to utilise the sexual symbol bisexually betrays an archaic trend, for in childhood a difference in the genitals is unknown, and the same genitals are attributed to both sexes.

These very incomplete suggestions may suffice to stimulate others to make a more careful collection.[60]

I shall now add a few examples of the application of such symbolisms in dreams, which will serve to show how impossible it becomes to interpret a dream without taking into account the symbolism of dreams, and how imperatively it obtrudes itself in many cases.

1. The hat as a symbol of the man (of the male genital):[61] (a fragment from the dream of a young woman who suffered from agoraphobia on account of a fear of temptation).

"I am walking in the street in summer, I wear a straw hat of peculiar shape, the middle piece of which is bent upwards and the side pieces of which hang downwards (the description became here obstructed), and in such a fashion that one is lower than the other. I am cheerful and in a confidential mood, and as I pass a troop of young officers I think to myself: None of you can have any designs upon me."

As she could produce no associations to the hat, I said to her: "The hat is really a male genital, with its raised middle piece and the two downward hanging side pieces." I intentionally refrained from interpreting those details concerning the unequal downward hanging of the two side pieces, although just such individualities in the determinations lead the way to the interpretation. I continued by saying that if she only had a man with such

228

a virile genital she would not have to fear the officers—that is, she would have nothing to wish from them, for she is mainly kept from going without protection and company by her fancies of temptation. This last explanation of her fear I had already been able to give her repeatedly on the basis of other material.

It is quite remarkable how the dreamer behaved after this interpretation. She withdrew her description of the hat, and claimed not to have said that the two side pieces were hanging downwards. I was, however, too sure of what I had heard to allow myself to be misled, and I persisted in it. She was quiet for a while, and then found the courage to ask why it was that one of her husband's testicles was lower than the other, and whether it was the same in all men. With this the peculiar detail of the hat was explained, and the whole interpretation was accepted by her. The hat symbol was familiar to me long before the patient related this dream. From other but less transparent cases I believe that the hat may also be taken as a female genital.

2. The little one as the genital—to be run over as a symbol of sexual intercourse (another dream of the same agoraphobic patient).

"Her mother sends away her little daughter so that she must go alone. She rides with her mother to the railroad and sees her little one walking directly upon the tracks, so that she cannot avoid being run over. She hears the bones crackle. (From this she experiences a feeling of discomfort but no real horror.) She then looks out through the car window to see whether the parts cannot be seen behind. She then reproaches her mother for allowing the little one to go out alone." Analysis. It is not an easy matter to give here a complete interpretation of the dream. It forms part of a cycle of dreams, and can be fully understood only in connection with the others. For it is not easy to get the necessary material sufficiently isolated to prove the symbolism. The patient at first finds that the railroad journey is to be interpreted historically as an allusion to a departure from a sanitorium for nervous diseases, with the superintendent of which she naturally was in love. Her mother took her away from this place, and the physician came to the railroad station and handed her a bouquet of flowers on leaving; she felt uncomfortable because her mother witnessed this homage. Here the mother, therefore, appears as a disturber of her love affairs, which is the rôle actually played by this strict woman during her daughter's girlhood. The next thought referred to the sentence: "She then looks to see whether the parts can be seen behind." In the dream façade one would naturally be compelled to think of the parts of the little daughter run over and ground up. The thought, however, turns in quite a different direction. She recalls that she once saw her father in the bath-room naked from behind; she then begins to talk about the sex differentiation, and asserts that in the man the genitals can be seen from behind, but in the woman they cannot. In this connection she now herself offers the interpretation that the little one is the genital, her little one (she has a four-year-old daughter) her own genital. She reproaches her mother for wanting her to live as though she had no genital, and recognises this reproach in the introductory sentence of the dream; the mother sends away her little one so that she must go alone. In her phantasy going alone on the street signifies to have no man and no sexual relations (coire = to go together), and this she does not like. According to all her statements she really suffered as a girl on account of the jealousy of her mother, because she showed a preference for her father.

The "little one" has been noted[62] as a symbol for the male or the female genitals by Stekel, who can refer in this connection to a very widespread usage of language.

The deeper interpretation of this dream depends upon another dream of the same night in which the dreamer identifies herself with her brother. She was a "tomboy," and was always being told that she should have been born a boy. This identification with the brother shows with special clearness that "the little one" signifies the genital. The mother threatened him (her) with castration, which could only be understood as a punishment for playing with the parts, and the identification, therefore, shows that she herself had masturbated as a child, though this fact she now retained only in a memory concerning her brother. An early knowledge of the male genital which she later lost she must have acquired at that time according to the assertions of this second dream. Moreover the second dream points to the infantile sexual theory that girls originate from boys through castration. After I had told her of this childish belief, she at once confirmed it with an anecdote in which the boy asks the girl: "Was it cut off?" to which the girl replied, "No, it's always been so."

The sending away of the little one, of the genital, in the first dream therefore also refers to the threatened castration. Finally she blames her mother for not having been born a boy.

230

That "being run over" symbolises sexual intercourse would not be evident from this dream if we were not sure of it from many other sources.

3. Representation of the genital by structures, stairways, and shafts. (Dream of a young man inhibited by a father complex.)

"He is taking a walk with his father in a place which is surely the Prater, for the *Rotunda* may be seen in front of which there is a small front structure to which is attached a captive balloon; the balloon, however, seems quite collapsed. His father asks him what this is all for; he is surprised at it, but he explains it to his father. They come into a court in which lies a large sheet of tin. His father wants to pull off a big piece of this, but first looks around to see if anyone is watching. He tells his father that all he needs to do is to speak to the watchman, and then he can take without any further difficulty as much as he wants to. From this court a stairway leads down into a shaft, the walls of which are softly upholstered something like a leather pocketbook. At the end of this shaft there is a longer platform, and then a new shaft begins. . . . "

Analysis

This dream belongs to a type of patient which is not favourable from a therapeutic point of view. They follow in the analysis without offering any resistances whatever up to a certain point, but from that point on they remain almost inaccessible. This dream he almost analysed himself. "The Rotunda," he said, "is my genital, the captive balloon in front is my penis, about the weakness of which I have worried. We must, however, interpret in greater detail; the Rotunda is the buttock which is regularly associated by the child with the genital, the smaller front structure is the scrotum. In the dream his father asks him what this is all for—that is, he asks him about the purpose and arrangement of the genitals. It is quite evident that this state of affairs should be turned around, and that he should be the questioner. As such a questioning on the side of the father has never taken place in reality, we must conceive the dream thought as a wish, or take it conditionally, as follows: "If I had only asked my father for sexual enlightenment." The continuation of this thought we shall soon find in another place.

The court in which the tin sheet is spread out is not to be conceived symbolically in the first instance, but originates from his father's place of business. For discretionary reasons I have inserted the tin for another material in which the father deals, without, however, changing anything in the verbal expression of the dream. The dreamer had entered his father's business, and had taken a terrible dislike to the questionable practices upon which profit mainly depends. Hence the continuation of the above dream thought ("if I had only asked him") would be: "He would have deceived me just as he does his customers." For the pulling off, which serves to represent commercial dishonesty, the dreamer himself gives a second explanation—namely, onanism. This is not only entirely familiar to us, but agrees very well with the fact that the secrecy of onanism is expressed by its opposite ("Why one can do it quite openly"). It, moreover, agrees entirely with our expectations that the onanistic activity is again put off on the father, just as was the questioning in the first scene of the dream. The shaft he at once interprets as the vagina by referring to the soft upholstering of the walls. That the act of coition in the vagina is described as a going down instead of in the usual way as a going up, I have also found true in other instances.[63]

The details that at the end of the first shaft there is a longer platform and then a new shaft, he himself explains biographically. He had for some time consorted with women sexually, but had then given it up because of inhibitions and now hopes to be able to take it up again with the aid of the treatment. The dream, however, becomes indistinct toward the end, and to the experienced interpreter it becomes evident that in the second scene of the dream the influence of another subject has begun to assert itself; in this his father's business and his dishonest practices signify the first vagina represented as a shaft so that one might think of a reference to the mother.

4. The male genital symbolised by persons and the female by a landscape.

(Dream of a woman of the lower class, whose husband is a policeman, reported by B. Dattner.)

. . . Then someone broke into the house and anxiously called for a policeman. But he went with two tramps by mutual consent into a church,[64] to which led a great many stairs;[65] behind the church there was a mountain,[66] on top of which a dense forest.[67] The policeman was furnished with a helmet, a gorget, and a cloak.[68] The two vagrants, who went along with the policeman quite peaceably, had tied to their loins sack-like aprons.[69] A road led from the church to the mountain. This road was overgrown on

232

each side with grass and brushwood, which became thicker and thicker as it reached the height of the mountain, where it spread out into quite a forest.

5. A stairway dream.

(Reported and interpreted by Otto Rank.)

For the following transparent pollution dream, I am indebted to the same colleague who furnished us with the dental-irritation dream reported on p. 280.

"I am running down the stairway in the stair-house after a little girl, whom I wish to punish because she has done something to me. At the bottom of the stairs some one held the child for me. (A grown-up woman?) I grasp it, but do not know whether I have hit it, for I suddenly find myself in the middle of the stairway where I practise coitus with the child (in the air as it were). It is really no coitus, I only rub my genital on her external genital, and in doing this I see it very distinctly, as distinctly as I see her head which is lying sideways. During the sexual act I see hanging to the left and above me (also as if in the air) two small pictures, landscapes, representing a house on a green. On the smaller one my surname stood in the place where the painter's signature should be; it seemed to be intended for my birthday present. A small sign hung in front of the pictures to the effect that cheaper pictures could also be obtained. I then see myself very indistinctly lying in bed, just as I had seen myself at the foot of the stairs, and I am awakened by a feeling of dampness which came from the pollution."

Interpretation. The dreamer had been in a book-store on the evening of the day of the dream, where, while he was waiting, he examined some pictures which were exhibited, which represented motives similar to the dream pictures. He stepped nearer to a small picture which particularly took his fancy in order to see the name of the artist, which, however, was quite unknown to him.

Later in the same evening, in company, he heard about a Bohemian servant-girl who boasted that her illegitimate child "was made on the stairs." The dreamer inquired about the details of this unusual occurrence, and learned that the servant-girl went with her lover to the home of her parents, where there was no opportunity for sexual relations, and that the excited man performed the act on the stairs. In witty allusion to the mischievous expression used about wine-adulterers, the dreamer remarked, "The child really grew on the cellar steps."

These experiences of the day, which are quite prominent in the dream content, were readily reproduced by the dreamer. But he just as readily reproduced an old fragment of infantile recollection which was also utilised by the dream. The stair-house was the house in which he had spent the greatest part of his childhood, and in which he had first become acquainted with sexual problems. In this house he used, among other things, to slide down the banister astride which caused him to become sexually excited. In the dream he also comes down the stairs very rapidly—so rapidly that, according to his own distinct assertions, he hardly touched the individual stairs, but rather "flew" or "slid down," as we used to say. Upon reference to this infantile experience, the beginning of the dream seems to represent the factor of sexual excitement. In the same house and in the adjacent residence the dreamer used to play pugnacious games with the neighbouring children, in which he satisfied himself just as he did in the dream.

If one recalls from Freud's investigation of sexual symbolism[70] that in the dream stairs or climbing stairs almost regularly symbolises coitus, the dream becomes clear. Its motive power as well as its effect, as is shown by the pollution, is of a purely libidinous nature. Sexual excitement became aroused during the sleeping state (in the dream this is represented by the rapid running or sliding down the stairs) and the sadistic thread in this is, on the basis of the pugnacious playing, indicated in the pursuing and overcoming of the child. The libidinous excitement becomes enhanced and urges to sexual action (represented in the dream by the grasping of the child and the conveyance of it to the middle of the stairway). Up to this point the dream would be one of pure sexual symbolism, and obscure for the unpractised dream interpreter. But this symbolic gratification, which would have insured undisturbed sleep, was not sufficient for the powerful libidinous excitement. The excitement leads to an orgasm, and thus the whole stairway symbolism is unmasked as a substitute for coitus. Freud lays stress on the rhythmical character of both actions as one of the reasons for the sexual utilisation of the stairway symbolism, and this dream especially seems to corroborate this, for, according to the express assertion of the dreamer, the rhythm of a sexual act was the most pronounced feature in the whole dream.

Still another remark concerning the two pictures, which, aside from their real significance, also have the value of "Weibsbilder" (literally *woman-*

234

pictures, but idiomatically *women*). This is at once shown by the fact that the dream deals with a big and a little picture, just as the dream content presents a big (grown up) and a little girl. That cheap pictures could also be obtained points to the prostitution complex, just as the dreamer's surname on the little picture and the thought that it was intended for his birthday, point to the parent complex (to be born on the stairway—to be conceived in coitus).

The indistinct final scene, in which the dreamer sees himself on the staircase landing lying in bed and feeling wet, seems to go back into childhood even beyond the infantile onanism, and manifestly has its prototype in similarly pleasurable scenes of bed-wetting.

6. A modified stair-dream.

To one of my very nervous patients, who was an abstainer, whose fancy was fixed on his mother, and who repeatedly dreamed of climbing stairs accompanied by his mother, I once remarked that moderate masturbation would be less harmful to him than enforced abstinence. This influence provoked the following dream:

"His piano teacher reproaches him for neglecting his piano-playing, and for not practising the *Études* of Moscheles and Clementi's *Gradus ad Parnassum.*" In relation to this he remarked that the *Gradus* is only a stairway, and that the piano itself is only a stairway as it has a scale.

It is correct to say that there is no series of associations which cannot be adapted to the representation of sexual facts. I conclude with the dream of a chemist, a young man, who has been trying to give up his habit of masturbation by replacing it with intercourse with women.

Preliminary statement.—On the day before the dream he had given a student instruction concerning Grignard's reaction, in which magnesium is to be dissolved in absolutely pure ether under the catalytic influence of iodine. Two days before, there had been an explosion in the course of the same reaction, in which the investigator had burned his hand.

Dream: I. *He is to make phenylmagnesiumbromid; he sees the apparatus with particular clearness, but he has substituted himself for the magnesium. He is now in a curious swaying attitude. He keeps repeating to himself, "This is the right thing, it is working, my feet are beginning to dissolve and my knees are getting soft." Then he reaches down and feels for his feet, and meanwhile (he does not know how) he takes his legs out of the crucible, and then again he says to himself, "That cannot be. . . . Yes, it must be so, it has been done correctly." Then he partially awakens, and repeats the dream to himself, because he wants to tell it to me. He is distinctly afraid of the analysis of the dream. He is much excited during this semi-sleeping state, and repeats continually, "Phenyl, phenyl."*

II. *He is in . . . ing with his whole family; at half-past eleven. He is to be at the Schottenthor for a rendezvous with a certain lady, but he does not wake up until half-past eleven. He says to himself, "It is too late now when you get there it will be half-past twelve." The next instant he sees the whole family gathered about the table—his mother and the servant girl with the soup-tureen with particular clearness. Then he says to himself, "Well, if we are eating already, I certainly can't get away."*

Analysis

He feels sure that even the first dream contains a reference to the lady whom he is to meet at the rendezvous (the dream was dreamed during the night before the expected meeting). The student to whom he gave the instruction is a particularly unpleasant fellow; he had said to the chemist: "That isn't right," because the magnesium was still unaffected, and the latter answered as though he did not care anything about it: "It certainly isn't right." He himself must be this student; he is as indifferent towards his analysis as the student is towards his synthesis; the *He* in the dream, however, who accomplishes the operation, is myself. How unpleasant he must seem to me with his indifference towards the success achieved!

Moreover, he is the material with which the analysis (synthesis) is made. For it is a question of the success of the treatment. The legs in the dream recall an impression of the previous evening. He met a lady at a dancing lesson whom he wished to conquer; he pressed her to him so closely that she once cried out. After he had stopped pressing against her legs, he felt her firm responding pressure against his lower thighs as far as just above his knees, at the place mentioned in the dream. In this situation, then, the woman is the magnesium in the retort, which is at last working. He is feminine towards me, as he is masculine towards the woman. If it will work with the woman, the treatment will also work. Feeling and becoming aware of himself in the region of his knees refers to masturbation, and corresponds to his fatigue of the previous day. . . . The rendezvous had actually been set for half-past eleven. His wish to over-sleep and to remain with his usual sexual objects (that is, with masturbation) corresponds with his resistance.

In relation to the repetition of the name phenyl, he gives the following thoughts: All these radicals ending in *yl* have always been

236

pleasing to him; they are very convenient to use: benzyl, azetyl, &c. That, however, explained nothing. But when I proposed the radical Schlemihl[71] he laughed heartily, and related that during the summer he had read a book by Prévost which contained a chapter: "Les exclus de l'amour," the description in which made him think of the Schlemihls, and he added, "That is my case." He would have again acted the Schlemihl if he had missed the rendezvous.

Notes:

1. It is clear that the conception of Robert, that the dream is intended to rid our memory of the useless impressions which it has received during the day, is no longer tenable, if indifferent memories of childhood appear in the dream with some degree of frequency. The conclusion would have to be drawn that the dream ordinarily performs very inadequately the duty which is prescribed for it.

2. As mentioned in the first chapter, p. 89, H. Swoboda applies broadly to the psychic activity, the biological intervals of twenty-three and twenty-eight days discovered by W. Fliess, and lays especial emphasis upon the fact that these periods are determinant for the appearance of the dream elements in dreams. There would be no material change in dream interpretation if this could be proven, but it would result in a new source for the origin of the dream material. I have recently undertaken some examination of my own dreams in order to test the applicability of the "Period Theory" to the dream material, and I have selected for this purpose especially striking elements of the dream content, whose origin could be definitely ascertained:—

   I.—*Dream from October* 1–2, 1910

   (Fragment) . . . Somewhere in Italy. Three daughters show me small costly objects, as if in an antiquity shop. At the same time they sit down on my lap. Of one of the pieces I remark: "Why, you got this from me." I also see distinctly a small profile mask with the angular features of Savonarola.

   When have I last seen a picture of Savonarola? According to my travelling diary, I was in Florence on the fourth and fifth of September, and while there thought of showing my travelling companion the plaster medallion of the features of the fanatical monk in the Piazza Signoria, the same place where he met his death by burning. I believe that I called his attention to it at 3 A.M. To be sure, from this impression, until its return in the dream, there was an interval of twenty-seven and one days—a "feminine period," according to Fliess. But, unfortunately for the demonstrative force of this example, I must add that on the very day of the dream I was visited (the first time after my return) by the able but melancholy-looking colleague whom I had already years before nicknamed "Rabbi Savonarola." He brought me a patient who had met with an accident on the Pottebba railroad, on which I had myself travelled eight days before, and my thoughts were thus turned to my last Italian journey. The appearance in the dream content of the striking element of Savonarola is explained by the visit of my colleague on the day of the dream; the twenty-eight day interval had no significance in its origin.

II.—*Dream from October* 10–11

I am again studying chemistry in the University laboratory. Court Councillor L. invites me to come to another place, and walks before me in the corridor carrying in front of him in his uplifted hand a lamp or some other instrument, and assuming a peculiar attitude, his head stretched forward. We then come to an open space . . . (rest forgotten).

In this dream content, the most striking part is the manner in which Court Councillor L. carries the lamp (or lupe) in front of him, his gaze directed into the distance. I have not seen L. for many years, but I now know that he is only a substitute for another greater person—for Archimedes near the Arethusa fountain in Syracuse, who stands there exactly like L. in the dream, holding the burning mirror and gazing at the besieging army of the Romans. When had I first (and last) seen this monument? According to my notes, it was on the seventeenth day of September, in the evening, and from this date to the dream there really passed 13 and 10, equals 23, days—according to Fliess, a "masculine period."

But I regret to say that here, too, this connection seems somewhat less inevitable when we enter into the interpretation of this dream. The dream was occasioned by the information, received on the day of the dream, that the lecture-room in the clinic in which I was invited to deliver my lectures had been changed to some other place. I took it for granted that the new room was very inconveniently situated, and said to myself, it is as bad as not having any lecture-room at my disposal. My thoughts must have then taken me back to the time when I first became a docent, when I really had no lecture-room, and when, in my efforts to get one, I met with little encouragement from the very influential gentlemen councillors and professors. In my distress at that time, I appealed to L., who then had the title of dean, and whom I considered kindly disposed. He promised to help me, but that was all I ever heard from him. In the dream he is the Archimedes, who gives me the [Greek] and leads me into the other room. That neither the desire for revenge nor the consciousness of one's own importance is absent in this dream will be readily divined by those familiar with dream interpretation. I must conclude, however, that without this motive for the dream, Archimedes would hardly have got into the dream that night. I am not certain whether the strong and still recent impression of the statue in Syracuse did not also come to the surface at a different interval of time.

III.—*Dream from October* 2–3, 1910

(Fragment) . . . Something about Professor Oser, who himself prepared the menu for me, which served to restore me to great peace of mind (rest forgotten).

The dream was a reaction to the digestive disturbances of this day, which made me consider asking one of my colleagues to arrange a diet for me. That in the dream I selected for this purpose Professor Oser, who had died in the summer, is based on the recent death (October 1) of another university teacher, whom I highly revered. But when did Oser die, and when did I hear of his death? According to the newspaper notice, he died on the 22nd of August, but as I was at the time in Holland, whither my Vienna newspapers were regularly pent me, I must have read the obituary notice on the 24th or 25th of August. This interval no longer corresponds to any period. It takes in 7 and 30 and 2, equals 39, days, or perhaps 38 days. I cannot recall having spoken or thought of Oser during this interval.

Such intervals as were not available for the "period theory" without further elaboration, were shown from my dreams to be far more frequent than the regular ones. As maintained in the text, the only thing constantly found is the relation to an impression of the day of the dream itself.

3. *Cf.* my essay, "Ueber Deckerinnerungen," in the *Monatschrift für Psychiatrie und Neurologie,* 1899.

4. Ger., *blühend.*

5. The tendency of the dream function to fuse everything of interest which is present into simultaneous treatment has already been noticed by several authors, for instance, by Delage," p. 41, Delboeuf, *Rapprochement Forcé,* p. 236.

6. The dream of Irma's injection; the dream of the friend who is my uncle.

7. The dream of the funeral oration of the young physician.

8. The dream of the botanical monograph.

9. The dreams of my patients during analysis are mostly of this kind.

10. *Cf.* Chap. 7. upon "Transference."

11. Substitution of the opposite, as will become clear to us after interpretation.

12. The Prater is the principal drive of Vienna. (Transl.)

13. I have long since learned that it only requires a little courage to fulfil even such unattainable wishes.

14. In the first edition there was printed here the name Hasdrubal, a confusing error, the explanation of which I have given in my *Psychopathologie des Alltagalebens.*

15. A street in Vienna.

16. *Fensterln* is the practice, now falling into disuse, found in rural districts of the German Schwarzwald, of lovers wooing at the windows of their sweethearts, bringing ladders with them, and becoming so intimate that they practically enjoy a system of trial marriages. The reputation of the young woman never suffers on account of *fensterln,* unless she becomes intimate with too many suitors. (Translator.)

17. Both the emotions which belong to these childish scenes— astonishment and resignation to the inevitable—had appeared in a dream shortly before, which was the first thing that brought back the memory of this childhood experience.

18. I do not elaborate plagiostomi purposely; they recall an occasion of angry disgrace before the same teacher.

19. *Cf.* Maury's dream about kilo-lotto, p. 59.

20. Popo = backside in German nursery language.

21. This repetition has insinuated itself into the text of the dream apparently through my absent-mindedness, and I allow it to remain because the analysis shows that it has its significance.

22. Not in *Germinal*, but in *La Terre*—a mistake of which I became aware only in the analysis. I may call attention also to the identity of the letters in *Huflattich* and *Flatus*.

23. Translator's note.

24. In his significant work ("Phantasie und Mythos," *Jahrbuch für Psychoanalyse*, Bd. ii., 1910), H. Silberer has endeavoured to show from this part of the dream that the dreamwork is able to reproduce not only the latent dream thoughts, but also the psychic processes in the dream formation "Das functionale Phänomen").

25. Another interpretation: He is one-eyed like Odin, the father of the gods . . . Odin's consolation. The consolation in the childish scene, that I will buy him a new bed.

26. I here add some material for interpretation. Holding the urinal recalls the story of a peasant who tries one glass after another at the opticians, but still cannot read (peasant-catcher, like girl-catcher in a portion of the dream). The treatment among the peasants of the father who has become weak-minded in Zola's *La Terre*. The pathetic atonement that in his last days the father soils his bed like a child; hence, also, I am his sick-attendant in the dream. Thinking and experiencing are here, as it were; the same thing recalls a highly revolutionary closet drama by Oscar Panizza, in which the Godhead is treated quite contemptuously, as though he were a paralytic old man. There occurs a passage: "Will and deed are the same thing with him, and he must be prevented by his archangel, a kind of Ganymede, from scolding and swearing, because these curses would immediately be fulfilled." Making plans is a reproach against my father, dating from a later period in the development of my critical faculty; just as the whole rebellious, sovereign-offending dream, with its scoff at high authority, originates in a revolt against my father. The sovereign is called father of the land (*Landesvater*), and the father is the oldest, first and only authority for the child, from the absolutism of which the other social authorities have developed in the course of the history of human civilisation (in so far as the "mother's right" does not force a qualification of this thesis). The idea in the dream, "thinking and experiencing are the same thing," refers to the explanation of hysterical symptoms, to which the male urinal (glass) also has a relation. I need not explain the principle of the "Gschnas" to a Viennese; it consists in constructing objects of

rare and costly appearance out of trifles, and preferably out of comical and worthless material—for example, making suits of armour out of cooking utensils, sticks and "salzstangeln" (elongated rolls), as our artists like to do at their jolly parties. I had now learned that hysterical subjects do the same thing; besides what has actually occurred to them, they unconsciously conceive horrible or extravagant fantastic images, which they construct from the most harmless and commonplace things they have experienced. The symptoms depend solely upon these phantasies, not upon the memory of their real experiences, be they serious or harmless. This explanation helped me to overcome many difficulties and gave me much pleasure. I was able to allude to it in the dream element "male urinal" (glass) because I had been told that at the last "Gschnas" evening a poison chalice of Lucretia Borgia had been exhibited, the chief constituent of which had consisted of a glass urinal for men, such as is used in hospitals.

27. Of. the passage in Griesinger and the remarks in my second essay on the "defence-neuropsychoses"—*Selected Papers on Hysteria,* translated by A. A. Brill.

28. In the two sources from which I am acquainted with this dream, the report of its contents do not agree.

29. An exception is furnished by those cases in which the dreamer utilises in the expression of his latent dream thoughts the symbols which are familiar to us.

30. "The Emperor's New Clothes."

31. The child also appears in the fairy tale, for there a child suddenly calls: "Why, he hasn't anything on at all."

32. Ferenczi has reported a number of interesting dreams of nakedness in women which could be traced to an infantile desire to exhibit, but which differ in some features from the "typical" dream of nakedness discussed above.

33. For obvious reasons the presence of the "whole family" in the dream has the same significance.

34. *Cf.* "Analyse der Phobie eines fünfjährigen Knaben" in the *Jahrbuch für psychoanalytische und psychopathologische Forschungen,* vol. i., 1909, and "Ueber infantile Sexualtheorien," in *Sexualprobleme,* vol. i., 1908.

35.

36. The three-and-a-half-year-old Hans, whose phobia is the subject of

analysis in the above-mentioned publication, cries during fever shortly after the birth of his sister: "I don't want a little sister." In his neurosis, one and a half years later, he frankly confesses the wish that the mother should drop the little one into the bath-tub while bathing it, in order that it may die. With all this, Hans is a good-natured, affectionate child, who soon becomes fund of his sister, and likes especially to take her under his protection.

37. The three-and-a-half-year old Hans embodies his crushing criticism of his little sister in the identical word (see previous notes). He assumes that she is unable to speak on account of her lack of teeth.

38. I heard the following idea expressed by a gifted boy of ten, after the sudden death of his father: "I understand that father is dead, but I cannot see why he does not come home for supper."

39. At least a certain number of mythological representations. According to others, emasculation is only practised by Kronos on his father.

With regard to mythological significance of this motive, *cf.* Otto Rank's "Der Mythus von der Geburt des Helden," fifth number of *Schriften zur angew. Seelenkunde,* 1909.

40. Act i. sc. 2. Translated by George Somers Clark.

41. Another of the great creations of tragic poetry, Shakespeare's *Hamlet,* is founded on the same basis as the *Oedipus.* But the whole difference in the psychic life of the two widely separated periods of civilisation—the age-long progress of repression in the emotional life of humanity—is made manifest in the changed treatment of the identical material. In *Oedipus* the basic wish-phantasy of the child is brought to light and realised as it is in the dream; in *Hamlet* it remains repressed, and we learn of its existence—somewhat as in the case of a neurosis—only by the inhibition which results from it. The fact that it is possible to remain in complete darkness concerning the character of the hero, has curiously shown itself to be consistent with the overpowering effect of the modern drama. The play is based upon Hamlet's hesitation to accomplish the avenging task which has been assigned to him; the text does not avow the reasons or motives of this hesitation, nor have the numerous attempts at interpretation succeeded in giving them. According to the conception which is still current to-day, and which goes back to Goethe, Hamlet represents the type of man whose prime energy is paralysed by over-development of thought activity. ("Sicklied

o'er with the pale cast of thought.") According to others the poet has attempted to portray a morbid, vacillating character who is subject to neurasthenia. The plot of the story, however, teaches us that Hamlet is by no means intended to appear as a person altogether incapable of action. Twice we see him asserting himself actively, once in headlong passion, where he stabs the eavesdropper behind the arras, and on another occasion where he sends the two courtiers to the death which has been intended for himself—doing this deliberately, even craftily, and with all the lack of compunction of a prince of the Renaissance. What is it, then, that restrains him in the accomplishment of the task which his father's ghost has set before him? Here the explanation offers itself that it is the peculiar nature of this task. Hamlet can do everything but take vengeance upon the man who has put his father out of the way, and has taken his father's place with his mother—upon the man who shows him the realisation of his repressed childhood wishes. The loathing which ought to drive him to revenge is thus replaced in him by self-reproaches, by conscientious scruples, which represent to him that he himself is no better than the murderer whom he is to punish. I have thus translated into consciousness what had to remain unconscious in the mind of the hero; if some one wishes to call Hamlet a hysteric subject I cannot but recognise it as an inference from my interpretation. The sexual disinclination which Hamlet expresses in conversation with Ophelia, coincides very well with this view—it is the same sexual disinclination which was to take possession of the poet more and more during the next few years of his life, until the climax of it is expressed in *Timon of Athens*. Of course it can only be the poet's own psychology with which we are confronted in *Hamlet;* from a work on Shakespeare by George Brandes (1896), I take the fact that the drama was composed immediately after the death of Shakespeare's father—that is to say, in the midst of recent mourning for him—during the revival, we may assume, of his childhood emotion towards his father. It is also known that a son of Shakespeare's, who died early, bore the name of Hamnet (identical with Hamlet). Just as *Hamlet* treats of the relation of the son to his parents, *Macbeth,* which appears subsequently, is based upon the theme of childlessness. Just as every neurotic symptom, just as the dream itself, is capable of re-interpretation, and even requires it in order to be perfectly intelligible, so every genuine poetical creation must have proceeded from more than one motive, more than one impulse in the mind of the poet, and must admit of more than one interpretation. I have here attempted to interpret only the most profound group of impulses in the mind of the creative poet. The conception of the *Hamlet* problem contained in these remarks has been later confirmed in a detailed work based on many new arguments by Dr. Ernest Jones, of Toronto (Canada). The connection of the *Hamlet* material with the "Mythus von der Geburt des Helden" has also been demonstrated by O. Rank.—"The *Oedipus* Complex as an Explanation of *Hamlet*'s Mystery: a Study in Motive" (*American Journal of Psychology,* January 1910, vol. xxi.).

42. Likewise, anything large, over-abundant, enormous, and exaggerated, may be a childish characteristic. The child knows no more intense wish than to become big, and to receive as much of everything as grown-ups; the child is hard to satisfy; it knows no *enough,* and insatiably demands the repetition of whatever has pleased it or tasted good to it. It learns to practise moderation, to be modest and resigned, only through culture and education. As is well known. the neurotic is also inclined toward immoderation and excess.

43. While Dr. Jones was delivering a lecture before an American scientific society, and speaking of egotism in dreams, a learned lady took exception to this unscientific generalisation. She thought that the lecturer could only pronounce such judgment on the dreams of Austrians, and had no right to include the dreams of Americans. As for herself she was sure that all her dreams were strictly altruistic.

44. According to C. G. Jung, dreams of dental irritation in the case of women have the significance of parturition dreams.

45. *Cf.* the "biographic" dream on p. 264.

46. As the dreams of pulling teeth, and teeth falling out, are interpreted in popular belief to mean the death of a close friend, and as psychoanalysis can at most only admit of such a meaning in the above indicated parodical sense, I insert here a dream of dental irritation placed at my disposal by Otto Rank 109.

"Upon the subject of dreams of dental irritation I have received the following report from a colleague who has for some time taken a lively interest in the problems of dream interpretation:

I recently dreamed that I went to the dentist who drilled out one of my back teeth in the lower jaw. He worked so long at it that the tooth became useless. He then grasped it with the forceps, and pulled it out with such perfect ease that it astonished me. He said that I should not care about it, as this wat not really the tooth that had been treated; and he put it on the table where the tooth (as it seems to me now an upper incisor) fell apart into many strata. I arose from the operating chair, stepped inquisitively nearer, and, full of interest, put a medical question. While the doctor separated the individual pieces of the strikingly white tooth and ground them up (pulverised them) with an instrument, he explained to me that this had some connection with puberty, and that the teeth come out so easily only before puberty; the decisive moment for this in women is the birth of a child. I then noticed (as I believe half awake) that this dream was accompanied by a pollution which I cannot however definitely place at a particular point in the dream; I am inclined to think that it began with the pulling out of the tooth.

I then continued to dream something which I can no longer remember, which ended with the fact that I had left my hat and coat somewhere (perhaps at the dentist's), hoping that they would be brought after me, and dressed only in my overcoat I hastened to catch a departing train, I succeeded at the last moment in jumping upon the last car, where someone was already standing. I could not, however, get inside the car, but was compelled to make the journey in an uncomfortable position, from which I attempted to escape with final success. We journeyed through a long tunnel, in which two trains from the opposite direction passed through our own train as if it were a tunnel. I looked in as from the outside through a car window.

As material for the interpretation of this dream, we obtained the following experiences and thoughts of the dreamer:—

I. For a short time I had actually been under dental treatment, and at the time of the dream I was suffering from continual pains in the tooth of my lower jaw, which was drilled out in the dream, and on which the dentist had in fact worked longer than I liked. On the forenoon of the day of the dream I had again gone to the doctor's on account of the pain, and he had suggested that I should allow him to pull out another tooth than the one treated in the same jaw, from which the pain probably came. It was a 'wisdom tooth' which was just breaking through. On this occasion, and in this connection, I had put a question to his conscience as a physician.

II. On the afternoon of the same day I was obliged to excuse myself to a lady for my irritable disposition on account of the toothache, upon which she told me that she was afraid to have one of her roots pulled, though the crown was almost completely gone. She thought that the pulling out of eye teeth was especially painful and dangerous, although some acquaintance had told her that this was much easier when it was a tooth of the lower jaw. It was such a tooth in her case. The same acquaintance also told her that while under an anaesthetic one of her false teeth had been pulled—a statement which increased her fear of the necessary operation. She then asked me whether by eye teeth one was to understand molars or canines, and what was known about them. I then called her attention to the vein of superstitions in all these meanings, without however, emphasising the real significance of some of the popular views. She knew from her own experience, a very old and general popular belief, according to which *if a pregnant woman has toothache she will give birth to a boy.*

III. This saying interested me in its relation to the typical significance of dreams of dental irritation as a substitute for onanism as maintained by Freud in his *Traumdeutung* (2nd edition, p. 193), for the teeth and the male genital (Bub-boy) are brought in certain relations even in the popular saying. On the evening of the same day I therefore read the passage in question in the *Traumdeutung,* and found there among other things the statements which will be quoted in a moment, the influence of which on my dream is as plainly recognisable as the influence of the two above-mentioned experiences. Freud writes concerning dreams of dental irritation that 'in the case of men nothing else than cravings for masturbation from the time of puberty furnishes the motive power for these dreams,' p. 193. Further, 'I am of the opinion that the frequent modifications of the typical dream of dental irritation—that *e.g.* of another person drawing the tooth from the dreamer's mouth—are made intelligible by means of the same explanation. It may seem problematic, however, how "dental irritation" can arrive at this significance. I here call attention to the transference from below to above (in the dream in question from the lower to the upper jaw), which occurs so frequently, which is at the service of sexual repression, and by means of which all kinds of sensations and intentions occurring in hysteria which ought to be enacted in the genitals can be realised upon less objectionable parts of the body,' p. 194. 'But I must also refer to another connection contained in an idiomatic expression. In our country there is in use an indelicate designation for the act of masturbation, namely: To pull one out, or to pull one down,' p. 195, 2nd edition. This expression had been familiar to me in early youth as a designation for onanism, and from here on it will not be difficult for the experienced dream interpreter to get access to the infantile material which may lie at the basis of this dream. I only wish to add that the facility with which the tooth in the dream came out, and the fact that it became transformed after coming out into an upper incisor, recalls to me an experience of childhood when I myself easily and painlessly pulled out one of my wobbling front teeth. This episode, which I can still to this day distinctly remember with all its details, happened at the same early period in which my first conscious attempts at onanism began— (Concealing Memory).

The reference of Freud to an assertion of C. G. Jung that dreams of dental irritation in women signify parturition (footnote p. 194), together with the popular belief in the significance of toothache in pregnant women, has established an opposition between the feminine significance and the masculine (puberty). In this connection I recall an earlier dream which I dreamed soon after i was discharged by the dentist after the treatment, that the gold crowns which had just been put in fell out, whereupon I was greatly chagrined in the dream on account of the considerable expense, concerning which I had not yet stopped worrying. In view of a certain experience this dream now becomes comprehensible as a commendation of the material advantages of masturbation when contrasted with every form of the economically less advantageous object-love (gold crowns are also Austrian gold coins).

Theoretically this case seems to show a double interest. First it verifies the connection revealed by Freud, inasmuch as the ejaculation in the dream takes place during the act of tooth-pulling. For no matter in what form a pollution may appear, we are obliged to look upon it as a masturbatic gratification which takes place without the help of mechanical excitation. Moreover the gratification by pollution in this case does not take place, as is usually the case, through an imaginary object, but it is without an object; and, if one may be allowed to say so, it is purely autoerotic, or at most it perhaps shows a slight homosexual thread (the dentist).

The second point which seems to be worth mentioning is the following: The objection is quite obvious that we are seeking here to validate the Freudian conception in a quite superfluous manner, for the experiences of the reading itself are perfectly sufficient to explain to us the content of the dream. The visit to the dentist, the conversation with the lady, and the reading of the *Traumdeutung* are sufficient to explain why the sleeper, who was also disturbed during the night by toothache, should dream this dream, it may even explain the removal of the sleep-disturbing pain (by means of the presentation of the removal of the painful tooth and simultaneous over-accentuation of the dreaded painful sensation through libido). But no matter how much of this assumption we may admit, we cannot earnestly maintain that the readings of Freud's explanations have produced in the dreamer the connection of the tooth-pulling with the act of masturbation; it could not even have been made effective had it not been for the fact, as the dreamer himself admitted ('to pull one off') that this association had already been formed long ago. What may have still more stimulated this association in connection with the conversation with the lady is shown by a later assertion of the dreamer that while reading the *Traumdeutung* he could not, for obvious reasons, believe in this typical meaning of dreams of dental irritation, and entertained the wish to know whether it held true for all dreams of this nature. Tin-dream now confirms this at least for his own person, and shows him why he had to doubt it. The dream is therefore also in this respect the fulfilment of a wish; namely, to be convinced of the importance and stability of this conception of Freud."

47. A young colleague, who is entirely free from nervousness, tells me in this connection: "I know from my own experience that while swinging, and at the moment at which the downward movement had the greatest impetus, I used to get a curious feeling in my genitals, which I must designate, although it was not really pleasant to me, as a voluptuous feeling." I have often heard from patients that their first erections accompanied by voluptuous sensations had occurred in boyhood while they were climbing. It is established with complete certainty by psychoanalyses that the first sexual impulses have often originated in the scufflings and wrestlings of childhood.

48. This naturally holds true only for German-speaking dreamers who are acquainted with the vulgarism *"vögeln."*
49. *Sammlung kl. Schriften zur Neurosenlehre,* zweite Folge, 1909.
50. *Cf.* the author's *Three Contributions to the Sexual Theory,* translated by A. A. Brill.
51. W. Stekel, *Die Sprache des Traumes,* 1911.
52. Alf. Adler, "Der Psychische Hermaphroditismus im Leben und in der Neurose," *Fortschrifte der Medizin,* 1910, No. 16, and later works in the *Zentralblatt für Psychoanalyse,* 1, 1910–1911.
53. I have published a typical example of such a veiled Oedipus dream in No. 1 of the *Zentralblatt für Psychoanalyse;* another with a detailed analysis was reported in the same journal, No. IV., by Otto Rank. Indeed the ancients were not unfamiliar with the symbolic interpretation of the open Oedipus dream (see O. Rank, p. 542); thus a dream of sexual relations with the mother has been transmitted to us by Julius Caesar which the oneiroscopists interpreted as a favourable omen for taking possession of the earth (Mother-earth). It is also known that the oracle declared to the Tarquinii that that one of them would become ruler of Rome who should first kiss the mother (*osculum matri tulerit*), which Brutus conceived as referring to the mother-earth (*terram osculo contigit, scilicet quod ea communia mater omnium mortalium esset,* Livius, 1., lxi.). These myths and interpretations point to a correct psychological knowledge. I have found that persons who consider themselves preferred or favoured by their mothers manifest in life that confidence in themselves and that firm optimism which often seems heroic and brings about real success by force.
54. It is only of late that I have learned to value the significance of fancies and unconscious thoughts about life in the womb. They contain the explanation of the curious fear felt by so many people of being buried alive, as well as the profoundest unconscious reason for the belief in a life after death which represents nothing but a projection into the future of this mysterious life before birth. *The act of birth, moreover, is the first experience with fear, and is thus the source and model of the emotion of fear.*
55. For such a dream see Pfister: "Ein Fall von Psychanalytischer Seelensorge und Seelenheilung," *Evangelische Freiheit,* 1909. Concerning the symbol of "saving" see my lecture, "Die Zukünftigen Chancen der psychoanalytischen Therapie," *Zentralblatt für Psychoanalyse,* No. I.,

1910. Also "Beiträge zur Psychologie des Liebeslebens, I. Ueber einen besonderen Typus der objektwahl beim Manne," *Jahrbuch,* Bleuler-Freud, vol. ii., 1910.

56. *Cf.* the works of Bleuler and of his pupils Maeder, Abraham, and others of the Zurich school upon symbolism, and of those authors who are not physicians (Kleinpaul and others), to which they refer.

57. In this country the President, the Governor, and the Mayor often represent the father in the dream. (Translator.)

58. I may here repeat what I have said in another place ("Die Zukünftigen Chancen der psychoanalytischen Therapie," *Zentralblatt für Psychoanalyse,* I., No. 1 and 2, 1910): "Some time ago I learned that a psychologist who is unfamiliar with our work remarked to one of my friends that we are surely over-estimating the secret sexual significance of dreams. He stated that his most frequent dream was of climbing a stairway, and that there was surely nothing sexual behind this. Our attention having been called to this objection, we directed our investigations to the occurrence of stairways, stairs, and ladders in the dream, and we soon ascertained that stairs (or anything analogous to them) represent a definite symbol of coitus. The basis for this comparison is not difficult to find; under rhythmic intervals and with increasing difficulty in breathing one reaches to a height, and may come down again in a few rapid jumps. Thus the rhythm of coitus is recognisable in climbing stairs. Let us not forget to consider the usage of language. It shows us that the "climbing" or "mounting" is, without further addition, used as a substitutive designation of the sexual act. In French the step of the stairway is called *"la marche"; "un vieux marcheur"* corresponds exactly to our "an old climber."

59. In this country where the word "necktie" is almost exclusively used, the translator has also found it to be a symbol of a burdensome woman from whom the dreamer longs to be freed—"necktie—something tied to my neck like a heavy weight—my fiancée," are the associations from the dream of a man who eventually broke his marriage engagement.

60. In spite of all the differences between Scherner's conception of dream symbolism and the one developed here, I must still assert that Scherner" should be recognised us the true discoverer of symbolism in dreams, and that the experience of psychoanalysis has brought his

book into honourable repute after it had been considered fantastic for about fifty years.

61. From "Nachträge zur Traumdeutung," *Zentralblatt für Psychoanalyse,* I., No. 5 and 6, 1911.

62. "Beiträge zur Traumdeutung," *Jahrbuch für Psychoanalyt. und psychop. Forsch.,* Bd. I., 1909, p. 473. Here also (p. 475) a dream is reported in which a hat with a feather standing obliquely in the middle symbolises the (impotent) man.

63. Cf. *Zentralblatt für Psychoanalyse,* I.

64. Or chapel—vagina.

65. Symbol of coitus.

66. Mons veneris.

67. Crines pubis.

68. Demons in cloaks and capucines are, according to the explanation of a man versed in the subject, of a phallic nature.

69. The two halves of the scrotum.

70. See *Zentralblatt für Psychoanalyse,* vol. i., p. 2.

71. This Hebrew word is well known in German-speaking countries, even among non-Jews, and signifies an unlucky, awkward person. (Translator.)

6

The Dream-Work

All previous attempts to solve the problems of the dream have been based directly upon the manifest dream content as it is retained in the memory, and have undertaken to obtain an interpretation of the dream from this content, or, if interpretation was dispensed with, to base a judgment of the dream upon the evidence furnished by this content. We alone are in possession of new data; for us a new psychic material intervenes between the dream content and the results of our investigations: and this is the *latent* dream content or the dream thoughts which are obtained by our method. We develop a solution of the dream from this latter, and not from the manifest dream content. We are also confronted for the first time with a problem which has not before existed, that of examining and tracing the relations between the latent dream thoughts and the manifest dream content, and the processes through which the former have grown into the latter.

We regard the dream thoughts and the dream content as two representations of the same meaning in two different languages; or to express it better, the dream content appears to us as a translation of the dream thoughts into another form of expression, whose signs and laws of composition we are to learn by comparing the original with the translation. The dream thoughts are at once intelligible to us as soon as we have ascertained them. The dream content is, as it were, presented in a picture-writing, whose signs are to be translated one by one into the language of the dream thoughts. It would of course be incorrect to try to read these signs according to their values as pictures instead of according to their significance as signs. For instance, I have before me a picture-puzzle (rebus): a house, upon whose roof there is a boat; then a running figure whose head has been apostrophised away, and the like. I might now be tempted as a critic to consider this composition and its elements nonsensical. A boat does not belong on the roof of a house and a person without a head cannot run; the person, too, is larger than the house, and if the whole thing is to represent a landscape, the single letters of the alphabet do not fit into it, for of course they do not occur in pure nature. A correct judgment of the picture-puzzle results only if I make no such objections to the whole and its parts, but if, on the contrary, I take pains to replace each picture by the syllable or word which it is capable of representing by means of any sort of reference, the words which are thus brought together are no longer meaningless, but may constitute a most beautiful and sensible expression. Now the dream is a picture-puzzle of this sort, and our predecessors in the field of dream interpretation have made the mistake of judging the rebus as an artistic composition. As such it appears nonsensical and worthless.

(A)   The Condensation Work

The first thing which becomes clear to the investigator in the comparison of the dream content with the dream thoughts is that a tremendous work of condensation has taken place. The dream is reserved, paltry, and laconic when compared with the range and copiousness of the dream thoughts. The dream when written down fills half a page; the analysis, in which the dream thoughts are contained, requires six, eight, twelve times as much space. The ratio varies with different dreams; it never changes its essential meaning, as far as I have been able to observe. As a rule the extent of the compression which has taken place is under-estimated, owing to the fact that the dream thoughts which are brought to light are considered the complete material, while continued work of interpretation may reveal new thoughts which are concealed behind the dream. We have already mentioned that one is really never sure of having interpreted a dream completely; even if the solution seems satisfying and flawless, it still always remains possible that there is a further meaning which is manifested by the same dream. Thus the *amount of condensation* is—strictly speaking—indeterminable. An objection, which at first sight seems very plausible, might be raised against the assertion that the disproportion between dream content and dream thought justifies the conclusion that an abundant condensation of psychic material has taken place in the formation of dreams. For we so often have the impression that we have dreamed a great deal throughout the night and then have forgotten the greater part. The dream which we recollect upon awakening would thus be only a remnant of the total dream-work, which would probably equal the dream thoughts in range if we were able to remember the former completely. In part this is certainly true; there can be no mistake about the observation that the dream is most accurately reproduced if one tries to remember it immediately after awakening, and that the recollection of it becomes more and more defective towards evening. On the other hand, it must be admitted that the impression that we have dreamed a good deal more than we are able to reproduce is often based upon an illusion, the cause of which will be explained later. Moreover, the assumption of condensation in the dream activity is not affected by the possibility of forgetting in dreams, for it is proved by groups of ideas belonging to those particular parts of the dream which have remained in the memory. If a large part of the dream has actually been lost to memory, we are probably deprived of access to a new series of dream thoughts. It is altogether unjustifiable to expect that those portions of the dream which have been lost also relate to the thoughts with which we are already acquainted from the analysis of the portions which have been preserved.

In view of the great number of ideas which analysis furnishes for each individual element of the dream content, the chief doubt with many readers will be whether it is permissible to count everything that subsequently comes to mind during analysis as a part of the dream thoughts—to assume, in other words, that all these thoughts have been active in the sleeping state and have taken part in the formation of the dream. Is it not more probable that thought connections are developed in the course of analysis which did not participate in the formation of the dream? I can meet this doubt only conditionally. It is true, of course, that particular thought connections first arise only during analysis; but one may always be sure that such new connections have been established only between thoughts which have already been connected in the dream thoughts by other means; the new connections are, so to speak, corollaries, short circuits, which are made possible by the existence of other more fundamental means of connection. It must be admitted that the huge number of trains of thought revealed by analysis have already been active in the formation of the dream, for if a chain of thoughts has been worked out, which seems to be without connection with the formation of the dream, a thought is suddenly encountered which, being represented in the dream, is indispensable to its interpretation—which nevertheless is inaccessible except through that chain of thoughts. The reader may here turn to the dream of the botanical monograph, which is obviously the result of an astonishing condensation activity, even though I have not given the analysis of it completely.

But how, then, is the psychic condition during sleep which precedes dreaming to be imagined? Do all the dream thoughts exist side by side, or do they occur one after another, or are many simultaneous trains of thought constructed from different centres, which meet later on? I am of the opinion that it is not yet necessary to form a plastic conception of the psychic condition of dream formation. Only let us not forget that we are concerned with unconscious thought, and that the process may easily be a different one from that which we perceive in ourselves in intentional contemplation accompanied by consciousness.

The fact, however, that dream formation is based on a process of condensation, stands indubitable. How, then, is this condensation brought about?

If it be considered that of those dream thoughts which are found only the smallest number are represented in the dream by means of one of its ideal elements, it might be concluded that condensation is accomplished by means of ellipsis, in that the dream is not an accurate translation or a projection point by point of the dream thoughts, but a very incomplete and defective reproduction of them. This view, as we shall soon find, is a very inadequate one. But let us take it as a starting point for the present, and ask ourselves: If only a few of the elements of the dream thoughts get into the dream content, what conditions determine their choice?

In order to gain enlightenment on this subject let us turn our attention to those elements of the dream content which must have fulfilled the conditions we are seeking. A dream to the formation of which an especially strong condensation has contributed will be the most suitable material for this investigation. I select the dream, cited on page 159, of the botanical monograph.

Dream content: *I have written a monograph upon a (obscure) certain plant. The book lies before me, I am just turning over a folded coloured plate. A dried specimen of the plant is bound with every copy as though from a herbarium.*

The most prominent element of this dream is the botanical monograph. This comes from the impressions received on the day of the dream; I had actually *seen a monograph on the genus "cyclamen"* in the show-window of a book-store. The mention of this genus is lacking in the dream content, in which only the monograph and its relation to botany have remained. The "botanical monograph" immediately shows its relation to the work on cocaine which I had once written; thought connections proceed from cocaine on the one hand to a "Festschrift," and on the other to my friend, the eye specialist, Dr. Koenigstein, who has had a share in the utilisation of cocaine. Moreover, with the person of this Dr. Koenigstein is connected the recollection of the interrupted conversation which I had had with him on the previous evening and of the manifold thoughts about remuneration for medical services among colleagues. This conversation, then, is properly the actual stimulus of the dream; the monograph about cyclamen is likewise an actuality but of an indifferent nature; as I soon see, the "botanical monograph" of the dream turns out to be a common mean between the two experiences of the day, and to have been taken over unchanged from an indifferent impression and bound up with the psychologically significant experience by means of the most abundant associations.

Not only the combined idea, "botanical monograph," however, but also each of the separate elements, "botanical" and "monograph," penetrates deeper and deeper into the confused tangle of the dream thoughts. To "botanical" belong the recollections of the person of Professor *Gartner* (German: Gärtner = gardener), of his *blooming* wife, of my patient whose name is *Flora,* and of a lady about whom I told the story of the forgotten *flowers. Gartner,* again, is connected with the laboratory and the conversation with *Koenigstein;* the mention of the two female patients also belongs to the same conversation. A chain of thoughts, one end of which is formed by the title of the hastily seen monograph, leads off in the other direction from the lady with the flowers to the *favourite flowers* of my wife. Besides this, "botanical" recalls not only an episode at the Gymnasium, but an examination taken while I was at the university; and a new subject matter—

my hobbies—which was broached in the conversation already mentioned, is connected by means of my humorously so-called *favourite flower,* the artichoke, with the chain of thoughts proceeding from the forgotten flowers; behind "artichoke" there is concealed on the one hand a recollection of Italy, and on the other a reminiscence of a childhood scene in which I first formed my connection with books which has since grown so intimate. "Botanical," then, is a veritable nucleus, the centre for the dream of many trains of thought, which, I may assure the reader, were correctly and justly brought into relation to one another in the conversation referred to. Here we find ourselves in a thought factory, in which, as in the "Weaver's Masterpiece":

> "One tread moves thousands of threads,
> The little shuttles fly back and forth,
> The threads flow on unseen,
> One stroke ties thousands of knots."

"Monograph" in the dream, again, has a bearing upon two subjects, the one-sidedness of my studies and the costliness of my hobbies.

The impression is gained from this first investigation that the elements "botanical" and "monograph" have been accepted in the dream content because they were able to show the most extensive connections with the dream thoughts, and thus represent nuclei in which a great number of dream thoughts come together, and because they have manifold significance for the dream interpretation. The fact upon which this explanation is based may be expressed in another form: Every element of the dream content turns out to be *over-determined*—that is, it enjoys a manifold representation in the dream thoughts.

We shall learn more by testing the remaining component parts of the dream as to their occurrence in the dream thoughts. *The coloured plate* refers (*cf.* the analysis on p. 162) to a new subject, the criticism passed upon my work by colleagues, and to a subject already represented in the dream—my hobbies—and also to a childish recollection in which I pull to pieces the book with the coloured plates; the dried specimen of the plant relates to an experience at the Gymnasium centering about and particularly emphasizing the herbarium. Thus I see what sort of relation exists between the dream content and dream thoughts: Not only do the elements of the dream have a manifold determination in the dream thoughts, but the individual dream thoughts are represented in the dream by many elements. Starting from an element of the dream the path of associations leads to a number of dream thoughts; and from a dream thought to several elements of the dream. The formation of the dream does not, therefore, take place in such fashion that a single one of the dream thoughts or a group of them furnishes the dream content with an abridgment as its representative therein, and that then another dream thought furnishes another abridgment as its representative—somewhat as popular representatives are elected from among the people—but the whole mass of the dream thoughts is subjected to a certain elaboration, in the course of which those elements that receive the greatest and completest support stand out in relief, analogous, perhaps, to election by *scrutins des listes*. Whatever dream I may subject to such dismemberment, I always find the same fundamental principle confirmed—that the dream elements are constructed from the entire mass of the dream thoughts and that every one of them appears in relation to the dream thoughts to have a multiple determination.

It is certainly not out of place to demonstrate this relation of the dream content to the dream thoughts by means of a fresh example, which is distinguished by a particularly artful intertwining of reciprocal relations. The dream is that of a patient whom I am treating for claustrophobia (fear in enclosed spaces). It will soon become evident why I feel myself called upon to entitle this exceptionally intellectual piece of dream activity in the following manner:

II. "A Beautiful Dream"

The dreamer is riding with much company to X-street, where there is a modest road-house (which is not the fact). *A theatrical performance is being given in its rooms. He is first audience, then actor. Finally the company is told to change their clothes, in order to get back into the city. Some of the people are assigned to the rooms on the ground floor, others to the first floor. Then a dispute arises. Those above are angry because those below have not yet finished, so that they cannot come down. His brother is upstairs, he is below, and he is angry at his brother because there is such crowding.* (This part obscure.) *Besides it has already been decided upon their arrival who is to be upstairs and who down. Then he goes alone over the rising ground, across which X-street leads toward the city, and he has such difficulty and hardship in walking that he cannot move from the spot. An elderly gentleman joins him and scolds about the King of Italy. Finally, towards the end of the rising ground walking becomes much easier.*

The difficulties experienced in walking were so distinct that for some time after waking he was in doubt whether they were dream or reality.

According to the manifest content, this dream can hardly be praised. Contrary to the rules, I shall begin with that portion which the dreamer referred to as the most distinct.

The difficulties which were dreamed of, and which were probably experienced during the dream—difficult climbing accompanied by dyspnoea—is one of the symptoms which the patient had actually shown years before, and which, in conjunction with other symptoms, was at that time attributed to tuberculosis (probably hysterically simulated). We are already from exhibition dreams acquainted with this sensation of being hindered, peculiar to the dream, and here again we find it used for the purpose of any kind of representation, as an ever-ready material. That part of the dream content which ascribes the climbing as difficult at first, and as becoming easier at the end of the hill, made me think while it was being told of the well-known masterful introduction to *Sappho* by A. Daudet. Here a young man carries the girl whom he loves upstairs— she is at first as light as a feather; but the higher he mounts the more heavily she weighs upon his arm, and this scene symbolises a course of events by recounting which Daudet tries to warn young men not to waste serious affection upon girls of humble origin or of questionable past.[1] Although I knew that my patient had recently had a love affair with a lady of the theatre, and had broken it off, I did not expect to find that the interpretation which had occurred to me was correct. Moreover, the situation in *Sappho* was the *reverse* of that in the dream; in the latter the climbing was difficult at the beginning and easy later on; in the novel the symbolism serves only if what was at first regarded as easy finally turns out to be a heavy load. To my astonishment, the patient remarked that the interpretation corresponded closely to the plot of a play which he had seen on the evening before at the theatre. The play was called *Round about Vienna,* and treated of the career of a girl who is respectable at first but later goes over to the *demi-monde,* who has affairs with persons in high places, thus "climbing," but finally "goes down" faster and faster. This play had reminded him of another entitled *From Step to Step,* in the advertisement of which had appeared a *stairway* consisting of several steps.

Now to continue the interpretation. The actress with whom he had had his most recent affair, a complicated one, had lived in X-street. There is no inn in this street. However, while he was spending a part of the summer in Vienna for the sake of the lady, he had lodged (German *abgestiegen* = stopped, literally *stepped off*) at a little hotel in the neighbourhood. As he was leaving the hotel he said to the cab-driver, "I am glad I didn't get any vermin anyway" (which incidentally is one of his phobias). Whereupon the cab-driver answered: "How could anybody stop there! It isn't a hotel at all, it's really nothing but a *road-house!*"

The road-house immediately suggests to the dreamer's recollection a quotation:

> "Of that marvellous host
> I was once a guest."

But the host in the poem by Uhland is an apple tree. Now a second quotation continues the train of thought:

> FAUST (*dancing with the young witch*).
> "A lovely dream once came to me;
> I then beheld an apple tree,
> And there two fairest apples shone:
> They lured me so, I climbed thereon."
>
> THE FAIR ONE.
> "Apples have been desired by you,
> Since first in Paradise they grew;
> And I am moved with joy to know
> That such within my garden grow."

> Translated by BAYARD TAYLOR.

There remains not the slightest doubt what is meant by the apple tree and the apples. A beautiful bosom stood high among the charms with which the actress had bewitched our dreamer.

According to the connections of the analysis we had every reason to assume that the dream went back to an impression from childhood. In this case it must have reference to the nurse of the patient, who is now a man of nearly fifty years of age. The bosom of the nurse is in reality a road-house for the child. The nurse as well as Daudet's Sappho appears as an allusion to his abandoned sweetheart.

The (elder) brother of the patient also appears in the dream content; he is upstairs, the dreamer himself is below. This again is an *inversion,* for the brother, as I happen to know, has lost his social position, my patient has retained his. In reporting the dream content the dreamer avoided saying that his brother was upstairs and that he himself was *down.* It would have been too frank an expression, for a person is said to be "down and out" when he has lost his fortune and position. Now the fact that at this point in the dream something is represented as inverted must have a meaning. The inversion must apply rather to some other relation between the dream thoughts and dream content. There is an indication which suggests how this inversion is to be taken. It obviously applies to the end of the dream, where the circumstances of climbing are the reverse of those in *Sappho.* Now it may easily be seen what inversion is referred to; in *Sappho* the man carries the woman who stands in a sexual relation to him; in the dream thoughts, *inversely,* a woman carries a man, and as this state of affairs can only occur during childhood, the reference is again to the nurse who carries the heavy child. Thus the final portion of the dream succeeds in representing *Sappho* and the nurse in the same allusion.

Just as the name *Sappho* has not been selected by the poet without reference to a Lesbian custom, so the elements of the dream in which persons act *above* and *below,* point to fancies of a sexual nature with which the dreamer is occupied and which as suppressed cravings are not without connection with his neurosis. Dream interpretation itself does not show that these are fancies and not recollections of actual happenings; it only furnishes us with a set of thoughts and leaves us to determine their value as realities. Real and fantastic occurrences at first appear here as of equal value—and not only here but also in the creation of more important psychic structures than dreams. Much company, as we already know, signifies a secret. The brother is none other than a representative, drawn into the childhood scene by "fancying backwards," of all of the later rivals for the woman. Through the agency of an experience which is indifferent in itself, the episode with the gentleman who scolds about the King of Italy again refers to the intrusion of people of low rank into aristocratic society. It is as though the warning which Daudet gives to youth is to be supplemented by a similar warning applicable to the suckling child.[2]

In order that we may have at our disposal a third example for the study of condensation in dream formation, I shall cite the partial analysis of another dream for which I am indebted to an elderly lady who is being psychoanalytically treated. In harmony with the condition of severe anxiety from which the patient suffered, her dreams contained a great abundance of sexual thought material, the discovery of which astonished as well as frightened her. Since I cannot carry the interpretation of the dream to completion, the material seems to fall apart into several groups without apparent connection.

III. Content of the dream

She remembers that she has two June bugs in a box, which she must set at liberty, for otherwise they will suffocate. She opens the box, and the bugs are quite exhausted; one of them flies out of the window, but the other is crushed on the casement while she is shutting the window, as some one or other requests her to do(expressions of disgust).

Analysis

Her husband is away travelling, and her fourteen-year-old daughter is sleeping in the bed next to her. In the evening the little one calls her attention to the fact that a moth has fallen into her glass of water; but she neglects to take it out, and feels sorry for the poor little creature in the morning. A story which she had read in the evening told of boys throwing a cat into boiling water, and the twitchings of the animal were described. These are the occasions for the dream, both of which are indifferent in themselves. She is further occupied with the subject of *cruelty to animals*. Years before, while they were spending the summer at a certain place, her daughter was very cruel to animals. She started a butterfly collection, and asked her for arsenic with which to kill the butterflies. Once it happened that a moth flew about the room for a long time with a needle through its body; on another occasion she found that some moths which had been kept for metamorphosis had died of starvation. The same child while still at a tender age was in the habit of pulling out the wings of beetles and butterflies; now she would shrink in horror from these cruel actions, for she has grown very kind.

Her mind is occupied with this contrast. It recalls another contrast, the one between appearance and disposition, as it is described in *Adam Bede* by George Eliot. There a beautiful but vain and quite stupid girl is placed side by side with an ugly but high-minded one. The aristocrat who seduces the little goose, is opposed to the working man who feels *aristocratic,* and behaves accordingly. It is impossible to tell character from people's *looks.* Who could tell from *her* looks that she is tormented by sensual desires?

In the same year in which the little girl started her butterfly collection, the region in which they were staying suffered much from a pest of June bugs. The children made havoc among the bugs, and *crushed* them cruelly. At that time she saw a person who tore the wings off the June bugs and ate them. She herself had been born in June and also married in June. Three days after the wedding she wrote a letter home, telling how happy she was. But she was by no means happy.

During the evening before the dream she had rummaged among her old letters and had read various ones, comical and serious, to her family—an extremely ridiculous letter from a piano-teacher who had paid her attention when she was a girl, as well as one from an aristocratic admirer.[3]

She blames herself because a bad book by de Maupassant had fallen into the hands of one of her daughters.[4] The arsenic which her little girl asks for recalls the arsenic pills which restored the power of youth to the Due de Mora in Nabab.

"Set at liberty" recalls to her a passage from the *Magic Flute:*

> "I cannot compel you to love,
> But I will not give you your liberty."

"June bugs "suggests the speech of Katie:[5]

> "I love you like a little beetle."

Meanwhile the speech from *Tannhäuser:* "For you are wrought with evil passion."

She is living in fear and anxiety about her absent husband. The dread that something may happen to him on the journey is expressed in numerous fancies of the day. A little while before, during the analysis, she had come upon a complaint about his "senility" in her unconscious thoughts. The wish thought which this dream conceals may perhaps best be conjectured if I say that several days before the dream she was suddenly astounded by a command which she directed to her husband in the midst of her work: *"Go hang yourself."* It was found that a few hours before she had read somewhere that a vigorous erection is induced when a person is hanged. It was for the erection which freed itself from repression in this terror-inspiring veiled form. "Go hang yourself" is as much as to say: "Get up an erection, at any cost." Dr. Jenkin's arsenic pills in *Nabab* belong in this connection; for it was known to the patient that the strongest aphrodisiac, cantharides, is prepared by *crushing bugs* (so-called Spanish flies). The most important part of the dream content has a significance to this effect.

Opening and shutting the *window* is the subject of a standing quarrel with her husband. She herself likes to sleep with plenty of air, and her husband does not. *Exhaustion* is the chief ailment of which she complains these days.

In all three of the dreams just cited I have emphasized by italics those phrases where one of the elements of the dream recurs in the dream thoughts in order to make the manifold references of the former obvious. Since, however, the analysis of none of these dreams has been carried to completion, it will be well worth while to consider a dream with a fully detailed analysis, in order to demonstrate the manifold determination of its content. I select the dream of Irma's injection for this purpose. We shall see without effort in this example that the condensation work has used more than one means for the formation of the dream.

The chief person in the content of the dream is my patient Irma, who is seen with the features which belong to her in waking life, and who therefore in the first instance represents herself. But her attitude as I examine her at the window is taken from the recollection of another person, of the lady for whom I should like to exchange my patient, as the dream thoughts show. In as far as Irma shows a diphtheritic membrane which recalls my anxiety about my eldest daughter, she comes to represent this child of mine, behind whom is concealed the person of the patient who died from intoxication and who is brought into connection by the identity of her name. In the further course of the dream the significance of Irma's personality changes (without the alteration of her image as it is seen in the dream); she becomes one of the children whom we examine in the public dispensaries for children's diseases, where my friends show the difference of their mental capabilities. The transference was obviously brought about through the idea of my infant daughter. By means of her unwillingness to open her mouth the same Irma is changed into an allusion to another lady who was once examined by me, and besides that to my wife, in the same connection. Furthermore, in the morbid transformations which I discover in her throat I have gathered allusions to a great number of other persons.

All these people whom I encounter as I follow the associations suggested by "Irma," do not appear personally in the dream; they are concealed behind the dream person "Irma," who is thus developed into a collective image, as might be expected, with contradictory features. Irma comes to represent these other persons, who are discarded in the work of condensation, in that I cause to happen to her all the things which recall these persons detail for detail.

I may also construct a collective person for the condensation of the dream in another manner, by uniting the actual features of two or more persons in one dream image. It is in this manner that Dr. M. in my dream was constructed, he bears the name of Dr. M., and speaks and acts as Dr. M. does, but his bodily characteristics and his suffering belong to another person, my eldest brother; a single feature, paleness, is doubly determined, owing to the fact that it is common to both persons. Dr. R. in my dream about my uncle is a similar composite person. But here the dream image is prepared in still another manner. I have not united features peculiar to the one with features of the other, and thereby abridged the remembered image of each by certain features, but I have adopted the method employed by Galton in producing family portraits, by which he projects both pictures upon one another, whereupon the common features stand out in stronger relief, while those which do not coincide neutralize one another and become obscure in the picture. In the dream of my uncle the *blond beard* stands out in relief, as an emphasized feature, from the physiognomy, which belongs to two persons, and which is therefore blurred; furthermore the beard contains an allusion to my father and to myself, which is made possible by its reference to the fact of growing grey.

The construction of collective and composite persons is one of the chief resources of the activity of dream condensation. There will soon be an occasion for treating of this in another connection.

The notion "dysentery" in the dream about the injection likewise has a manifold determination, on the one hand because of its paraphasic assonance with diphtheria, and on the other because of its reference to the patient, whom I have sent to the Orient, and whose hysteria has been wrongly recognised.

The mention of "propyls" in the dream also proves to be an interesting case of condensation. Not "propyls" but "amyls" were contained in the dream thoughts. One might think that here a simple displacement had occurred in the dream formation. And this is the case, but the displacement serves the purposes of condensation, as is shown by the following supplementary analysis. If I dwell for a moment upon the word "propyls," its assonance to the word "propylaeum" suggests itself to me. But the propylaeum is to be found not only in Athens but also in Munich. In the latter city I visited a friend the year before who was seriously ill, and the reference to him becomes unmistakable on account of *trimethylamin,* which follows closely upon *propyls.*

I pass over the striking circumstance that here, as elsewhere in the analysis of dreams, associations of the most widely different values are employed for the establishment of thought connections as though they were equivalent, and I yield to the temptation to regard the process by which amyls in the dream thoughts are replaced by *propyls,* as though it were plastic in the dream content.

On the one hand is the chain of ideas about my friend Otto, who does not understand me, who thinks I am in the wrong, and who gives me the cordial that smells like amyls; on the other the chain of ideas—connected with the first by contrast—about my friend William, who understands me and who would always think I was in the right, and to whom I am indebted for so much valuable information about the chemistry of the sexual processes.

Those characteristics of the associations centering about Otto which ought particularly to attract my attention are determined by the recent occasions which are responsible for the dream; *amyls* belong to these elements so determined which are destined to get into the dream content. The group of associations "William" is distinctly vivified by the contrast to Otto, and the elements in it which correspond to those already excited in the "Otto" associations are thrown into relief. In this whole dream I am continually referring to a person who excites my displeasure and to another person whom I can oppose to him or her at will, and I conjure up the friend as against the enemy, feature for feature. Thus amyls in the Otto-group suggests recollections in the other group belonging to chemistry; trimethylamin, which receives support from several quarters, finds its way into the dream content. "Amyls," too, might have got into the dream content without undergoing change, but it yields to the influence of the

"William" group of associations, owing to the fact that an element which is capable of furnishing a double determination for amyls is sought out from the whole range of recollections which the name "William" covers. The association "propyls" lies in the neighbourhood of *amyls;* Munich with the propylaeum comes to meet amyls from the series of associations belonging to "William." Both groups are united in *propyls—propylaeum.* As though by a compromise, this intermediary element gets into the dream content. Here a *common mean* which permits of a manifold determination has been created. It thus becomes perfectly obvious that manifold determination must facilitate penetration into the dream content. A displacement of attention from what is really intended to something lying near in the associations has thoughtlessly taken place, for the sake of this mean-formation.

The study of the injection dream has now enabled us to get some insight into the process of condensation which takes place in the formation of dreams. The selection of those elements which occur in the dream content more than once, the formation of new unities (collective persons, composite images), and the construction of the common mean, these we have been able to recognise as details of the condensing process. The purpose which is served by condensation and the means by which it is brought about will be investigated when we come to study the psychic processes in the formation of dreams as a whole. Let us be content for the present with establishing dream *condensation* as an important relation between the dream thoughts and the dream content.

The condensing activity of the dream becomes most tangible when it has selected words and names as its object. In general words are often treated as things by the dream, and thus undergo the same combinations, displacements, and substitutions, and therefore also condensations, as ideas of things. The results of such dreams are comical and bizarre word formations. Upon one occasion when a colleague had sent me one of his essays, in which he had, in my judgment, overestimated the value of a recent physiological discovery and had expressed himself in extravagant terms, I dreamed the following night a sentence which obviously referred to this treatise: *"That is in true norekdal style."* The solution of this word formation at first gave me difficulties, although it was unquestionably formed as a parody after the pattern of the superlatives "colossal," "pyramidal"; but to tell where it came from was not easy. At last the monster fell apart into the two names Nora and Ekdal from two well-known plays by Ibsen. I had previously read a newspaper essay on Ibsen by the same author, whose latest work I was thus criticising in the dream.

II[6]

One of my female patients dreams that *a man with a light beard and a peculiar glittering eye is pointing to a sign board attached to a tree which reads: uclamparia—wet.*

Analysis

The man was rather authoritative looking, and his peculiar glittering eye at once recalled St. Paul's Cathedral, near Rome, where she saw in mosaics the Popes that have so far ruled. One of the early Popes had a golden eye (this was really an optical illusion which the guides usually call attention to). Further associations showed that the general physiognomy corresponded to her own clergyman (Pope), and the shape of the light beard recalled her doctor (myself), while the stature of the man in the dream recalled her father. All these persons stand in the same relation to her; they are all guiding and directing her course of life. On further questioning, the golden eye recalled gold—money—the rather expensive psychoanalytic treatment which gives her a great deal of concern. Gold, moreover, recalls the gold cure for alcoholism—Mr. D., whom she would have married if it had not been for his clinging to the disgusting alcohol habit—she does not object to a person taking an occasional drink; she herself sometimes drinks beer and cordials—this again brings her back to her visit to St. Paul's without the walls and its surroundings. She remembers that in the neighbouring monastery of the Three Fountains she drank a liquor made of eucalyptus by the Trappist monks who inhabit this monastery. She then relates how the monks transformed this malarial and swampy region into a dry and healthful neighbourhood by planting there many eucalyptus trees. The word "uclamparia" then resolves itself into eucalyptus and malaria, and the word "wet" refers to the former swampy nature of the place. Wet also suggests dry. Dry is actually the name of the man whom she would have married except for his over-indulgence in alcohol. The peculiar name of Dry is of Germanic origin (drei = three) and hence alludes to the Abbey of the Three (drei) Fountains above mentioned. In talking about Mr. Dry's habit she used the strong words, "He could drink a fountain." Mr. Dry jocosely refers to his habit by saying, "You know I must drink because I am always *dry*" (referring to his name). The eucalyptus also refers to her neurosis, which was at first diagnosed as malaria. She went to Italy because her attacks of anxiety, which were accompanied by marked trembling and shivering, were thought to be of malarial origin. She bought some eucalyptus oil from the monks, and she maintains that it has done her much good.

The condensation *uclamparia—wet* is therefore the point of junction for the dream as well as for the neurosis.[7]

III

In a somewhat long and wild dream of my own, the chief point of which is apparently a sea voyage, it happens that the next landing is called *Hearsing* and the one farther on *Fliess*. The latter is the name of my friend living in B., who has often been the objective point of my travels. But Hearsing is put together from the names of places in the local environment of Vienna, which so often end in *ing: Hietzing, Liesing, Moedling* (Medelitz, "meae deliciae," my own name, *"my joy"*) (joy = German Freude), and the English *hearsay*, which points to libel and establishes the relation to the indifferent dream excitement of the day—a poem in the *Fliegende Blaetter* about a slanderous dwarf, "Saidhe Hashesaid." By connecting the final syllable *"ing"* with the name *Fliess, "Vlissingen"* is obtained, which is a real port on the sea-voyage which my brother passes when he comes to visit us from England. But the English for *Vlissingen* is *Flushing,* which signifies blushing and recalls erythrophobia (fear of blushing), which I treat, and also reminds me of a recent publication by Bechterew about this neurosis, which has given occasion for angry feelings in me.

IV

Upon another occasion I had a dream which consisted of two parts. The first was the vividly remembered word "Autodidasker," the second was truthfully covered by a short and harmless fancy which had been developed a few days before, and which was to the effect that I must tell Professor N., when I saw him next: "The patient about whose condition I last consulted you is really suffering from a neurosis, just as you suspected." The coinage *"Autodidasker"* must, then, not only satisfy the requirement that it should contain or represent a compressed meaning, but also that this meaning should have a valid connection with my purpose, which is repeated from waking life, of giving Professor N. his due credit.

Now *Autodidasker* is easily separated into *author* (German *Autor*), *autodidact*, and *Lasker*, with whom is associated the name Lasalle. The first of these words leads to the occasion of the dream—which this time is significant. I had brought home to my wife several volumes by a well-known author, who is a friend of my brother's, and who, as I have learned, comes from the same town as I (J. J. David). One evening she spoke to me about the profound impression which the touching sadness of a story in one of David's novels, about a talented but degenerate person, had made upon her, and our conversation turned upon the indications of talent which we perceive in our own children. Under the influence of what she had just read, my wife expressed a concern relative to our children, and I comforted her with the remark that it is just such dangers that can be averted by education. During the night my train of thoughts proceeded further, took up the concern of my wife, and connected with it all sorts of other things. An opinion which the poet had expressed to my brother upon the subject of marriage showed my thoughts a by-path which might lead to a representation in the dream. This path led to Breslau, into which city a lady who was a very good friend of ours had married. I found in Breslau Lasker and Lasalle as examples realising our concern about being ruined at the hands of a woman, examples which enabled me to represent both manifestations of this influence for the bad at once.[8] The "Cherchez la femme," in which these thoughts may be summed up, when taken in another sense, brings me to my brother, who is still unmarried and whose name is Alexander. Now I see that Alex, as we abbreviate the name, sounds almost like inversion of Lasker and that this factor must have taken part in giving my thoughts their detour by way of Breslau.

But this playing with names and syllables in which I am here engaged contains still another meaning. The wish that my brother may have a happy family life is represented by it in the following manner. In the artistic romance *L'Oeuvre*, the writer, as is well known, has incidentally given an episodic account of himself and of his own family happiness, and he appears under the name of *Sandoz*. Probably he has taken the following course in the name transformation. *Zola* when inverted (as children like so much to do) gives *Aloz*. But that was still too undisguised for him; therefore he replaced the syllable *Al*, which stands at the beginning of the name Alexander, by the third syllable of the same name, *sand*, and thus *Sandoz* came about. In a similar manner my *autodidasker* originated.

My fancy, that I am telling Professor N. that the patient whom we had both seen is suffering from a neurosis, got into the dream in the following manner. Shortly before the close of my working year I received a patient in whose case my diagnosis failed me. A serious organic affliction—perhaps some changes in the spine—was to be assumed, but could not be proved. It would have been tempting to diagnose the trouble as a neurosis, and this would have put an end to all difficulties, had it not been for the fact that the sexual anamnesis, without which I am unwilling to admit a neurosis, was so energetically denied by the patient. In my embarrassment I called to my assistance the physician whom I respect most of all men (as others do also), and to whose authority I surrender most completely. He listened to my doubts, told me he thought them justified, and then said: "Keep on observing the man, it is probably a neurosis." Since I know that he does not share my opinions about the etiology of neuroses, I suppressed my disagreement, but I did not conceal my scepticism. A few days after I informed the patient that I did not know what to do with him, and advised him to go to some one else. Thereupon, to my great astonishment, he began to beg my pardon for having lied to me, saying that he had felt very much ashamed; and now he revealed to me just that piece of sexual etiology which I had expected, and which I found necessary for assuming the existence of a neurosis. This was a relief to me, but at the same time a humiliation; for I had to admit that my consultant, who was not disconcerted by the absence of anamnesis, had made a correct observation. I made up my mind to tell him about it when I saw him again, and to say to him that he had been in the right and I in the wrong.

This is just what I do in the dream. But what sort of a wish is supposed to be fulfilled if I acknowledge that I am in the wrong? This is exactly my wish; I wish to be in the wrong with my apprehensions—that is to say, I wish that my wife whose fears I have appropriated in the dream thoughts may remain in the wrong. The subject to which the matter of being in the right or in the wrong is related in the dream is not far distant from what is really interesting to the dream thoughts. It is the same pair of alternatives of either organic or functional impairment through a woman, more properly through the sexual life—either tabetic paralysis or a neurosis—with which the manner of Lasalle's ruin is more or less loosely connected.

In this well-joined dream (which, however, is quite transparent with the help of careful analysis) Professor N. plays a part not merely on account of this analogy and of my wish to remain in the wrong, or on account of the associated references to Breslau and to the family of our friend who is married there—but also on account of the following little occurrence which was connected with our consultation. After he had attended to our medical task by giving the above mentioned suggestion, his interest was directed to personal matters. "How many children have you now?"—"Six."—A gesture of respect and reflection.—"Girls, boys?"—"Three of each. They are my pride and my treasure."—"Well, there is no difficulty about the girls, but the boys give trouble later on in their education." I replied that until now they had been very tractable; this second diagnosis concerning the future of my boys of course pleased me as little as the one he had made earlier, namely, that my patient had only a neurosis. These two impressions, then, are bound together by contiguity, by being successively received, and if I incorporate the story of the neurosis into the dream, I substitute it for the conversation upon education which shows itself to be even more closely connected with the dream thoughts owing to the fact that it has such an intimate bearing upon the subsequently expressed concerns of my wife. Thus even my fear that N. may turn out to be right in his remarks on the educational difficulties in the case of boys is admitted into the dream content, in that it is concealed behind the representation of my wish that I may be wrong in such apprehensions. The same fancy serves without change to represent both conflicting alternatives.

The verbal compositions of the dream are very similar to those which are known to occur in paranoia, but which are also found in hysteria and in compulsive ideas. The linguistic habits of children, who at certain periods actually treat words as objects and invent new languages and artificial syntaxes, are in this case the common source for the dream as well as for psychoneuroses.

When speeches occur in the dream, which are expressly distinguished from thoughts as such, it is an invariable rule that the dream speech has originated from a remembered speech in the dream material. Either the wording has been preserved in its integrity, or it has been slightly changed in the course of expression; frequently the dream speech is pieced together from various recollections of speeches, while the wording has remained the same and the meaning has possibly been changed so as to have two or more significations. Not infrequently the dream speech serves merely as an allusion to an incident, at which the recollected speech occurred.[9]

(B)   The Work of Displacement

Another sort of relation, which is no less significant, must have come to our notice while we were collecting examples of dream condensation. We have seen that those elements which obtrude themselves in the dream content as its essential components play a part in the dream thoughts which is by no means the same. As a correlative to this the converse of this thesis is also true. That which is clearly the essential thing in the dream thoughts need not be represented in the dream at all. The dream, as it were, is *eccentric;* its contents are grouped about other elements than the dream thoughts as a central point. Thus, for example, in the dream about the botanical monograph the central point of the dream content is apparently the element "botanical"; in the dream thoughts we are concerned with the complications and conflicts which result from services rendered among colleagues which put them under obligations to one another, subsequently with the reproach that I am in the habit of sacrificing too much to my hobbies, and the element "botanical" would in no case find a place in this nucleus of the dream thoughts if it were not loosely connected with it by an antithesis, for botany was never among my favourite studies. In the Sappho dream of my patient the ascending and descending, being upstairs and down, is made the central point; the dream, however, is concerned with the danger of sexual relations with persons of low *degree,* so that only one of the elements of the dream thoughts seems to have been taken over into the dream content, albeit with unseemly elaboration. Similarly in the dream about June bugs, whose subject is the relation of sexuality to cruelty, the factor of cruelty has indeed reappeared but in a different connection and without the mention of the sexual, that is to say, it has been torn from its context and transformed into something strange. Again, in the dream about my uncle, the blond beard, which seems to be its central point, appears to have no rational connection with the wishes for greatness which we have recognised as the nucleus of the dream thoughts. It is only to be expected if such dreams give a displaced impression. In complete contrast to these examples, the dream of Irma's injection shows that individual elements can claim the same place in the formation of dreams which they occupy in the dream thoughts. The recognition of these new and entirely variable relations between the dream thoughts and the dream content is at first likely to excite our astonishment. If we find in a psychic process of normal life that an idea has been culled from among a number of others, and has acquired particular vividness in our consciousness, we are in the habit of regarding this result as a proof that the victorious idea is endowed with a peculiarly high degree of psychic value—a certain degree of interest. We now discover that this value of the individual elements in the dream thoughts is not preserved in the formation of the dream, or does not come into consideration. For there is no doubt as to the elements of the dream thoughts which are of the highest value; our judgment tells us immediately. In the formation of dreams those elements which are emphasized with intense interest may be treated as though they were inferior, and other elements are put in their place which certainly were inferior in the dream thoughts. We are at first given the impression that the psychic intensity[10] of the individual ideas does not come into consideration at all for the selection made by the dream, but only their greater or smaller multiplicity of determination. Not what is important in the dream thoughts gets into the dream, but what is contained in them several times over, one might be inclined to think; but our understanding of the formation of dreams is not much furthered by this assumption, for at the outset it will be impossible to believe that the two factors of manifold determination and of integral value do not tend in the same direction in the influence they exert on the selection made by the dream. Those ideas in the dream thoughts which are most important are probably also those which recur most frequently, for the individual dream thoughts radiate from them as from central points. And still the dream may reject those elements which are especially emphasized and which receive manifold support, and may take up into its content elements which are endowed only with the latter property.

This difficulty may be solved by considering another impression received in the investigation of the manifold determination of the dream content. Perhaps many a reader has already passed his own judgment upon this investigation by saying that the manifold determination of the elements of the dream is not a significant discovery, because it is a self-evident one. In the analysis one starts from the dream elements, and registers all the notions which are connected with them; it is no wonder, then, that these elements should occur with particular frequency in the thought material which is obtained in this manner. I cannot acknowledge the validity of this objection, but shall say something myself which sounds like it. Among the thoughts which analysis brings to light, many can be found which are far removed from the central idea of the dream, and which appear distinguished from the rest as artificial interpolations for a definite purpose. Their purpose may easily be discovered; they are just the ones which establish a connection, often a forced and far-fetched one, between the dream content and the dream thoughts, and if these elements were to be weeded out, not only over-determination but also a sufficient determination by means of the dream thoughts would often be lacking for the dream content. We are thus led to the conclusion that manifold determination, which decides the selection made by the dream, is perhaps not always a primary factor in dream formation, but is often the secondary manifestation of a psychic power which is still unknown to us. But in spite of all this, manifold determination must nevertheless control the entrance of individual elements into the dream, for it is possible to observe that it is established with considerable effort in cases where it does not result from the dream material without assistance.

The assumption is not now far distant that a psychic force is expressed in dream activity which on the one hand strips elements of high psychic value of their intensity, and which on the other hand creates new values, *by way of over-determination,* from elements of small value, these new values subsequently getting into the dream content. If this is the method of procedure, there has taken place in the formation of the dream a transference and displacement of the psychic intensities of the individual elements, of which the textual difference between the dream and the thought content appears as a result. The process which we assume here is nothing less than the essential part of the dream activity; it merits the designation of *dream displacement. Dream displacement* and *dream condensation* are the two craftsmen to whom we may chiefly attribute the moulding of the dream.

I think we also have an easy task in recognising the psychic force which makes itself felt in the circumstances of dream displacement. The result of this displacement is that the dream content no longer resembles the core of the dream thoughts at all, and that the dream reproduces only a disfigured form of the dream-wish in the unconscious. But we are already acquainted with dream disfigurement; we have traced it back to the censorship which one psychic instance in the psychic life exercises upon the other. Dream displacement is one of the chief means for achieving this disfigurement. *Is fecit, cui profuit.* We may assume that dream displacement is brought about by the influence of this censor, of the endopsychic repulsion.[11]

The manner in which the factors of displacement, condensation, and over-determination play into one another in the formation of the dream, which is the ruling factor and which the subordinate one, all this will be reserved as the subject of later investigations. For the present we may state, as a second condition which the elements must satisfy in order to get into the dream, *that they must be withdrawn from the censor of resistance.* From now on we shall take account of dream displacement as an unquestionable fact in the interpretation of dreams.

(C) Means of Representation in the Dream

Besides the two factors of dream condensation and dream displacement which we have found to be active in the transformation of the latent dream material into the manifest content, we shall come in the course of this investigation upon two other conditions which exercise an unquestionable influence upon the selection of the material which gets into the dream. Even at the risk of seeming to stop our progress, I should like to glance at the processes by which the interpretation of dreams is accomplished. I do not deny that I should succeed best in making them clear, and in showing that they are sufficiently reliable to insure them against attack, by taking a single dream as a paradigm and developing its interpretation, as I have done in Chapter II. in the dream of "Irma's Injection," and then putting together the dream thoughts which I have discovered, and reconstructing the formation of the dream from them—that is to say, by supplementing the analysis of dreams by a synthesis of them. I have accomplished this with several specimens for my own instruction; but I cannot undertake to do it here because I am prevented by considerations, which every right-minded person must approve of, relative to the psychic material necessary for such a demonstration. In the analysis of dreams these considerations present less difficulty, for an analysis may be incomplete and still retain its value even if it leads only a short way into the thought labyrinth of the dream. I do not see how a synthesis could be anything short of complete in order to be convincing. I could give a complete synthesis only of the dreams of such persons as are unknown to the reading public. Since, however, only neurotic patients furnish me with the means for doing this, this part of the description of the dream must be postponed until I can carry the psychological explanation of neuroses far enough—elsewhere—to be able to show their connection with the subject matter under consideration.[12]

From my attempts synthetically to construct dreams from the dream thoughts, I know that the material which is obtained from interpretation varies in value. For a part of it consists of the essential dream thoughts which would, therefore, completely replace the dream, and which would in themselves be sufficient for this replacement if there were no censor for the dream. The other part may be summed up under the term "collaterals"; taken as a whole they represent the means by which the real wish that arises from the dream thoughts is transformed into the dream-wish. A first part of these "collaterals" consists of allusions to the actual dream thoughts, which, considered schematically, correspond to displacements from the essential to the non-essential. A second part comprises the thoughts which connect these non-essential elements, that have become significant through displacement with one another, and which reach from them into the dream content. Finally a third part contains the ideas and thought connections which (in the work of interpretation) conduct us from the dream content to the intermediary collaterals, *all of which* need not *necessarily* have participated in the formation of the dream.

At this point we are interested exclusively in the essential dream thoughts. These are usually found to be a complex of thoughts and memories of the most intricate possible construction, and to possess all the properties of the thought processes which are known to us from waking life. Not infrequently they are trains of thought which proceed from more than one centre, but which do not lack points of connection; almost regularly a chain of thought stands next to its contradictory correlative, being connected with it by contrast associations.

The individual parts of this complicated structure naturally stand in the most manifold logical relations to one another. They constitute a foreground or background, digressions, illustrations, conditions, chains of argument, and objections. When the whole mass of these dream thoughts is subjected to the pressure of the dream activity, during which the parts are turned about, broken up, and pushed together, something like drifting ice, there arises the question, what becomes of the logical ties which until now had given form to the structure? What representation do "if," "because," "as though," "although," "either—or," and all the other conjunctions, without which we cannot understand a phrase or a sentence, receive in the dream?

At first we must answer that the dream has at its disposal no means for representing these logical relations among the dream thoughts. In most cases it disregards all these conjunctions, and undertakes the elaboration only of the objective content of the dream thoughts. It is left to the interpretation of the dream to restore the coherence which the activity of the dream has destroyed.

If the dream lacks ability to express these relations, the psychic material of which the dream is wrought must be responsible. The descriptive arts are limited in the same manner—painting and the plastic arts in comparison with poetry, which can employ speech; and here too the reason for this impotence is to be found in the material in the treatment of which the two arts strive to give expression to something. Before the art of painting had arrived at an understanding of the laws of expression by which it is bound, it attempted to escape this disadvantage. In old paintings little tags were hung from the mouths of the persons represented giving the speech, the expression of which in the picture the artist despaired of.

Perhaps an objection will here be raised challenging the assertion that the dream dispenses with the representation of logical relations. There are dreams in which the most complicated intellectual operations take place, in which proof and refutation are offered, puns and comparisons made, just as in waking thoughts. But here, too, appearances are deceitful; if the interpretation of such dreams is pursued, it is found that all of this is *dream material, not the representation of intellectual activity in the dream.* The *content* of the dream thoughts is reproduced by the apparent thinking of the dream, not *the relations of the dream thoughts to one another,* in the determination of which relations thinking consists. I shall give examples of this. But the thesis which is most easily established is that all speeches which occur in the dream, and which are expressly designated as such, are unchanged or only slightly modified copies of speeches which are likewise to be found in the recollections of the dream material. Often the speech is only an allusion to an event contained in the dream thoughts; the meaning of the dream is a quite different one.

I shall not deny, indeed, that there is also critical thought activity which does not merely repeat material from the dream thoughts and which takes part in the formation of the dream. I shall have to explain the influence of this factor at the close of this discussion. It will then become clear that this thought activity is evoked not by the dream thoughts, but by the dream itself after it is already finished in a certain sense.

We shall, therefore, consider it settled for the present that the logical relations among the dream thoughts do not enjoy any particular representation in the dream. For instance, where there is a contradiction in the dream, this is either a contradiction directed against the dream itself or a contradiction derived from the content of one of the dream thoughts; a contradiction in the dream corresponds to a contradiction *among* the dream thoughts only in a highly indirect manner.

But just as the art of painting finally succeeded in depicting in the represented persons, at least their intention in speaking—their tenderness, threatening attitude, warning mien, and the like—by other means than the dangling tag, Bo also the dream has found it possible to render account of a few of the logical relations among its dream thoughts by means of an appropriate modification of the peculiar method of dream representation. It will be found by experience that different dreams go to different lengths in taking this into consideration; while one dream entirely disregards the logical coherence of its material, another attempts to indicate it as completely as possible. In so doing the dream departs more or less widely from the subject-matter which it is to elaborate. The dream also takes a similarly varying attitude towards the temporal coherence of the dream thoughts, if such coherence has been established in the unconscious (as for example in the dream of Irma's injection).

But what are the means by which the dream activity is enabled to indicate these relations in the dream material which are so difficult to represent? I shall attempt to enumerate these separately.

In the first place, the dream renders account of the connection which is undeniably present between all the parts of the dream thoughts by uniting this material in a single composition as a situation or process. It reproduces *logical connection in the form of simultaneousness;* in this case it acts something like the painter who groups together all the philosophers or poets into a picture of the school of Athens or of Parnassus, although these were never at once present in any hall or on any mountain top—though they do, however, form a unity from the point of view of reflective contemplation.

The dream carries out this method of representation in detail. Whenever it shows two elements close together, it vouches for a particularly intimate connection between those elements which correspond to them in the dream thoughts. It is as in our method of writing: *to* signifies that the two letters are to be pronounced as one syllable, while *t* with *o* after a free space shows that *t* is the last letter of one word and *o* the first letter of another. According to this, dream combinations are not made of arbitrary, completely incongruent elements of the dream material, but of elements that also have a somewhat intimate relation to one another in the dream thoughts.

For representing causal relation the dream has two methods, which are essentially reducible to one. The more frequent method, in cases, for example, where the dream thoughts are to the effect: "Because this was so and so, this and that must happen," consists in making the premise an introductory dream and joining the conclusion to it in the form of the main dream. If my interpretation is correct, the sequence may also be reversed. That part of the dream which is more completely worked out always corresponds to the conclusion.

A female patient, whose dream I shall later give in full, once furnished me with a neat example of such a representation of causal relationship. The dream consisted of a short prologue and of a very elaborate but well organised dream composition, which might be entitled: "A flower of speech." The prologue of the dream is as follows: *She goes to the two maids in the kitchen and scolds them for taking so long to prepare "a little bite of food." She also sees a great many coarse dishes standing in the kitchen, inverted so that the water may drop off them, and heaped up in a pile. The two maids go to fetch water, and must, as it were, step into a river, which reaches up to the house or into the yard.*

Then follows the main dream, which begins as follows: *She is descending from a high place, over balustrades that are curiously fashioned, and she is glad that her dress doesn't get caught anywhere,* &c. Now the introductory dream refers to the house of the lady's parents. Probably she has often heard from her mother the words which are spoken in the kitchen. The piles of unwashed dishes are taken from an unpretentious earthenware shop which was located in the same house. The second part of this dream contains an allusion to the dreamer's father, who always had a great deal to do with servant girls, and who later contracted a fatal disease during a flood—the house stood near the bank of a river. The thought which is concealed behind the introductory dream, then, is to this effect: "Because I was born in this house, under such limited and unlovely circumstances." The main dream takes up the same thought, and presents it in a form that has been altered by the tendency to wish-fulfilment: "I am of exalted origin." Properly then: "Because I was born in such low circumstances, my career has been so and so."

As far as I can see, the partition of a dream into two unequal portions does not always signify a causal relation between the thoughts of the two portions. It often appears as though the same material were being presented in the two dreams from different points of view; or as though the two dreams have proceeded from two separated centres in the dream material and their contents overlap, so that the object which is the centre of one dream has served in the other as an allusion, and *vice versa*. But in a certain number of cases a division into shorter fore-dreams and longer subsequent dreams actually signifies a causal relation between the two portions. The other method of representing causal relation is used with less abundant material and consists in the change of one image in the dream, whether a person or a thing, into another. It is only in cases where we witness this change taking place in the dream that any causal relation is asserted to exist, not where we merely notice that one thing has taken the place of another. I said that both methods of representing causal relation are reducible to the same thing; in both cases *causation* is represented by a *succession,* now by the sequence of the dreams, now by the immediate transformation of one image into another. In the great majority of cases, of course, causal relation is not expressed at all, but is obliterated by the sequence of elements which is unavoidable in the dream process.

The dream is altogether unable to express the alternative, "either—or"; it is in the habit of taking both members of this alternative into one context, as though they were equally privileged. A classic example of this is contained in the dream of Irma's injection. Its latent thoughts obviously mean: I am innocent of the continued presence of Irma's pains; the fault rests either with her resistance to accepting the solution, *or* with the fact that she is living under unfavourable sexual conditions, which I am unable to change, *or* her pains are not of a hysteric nature at all, but organic. The dream, however, fulfils all these possibilities, which are almost exclusive, and is quite ready to extract from the dream-wish an additional fourth solution of this kind. After interpreting the dream I have therefore inserted the *either-or* in the sequence of the dream thoughts.

In the case where the dreamer finds occasion in telling the dream to use *either-or:* "It was either a garden or a living-room," &c., it is not really an alternative which occurs in the dream thoughts, but an "and," a simple addition. When we use *either-or* we are usually describing a characteristic of indistinctness belonging to an element of the dream which is still capable of being cleared up. The rule of interpretation for this case is as follows: The separate members of the alternative are to be treated as equals and connected by "and." For instance, after waiting for a long time in vain for the address of my friend who is living in Italy, I dream that I receive a telegram which tells me this address. Upon the strip of telegraph paper I see printed in blue the following; the first word is blurred:

perhaps *via,*
or *villa,*   the second is distinctly: *Sezerno*
or perhaps (*Casa*).

The second word, which sounds like an Italian name and which reminds me of our etymological discussions, also expresses my displeasure on account of the fact that my friend has kept his place of residence secret from me for so long a time; every member of the triple suggestion for the first word may be recognised in the course of analysis as a self-sufficient and equally well-justified starting point in the concatenation of ideas.

During the night before the funeral of my father I dreamed of a printed placard, a card or poster—perhaps something like signs in railway waiting-rooms which announce the prohibition of smoking—which reads either:

It is requested to shut the eyes

or

It is requested to shut an eye

which I am in the habit of representing in the following form:

It is requested to shut the/ an eye (s).

Each of the two variations has its own particular meaning, and leads us along particular paths in the interpretation of the dream. I had made the simplest kind of funeral arrangements, for I knew how the deceased thought about such matters. Other members of the family, however, did not approve of such puritanic simplicity; they thought we would have to be ashamed before the mourners. Hence one of the wordings of the dream requests the "shutting of one eye," that is to say, that people should show consideration. The significance of the blurring, which we describe with an *either-or,* may here be seen with particular ease. The dream activity has not succeeded in constructing a unified but at the same time ambiguous wording for the dream thoughts. Thus the two main trains of thought are already distinguished even in the dream content.

In a few cases the division of the dream into two equal parts expresses the alternative which the dream finds it so difficult to represent.

The attitude of the dream towards the category of antithesis and contradiction is most striking. This category is unceremoniously neglected; the word "No" does not seem to exist for the dream. Antitheses are with peculiar preference reduced to unity or represented as one. The dream also takes the liberty of representing any element whatever by its desired opposite, so that it is at first impossible to tell about any element capable of having an opposite, whether it is to be taken negatively or positively, in the dream thoughts.[13] In one of the last-mentioned dreams, whose introductory portion we have already interpreted ("because my parentage is such"), the dreamer descends over a balustrade and holds a blossoming twig in her hands. Since this picture suggests to her the angel in paintings of the Annunciation (her own name is Mary) carrying a lily stem in his hand, and the white-robed girls marching in the procession on Corpus Christi Day when the streets are decorated with green bows, the blossoming twig in the dream is very certainly an allusion to sexual innocence. But the twig is thickly studded with red blossoms, each one of which resembles a camelia. At the end of her walk, so the dream continues, the blossoms have already fallen considerably apart; then unmistakable allusions to menstruation follow. But this very twig which is carried like a lily and as though by an innocent girl, is also an allusion to Camille, who, as is known, always wore a white camelia, but a red one at the time of her menstruation. The same blossoming twig ("the flower of maidenhood" in the songs about the miller's daughter by Goethe) represents at once sexual innocence and its opposite. The same dream, also, which expresses the dreamer's joy at having succeeded in passing through life unsullied, hints in several places (as at the falling-off of the blossom), at the opposite train of thought—namely, that she had been guilty of various sins against sexual purity (that is in her childhood). In the analysis of the dream we may clearly distinguish the two trains of thought, of which the comforting one seems to be superficial, the reproachful one more profound. The two are diametrically opposed to each other, and their like but contrasting elements have been represented by the identical dream elements.

The mechanism of dream formation is favourable in the highest degree to only one of the logical relations. This relation is that of similarity, correspondence, contiguity, "as though," which is capable of being represented in the dream as no other can be, by the most varied expedients. The correspondences occurring in the dream, or cases of "as though," are the chief points of support for the formation of dreams, and no inconsiderable part of the dream activity consists in creating new correspondences of this sort in cases where those which are already at hand are prevented by the censor of resistance from getting into the dream. The effort towards condensation shown by the dream activity assists in the representation of the relation of similarity.

Similarity, agreement, community, are quite generally expressed in the dream by concentration into a *unity,* which is either already found in the dream material or is newly created. The first case may be referred to as *identification,* the second as *composition.* Identification is used where the dream is concerned with persons, composition where things are the objects of unification; but compositions are also made from persons. Localities are often treated as persons.

Identification consists in giving representation in the dream content to only one of a number of persons who are connected by some common feature, while the second or the other persons seem to be suppressed as far as the dream is concerned. This one representative person in the dream enters into all the relations and situations which belong to itself or to the persons who are covered by it. In cases of composition, however, when this has to do with persons, there are already present in the dream image features which are characteristic of, but not common to, the persons in question, so that a new unity, a composite person, appears as the result of the union of these features. The composition itself may be brought about in various ways. Either the dream person bears the name of one of the persons to whom it refers—and then we know, in a manner which is quite analogous to knowledge in waking life, that this or that person is the one who is meant—while the visual features belong to another person; or the dream image itself is composed of visual features which in reality are shared by both. Instead of visual features, also, the part played by the second person may be represented by the mannerisms which are usually ascribed to him, the words which he usually speaks, or the situations in which he is usually imagined. In the latter method of characterisation the sharp distinction between identification and composition of persons begins to disappear. But it may also happen that the formation of such a mixed personality is unsuccessful. The situation of the dream is then attributed to one person, and the other—as a rule the more important one—is introduced as an inactive and unconcerned spectator. The dreamer relates something like "My mother was also there" (Stekel).

The common feature which justifies the union of the two persons—that is to say, which is the occasion for it—may either be represented in the dream or be absent. As a rule, identification or composition of persons simply serves the purpose of dispensing with the representation of this common feature. Instead of repeating: "A is ill disposed towards me, and B is also," I make a composite person of A and B in the dream, or I conceive A as doing an unaccustomed action which usually characterises B. The dream person obtained in this way appears in the dream in some new connection, and the fact that it signifies both A and B justifies me in inserting that which is common to both—their hostility towards me—at the proper place in the interpretation of the dream. In this manner I often achieve a very extraordinary degree of condensation of the dream content; I can save myself the direct representation of very complicated relations belonging to a person, if I can find a second person who has an equal claim to a part of these relations. It is also obvious to what extent this representation by means of identification can circumvent the resisting censor, which makes the dream activity conform to such harsh conditions. That which offends the censor may lie in those very ideas which are connected in the dream material with the one person; I now find a second person, who likewise has relation to the objectionable material, but only to a part of it. The contact in that one point which offends the censor now justified me in forming a composite person, which is characterised on either hand by indifferent features. This person resulting from composition or identification, who is unobjectionable to the censor, is now suited for incorporation in the dream content, and by the application of dream condensation I have satisfied the demands of the dream censor.

In dreams where a common feature of two persons is represented, this is usually a hint to look for another concealed common feature, the representation of which is made impossible by the censor. A displacement of the common feature has here taken place partly in order to facilitate representation. From the circumstance that the composite person appears to me with an indifferent common feature, I must infer that another common feature which is by no means indifferent exists in the dream thoughts.

According to what has been said, identification or composition of persons serves various purposes in the dream; in the first place, to represent a feature common to the two persons; secondly, to represent a displaced common feature; and thirdly, even to give expression to a community of features that is merely *wished for.* As the wish for a community between two persons frequently coincides with the exchanging of these persons, this relation in the dream is also expressed through identification. In the dream of Irma's injection I wish to exchange this patient for another—that is to say, I wish the latter to be my patient as the former has been; the dream takes account of this wish by showing me a person who is called Irma, but who is examined in a position such as I have had the opportunity of seeing only when occupied with the other person in question. In the dream about my uncle this substitution is made the centre of the dream; I identify myself with the minister by judging and treating my colleague as shabbily as he does.

It has been my experience—and to this I have found no exception—that every dream treats of one's own person. Dreams are absolutely egotistic. In cases where not my ego, but only a strange person occurs in the dream content, I may safely assume that my ego is concealed behind that person by means of identification. I am permitted to supplement my ego. On other occasions when my ego appears in the dream, I am given to understand by the situation in which it is placed that another person is concealing himself behind the ego. In this case the dream is intended to give me notice that in the interpretation I must transfer something which is connected with this person—the hidden common feature—to myself. There are also dreams in which my ego occurs along with other persons which the resolution of the identification again shows to be my ego. By means of this identification I am instructed to unite in my ego certain ideas to whose acceptance the censor has objected. I may also give my ego manifold representation in the dream, now directly, now by means of identification with strangers. An extraordinary amount of thought material may be condensed by means of a few such identifications.[14]

The resolution of the identification of localities designated under their own names is even less difficult than that of persons, because here the disturbing influence of the ego, which is all-powerful in the dream, is lacking. In one of my dreams about Rome (p. 181) the name of the place in which I find myself is Rome; I am surprised, however, at the great number of German placards at a street corner. The latter is a wish-fulfilment, which immediately suggests Prague; the wish itself probably originated at a period in my youth when I was imbued with a German nationalistic spirit which is suppressed to-day. At the time of my dream I was looking forward to meeting a friend in Prague; the identification of Rome and Prague is thus to be explained by means of a desired common feature; I would rather meet my friend in Rome than in Prague, I should like to exchange Prague for Rome for the purpose of this meeting.

The possibility of creating compositions is one of the chief causes of the phantastic character so common in dreams, in that it introduces into the dream elements which could never have been the objects of perception. The psychic process which occurs in the formation of compositions is obviously the same which we employ in conceiving or fashioning a centaur or a dragon in waking life. The only difference is that in the phantastic creations occurring in waking life the intended impression to be made by the new creation is itself the deciding factor, while the composition of the dream is determined by an influence—the common feature in the dream thoughts—which is independent of the form of the image. The composition of the dream may be accomplished in a great many different ways. In the most artless method of execution the properties of the one thing are represented, and this representation is accompanied by the knowledge that they also belong to another object. A more careful technique unites the features of one object with those of the other in a new image, while it makes skilful use of resemblance between the two objects which exist in reality. The new creation may turn out altogether absurd or only phantastically ingenious, according to the subject-matter and the wit operative in the work of composition. If the objects to be condensed into a unity are too incongruous, the dream activity is content with creating a composition with a comparatively distinct nucleus, to which are attached less distinct modifications. The unification into one image has here been unsuccessful, as it were; the two representations overlap and give rise to something like a contest between visual images. If attempt were made to construct an idea out of individual images of perception, similar representations might be obtained in a drawing.

Dreams naturally abound in such compositions; several examples of these I have given in the dreams already analysed; I shall add more. In the dream on p. 321, which describes the career of my patient "in flowery language," the dream ego carries a blossoming twig in her hand, which, as we have seen, signifies at once innocence and sexual transgression. Moreover, the twig recalls cherry-blossoms on account of the manner in which the blossoms are clustered; the blossoms themselves, separately considered, are camelias, and finally the whole thing also gives the impression of an *exotic* plant. The common feature in the elements of this composition is shown by the dream thoughts. The blossoming twig is made up of allusions to presents by which she was induced or should have been induced to show herself agreeable. So it was with the cherries in her childhood and with the stem of camelias in her later years; the exotic feature is an illusion to a much-travelled naturalist, who sought to win her favour by means of a drawing of a flower. Another female patient creates a middle element out of bath-houses at a bathing resort, rural outside water-closets, and the garrets of our city dwellings. The reference to human nakedness and exposure is common to the two first elements; and we may infer from their connection with the third element that (in her childhood) the garret was likewise the scene of exposure. A dreamer of the male sex makes a composite locality out of two places in which "treatment" is given—my office and the public hall in which he first became acquainted with his wife. Another female patient, after her elder brother has promised to regale her with caviare, dreams that his legs are covered thick with black caviare pearls. The two elements, "contagion" in a moral sense and the recollection of a cutaneous eruption in childhood which made her legs look as though studded over with red dots instead of black ones, have here been united with the caviare pearls to form a new idea—the idea of "what she has inherited from her brother." In this dream parts of the human body are treated as objects, as is usually the case in dreams. In one of the dreams reported by Ferenczi there occurred a composition made up of the person of a physician and a horse, over which was spread a nightshirt. The common feature in these three components was shown in the analysis after the nightshirt had been recognised as an allusion to the father of the dreamer in an infantile scene. In each of the three cases there was some object of her sexual inquisitiveness. As a child she had often been taken by her nurse to the military breeding station, where she had the amplest opportunity to satisfy her curiosity, which was at that time uninhibited.

I have already asserted that the dream has no means for expressing the relation of contradiction, of contrast, of negation. I am about to contradict this assertion for the first time. A part of the cases, which may be summed up under the word "contrast," finds representation, as we have seen, simply by means of identification—that is, when an interchange or replacement can be connected with the contrast. We have given repeated examples of this. Another part of the contrasts in the dream thoughts, which perhaps falls into the category "turned into the opposite," is represented in the dream in the following remarkable manner, which may almost be designated as witty. The *"inversion"* does not itself get into the dream content, but manifests its presence there by means of the fact that a part of the already formed dream content which lies at hand for other reasons, is—as it were subsequently— inverted. It is easier to illustrate this process than to describe it. In the beautiful "Up and Down" dream (p. 296) the representation of ascending is an inversion of a prototype in the dream thoughts, that is to say, of the introductory scene of Daudet's *Sappho;* in the dream climbing is difficult at first, and easy later on, while in the actual scene it is easy at first, and later becomes more and more difficult. Likewise "above" and "below" in relation to the dreamer's brother are inverted in the dream. This points to a relation of contraries or contrasts as obtaining between two parts of the subject-matter of the dream thoughts and the relation we have found in the fact that in the childish fancy of the dreamer he is carried by his nurse, while in the novel, on the contrary, the hero carries his beloved. My dream about Goethe's attack upon Mr. M. (p. 372) also contains an "inversion" of this sort, which must first be set right before the interpretation of the dream can be accomplished. In the dream Goethe attacks a young man, Mr. M.; in reality, according to the dream thoughts, an eminent man, my friend, has been attacked by an unknown young author. In the dream I reckon time from the date of Goethe's death; in reality the reckoning was made from the year in which the paralytic was born. The thought determining the dream material is shown to be an objection to the treatment of Goethe as a lunatic. "The other way around," says the dream; "if you cannot understand the book, it is you who are dull-witted, not the author." Furthermore, all these dreams of inversion seem to contain a reference to the contemptuous phrase, "to turn one's back upon a person" (German: "einen die Kehrseite zeigen"; *cf.* the inversion in respect to the dreamer's brother in the *Sappho* dream). It is also remarkable how frequently inversion becomes necessary in dreams which are inspired by repressed homosexual feelings.

Moreover, inversion or transformation into an opposite is one of the favourite methods of representation, and one of the methods most capable of varied application which the dream activity possesses. Its first function is to create the fulfilment of a wish with reference to a definite element of the dream-thoughts. "If it were only just the other way!" is often the best expression of the relation of the ego to a disagreeable recollection. But inversion becomes extraordinarily useful for the purposes of the censor, for it brings about in the material represented a degree of disfiguration which all but paralyses our understanding of the dream. For this reason it is always permissible, in cases where the dream stubbornly refuses to yield its meaning, to try the inversion of definite portions of its manifest content, whereupon not infrequently everything becomes clear.

Besides this inversion, the subject-matter inversion in temporal relation is not to be overlooked. A frequent device of dream disfigurement consists in presenting the final issue of an occurrence or the conclusion of an argument at the beginning of the dream, or in supplying the premises of a conclusion or the causes of an effect at the end of it. Anyone who has not considered this technical method of dream disfigurement stands helpless before the problem of dream interpretation.[15]

Indeed in some cases we can obtain the sense of the dream only by subjecting the dream content to manifold inversion in different directions. For example, in the dream of a young patient suffering from a compulsion neurosis, the memory of an infantile death-wish against a dreaded father was hidden behind the following words: *His father upbraids him because he arrives so late.* But the context in the psychoanalytic treatment and the thoughts of the dreamer alike go to show that the sentence must read as follows: *He is angry at his father,* and, further, that his father is always coming home *too early* (*i.e.* too soon). He would have preferred that his father should not come home at all, which is identical with the wish (see page 232) that his father should die. As a little boy the dreamer was guilty of sexual aggression against another person while his father was away, and he was threatened with punishment in the words: "Just wait until father comes home."

If we attempt to trace the relations between dream content and dream thoughts further, we shall do this best by making the dream itself our starting-point and by asking ourselves the question: What do certain formal characteristics of dream representation signify with reference to the dream thoughts? The formal characteristics which must attract our attention in the dream primarily include variations in the distinctness of individual parts of the dream or of whole dreams in relation to one another. The variations in the intensity of individual dream images include a whole scale of degrees ranging from a distinctness of depiction which one is inclined to rate as higher—without warrant, to be sure—than that of reality, to a provoking indistinctness which is declared to be characteristic of the dream, because it cannot altogether be compared to any degree of indistinctness which we ever see in real objects. Moreover, we usually designate the impression which we get from an indistinct object in the dream as "fleeting," while we think of the more distinct dream images as remaining intact for a longer period of perception. We must now ask ourselves by what conditions in the dream material these differences in the vividness of the different parts of the dream content are brought about.

There are certain expectations which will inevitably arise at this point and which must be met. Owing to the fact that real sensations during sleep may form part of the material of the dream, it will probably be assumed that these sensations or the dream elements resulting from them are emphasized by peculiar intensity, or conversely, that what turns out to be particularly vivid in the dream is probably traceable to such real sensations during sleep. My experience has never confirmed this. It is incorrect to say that those elements of the dream which are the derivatives of impressions occurring in sleep (nervous excitements) are distinguished by their vividness from others which are based on recollections. The factor of reality is of no account in determining the intensity of dream images.

Furthermore, the expectation will be cherished that the sensory intensity (vividness) of individual dream images has a relation to the psychic intensity of the elements corresponding to them in the dream-thoughts. In the latter intensity is identical with psychic value; the most intense elements are in fact the most significant, and these are the central point of the dream. We know, however, that it is just these elements which are usually not accepted in the dream content owing to the censor. But still it might be possible that the elements immediately following these and representing them might show a higher degree of intensity, without, however, for that reason constituting the centre of the dream representation. This expectation is also destroyed by a comparison of the dream and the dream material. The intensity of the elements in the one has nothing to do with the intensity of the elements in the other; a complete "transvaluation of all psychic values" takes place between the dream-material and the dream. The very element which is transient and hazy and which is pushed into the background by more vigorous images is often the single and only element in which may be traced any direct derivative from the subject which entirely dominated the dream-thoughts.

The intensity of the elements of the dream shows itself to be determined in a different manner—that is, by two factors which are independent of each other. It is easy to see at the outset that those elements by means of which the wish-fulfilment is expressed are most distinctly represented. But then analysis also teaches us that from the most vivid elements of the dream, the greatest number of trains of thought start, and that the most vivid are at the same time those which are best determined. No change of sense is involved if we express the latter empirical thesis in the following form: the greatest intensity is shown by those elements of the dream for which the most abundant condensation activity was required. We may therefore expect that this condition and the others imposed by the wish-fulfilment can be expressed in a single formula.

The problem which I have just been considering—the causes of greater or less intensity or distinctness of individual elements of the dream—is one which I should like to guard against being confused with another problem, which has to do with the varying distinctness of whole dreams or sections of dreams. In the first case, the opposite of distinctness is blurredness; in the second, confusion. It is of course unmistakable that the intensities rise and fall in the two scales in unison. A portion of the dream which seems clear to us usually contains vivid elements; an obscure dream is composed of less intense elements. But the problem with which we are confronted by the scale, ranging from the apparently clear to the indistinct or confused, is far more complicated than that formed by variations in the vividness of the dream elements; indeed the former will be dropped from the discussion for reasons which will be given later. In isolated cases we are astonished to find that the impression of clearness or indistinctness produced by the dream is altogether without significance for its structure, and that it originates in the dream material as one of its constituents. Thus I remember a dream which seemed particularly well constructed, flawless, and clear, so that I made up my mind, while I was still in the somnolent state, to recognise a new class of dreams—those which had not been subject to the mechanism of condensation and displacement, and which might thus be designated "Fancies while asleep." A closer examination proved that this rare dream had the same breaches and flaws in its construction as every other; for this reason I abandoned the category of dream fancies. The content of the dream, reduced to its lowest terms, was that I was reciting to a friend a difficult and long-sought theory of bisexuality, and the wish-fulfilling power of the dream was responsible for the fact that this theory (which, by the way, was not stated in the dream) appeared so clear and flawless. What I considered a judgment upon the finished dream was thus a part of the dream content, and the essential one at that. The dream activity had extended its operations, as it were, into waking thought, and had presented to me in the form of a judgment that part of the dream material which it had not succeeded in reproducing with exactness. The exact opposite of this once came to my attention in the case of a female patient who was at first altogether unwilling to tell a dream which was necessary for the analysis, "because it was so obscure and confused," and who declared, after repeatedly denying the accuracy of her description, that several persons, herself, her husband, and her father, had occurred in the dream, and that it seemed as though she did not know whether her husband was her father, or who her father was anyway, or something of that sort. Upon considering this dream in connection with the ideas that occurred to the dreamer in the course of the sitting, it was found unquestionably to be concerned with the story of a servant girl who had to confess that she was expecting a child, and who was now confronted with doubts as to "who was really the father."[16] The obscurity manifested by the dream, therefore, is again in this case a portion of the material which excited it. A part of this material was represented in the form of the dream. The form of the dream or of dreaming is used with astonishing frequency to represent the concealed content.

Comments on the dream and seemingly harmless observations about it often serve in the most subtle manner to conceal—although they usually betray—a part of what is dreamed. Thus, for example, when the dreamer says: *Here the dream is vague,* and the analysis gives an infantile reminiscence of listening to a person cleaning himself after defecation. Another example deserves to be recorded in detail. A young man has a very distinct dream which recalls to him phantasies from his infancy which have remained conscious to him: he was in a summer hotel one evening, he mistook the number of his room, and entered a room in which an elderly lady and her two daughters were undressing to go to bed. He continues: *"Then there are some gaps in the dream; then something is missing;* and at the end there was a man in the room who wished to throw me out with whom I had to wrestle." He endeavoured in vain to recall the content and purpose of the boyish fancy to which the dream apparently alludes. But we finally become aware that the required content had already been given in his utterances concerning the indistinct part of the dream. The "gaps" were the openings in the genitals of the women who were retiring: "Here something is missing" described the chief character of the female genitals. In those early years he burned with curiosity to see a female genital, and was still inclined to adhere to the infantile sexual theory which attributes a male genital to the woman.

All the dreams which have been dreamed in the same night belong to the same whole when considered with respect to their content; their separation into several portions, their grouping and number, all these details are full of meaning. and may be considered as information coming from the latent dream content. In the interpretation of dreams consisting of many principal sections, or of dreams belonging to the same night, one must not fail to think of the possibility that these different and succeeding dreams bring to expression the same feelings in different material. The one that comes first in time of these homologous dreams is usually the most disfigured and most bashful, while the succeeding is bolder and more distinct.

Even Pharaoh's dream in the Bible of the ears and the kine, which Joseph interpreted, was of this kind. It is reported by Josephus (*Antiquities of the Jews,* bk. ii. chap. iii.) in greater detail than in the Bible. After relating the first dream, the King said: "When I had seen this vision I awaked out of my sleep, and being in disorder, and considering with myself what this appearance should be, I fell asleep again, and saw another dream much more wonderful than the first, which did still more affright and disturb me." After listening to the report of the dream, Joseph said, "This dream, O King, although seen under two forms, signifies one and the same issue of things."

Jung, who, in his *Beitrag zur Psychologie des Gerüchtes* relates how the veiled erotic dream of a school-girl was understood by her friends without interpretation and continued by them with variations, remarks in connection with reports of this dream, "that the last of a long series of dream pictures contained precisely the same thought whose representation had been attempted in the first picture of the series. The censor pushed the complex out of the way as long as possible, through constantly renewed symbolic concealments, displacements, deviations into the harmless, &c." (*l.c.* p. 87). Scherner was well acquainted with the peculiarities of dream disfigurement and describes them at the end of his theory of organic stimulation as a special law, p. 166:

"But, finally, the phantasy observes the general law in all nerve stimuli emanating from symbolic dream formations, by representing at the beginning of the dream only the remotest and freest allusions to the stimulating object; but towards the end, when the power of representation becomes exhausted, it presents the stimulus or its concerned organ or its function in unconcealed form, and in the way this dream designates its organic motive and reaches its end."

A new confirmation of Scherner's law has been furnished by Otto Rank in his work, *A Self Interpretation Dream.* This dream of a girl reported by him consisted of two dreams, separated in time of the same night, the second of which ended with pollution. This pollution dream could be interpreted in all its details by disregarding a great many of the ideas contributed by the dreamer, and the profuse relations between the two dream contents indicated that the first dream expressed in bashful language the same thing as the second, so that the latter—the pollution dream—helped to a full explanation of the former. From this example, Rank, with perfect justice, draws conclusions concerning the significance of pollution dreams in general.

But in my experience it is only in rare cases that one is in a position to interpret clearness or confusion in the dream as certainty or doubt in the dream material. Later I shall try to discover the factor in the formation of dreams upon whose influence this scale of qualities essentially depends.

In some dreams, which adhere for a time to a certain situation and scenery, there occur interruptions described in the following words: "But then it seemed as though it were at the same time another place, and there such and such a thing happened." What thus interrupts the main trend of the dream, which after a while may be continued again, turns out to be a subordinate idea, an interpolated thought in the dream material. A conditional relation in the dream-thoughts is represented by simultaneousness in the dream (wenn—wann; if—when).

What is signified by the sensation of impeded movement, which so often occurs in the dream, and which is so closely allied to anxiety? One wants to move, and is unable to stir from the spot; or one wants to accomplish something, and meets one obstacle after another. The train is about to start, and one cannot reach it; one's hand is raised to avenge an insult, and its strength fails, &c. We have already encountered this sensation in exhibition dreams, but have as yet made no serious attempt to interpret it. It is convenient, but inadequate, to answer that there is motor paralysis in sleep, which manifests itself by means of the sensation alluded to. We may ask: "Why is it, then, that we do not dream continually of these impeded motions?" And we are justified in supposing that this sensation, constantly appearing in sleep, serves some purpose or other in representation, and is brought about by a need occurring in the dream material for this sort of representation.

Failure to accomplish does not always appear in the dream as a sensation, but also simply as a part of the dream content. I believe that a case of this sort is particularly well suited to enlighten us about the significance of this characteristic of the dream. I shall give an abridged report of a dream in which I seem to be accused of dishonesty. *The scene is a mixture, consisting of a private sanatorium and several other buildings. A lackey appears to call me to an examination. I know in the dream that something has been missed, and that the examination is taking place because I am suspected of having appropriated the lost article. Analysis shows that examination is to be taken in two senses, and also means medical examination. Being conscious of my innocence, and of the fact that I have been called in for consultation, I calmly follow the lackey. We are received at the door by another lackey, who says, pointing to me, "Is that the person whom you have brought? Why, he is a respectable man." Thereupon, without any lackey, I enter a great hall in which machines are standing, and which reminds me of an Inferno with its hellish modes of punishment. I see a colleague strapped on to one apparatus who has every reason to be concerned about me; but he takes no notice of me. Then I am given to understand that I may now go. Then I cannot find my hat, and cannot go after all.*

The wish which the dream fulfils is obviously that I may be acknowledged to be an honest man, and may go; all kinds of subject-matter containing a contradiction of this idea must therefore be present in the dream-thoughts. The fact that I may go is the sign of my absolution; if, then, the dream furnishes at its close an event which prevents me from going, we may readily conclude that the suppressed subject-matter of the contradiction asserts itself in this feature. The circumstance that I cannot find my hat therefore means: "You are not an honest man after all." Failure to accomplish in the dream is the expression of a contradiction, a "No"; and therefore the earlier assertion, to the effect that the dream is not capable of expressing a negation, must be revised accordingly.[17]

In other dreams which involve failure to accomplish a thing not only as a situation but also as a sensation, the same contradiction is more emphatically expressed in the form of a volition, to which a counter volition opposes itself. Thus the sensation of impeded motion represents a *conflict of will.* We shall hear later that this very motor paralysis belongs to the fundamental conditions of the psychic process in dreaming. Now the impulse which is transferred to motor channels is nothing else than the will, and the fact that we are sure to find this impulse impeded in the dream makes the whole process extraordinarily well suited to represent volition and the "No" which opposes itself thereto. From my explanation of anxiety, it is easy to understand why the sensation of thwarted will is so closely allied to anxiety, and why it is so often connected with it in the dream. Anxiety is a libidinous impulse which emanates from the unconscious, and is inhibited by the fore-conscious. Therefore, when a sensation of inhibition in the dream is accompanied by anxiety, there must also be present a volition which has at one tune been capable of arousing a *libido;* there must be a sexual impulse.

What significance and what psychic force is to be ascribed to such manifestations of judgment as "For that is only a dream," which frequently comes to the surface in dreams, I shall discuss in another place (*vide infra,* p. 417). For the present I shall merely say that they serve to depreciate the value of the thing dreamed. An interesting problem allied to this, namely, the meaning of the fact that sometimes a certain content is designated in the dream itself as "dreamed"—the riddle of the "dream within the dream"—has been solved in a similar sense by W. Stekel through the analysis of some convincing examples. The part of the dream "dreamed" is again to be depreciated in value and robbed of its reality; that which the dreamer continues to dream after awakening from the dream within the dream, is what the dream-wish desires to put in place of the extinguished reality. It may therefore be assumed that the part "dreamed" contains the representation of the reality and the real reminiscence, while, on the other hand, the continued dream contains the representation of what the dreamer wished. The inclusion of a certain content in a "dream within the dream" is therefore equivalent to the wish that what has just been designated as a dream should not have occurred. The dream-work utilises the dream itself as a form of deflection.

(D) Regard for Presentability

So far we have been attempting to ascertain how the dream represents the relations among the dream-thoughts, but we have several times extended our consideration to the further question of what alterations the dream material undergoes for the purposes of dream formation. We now know that the dream material, after being stripped of the greater parts of its relations, is subjected to compression, while at the same time displacements of intensity among its elements force a psychic revaluation of this material. The displacements which we have considered were shown to be substitutions of one idea for another, the substitute being in some way connected with the original by associations, and the displacements were put to the service of condensation by virtue of the fact that in this manner a common mean between two elements took the place of these two elements in the formation of the dream. We have not yet mentioned any other kind of displacement. But we learn from the analyses that another exists, and that it manifests itself in a change of the verbal expression employed for the thought in question. In both cases we have displacement following a chain of associations, but the same process takes place in different psychic spheres, and the result of this displacement in the one case is that one element is substituted for another, while in the other case an element exchanges its verbal expression for another.

This second kind of displacement occurring in dream formation not only possesses great theoretical interest, but is also peculiarly well fitted to explain the semblance of phantastic absurdity in which the dream disguises itself. Displacement usually occurs in such a way that a colourless and abstract expression in the dream-thought is exchanged for one that is visual and concrete. The advantage, and consequently the purpose, of this substitution is obvious. Whatever is visual is *capable of representation* in the dream, and can be wrought into situations where the abstract expression would confront dream representation with difficulties similar to those which would arise if a political editorial were to be represented in an illustrated journal. But not only the possibility of representation, but also the interests of condensation and of the censor, can be furthered by this change. If the abstractly expressed and unwieldy dream-thought is recast into figurative language, this new expression and the rest of the dream material are more easily furnished with those identities and cross references, which are essential to the dream activity and which it creates whenever they are not at hand, for the reason that in every language concrete terms, owing to their evolution, are more abundant in associations than conceptual ones. It may be imagined that in dream formation a good part of the intermediary activity, which tries to reduce the separate dream-thoughts to the tersest and simplest possible expression in the dream, takes place in the manner above described—that is to say, in providing suitable paraphrase for the individual thoughts. One thought whose expression has already been determined on other grounds will thus exert a separating and selective influence upon the means available for expressing the other, and perhaps it will do this constantly throughout, somewhat after the manner of the poet. If a poem in rhyme is to be composed, the second rhyming line is bound by two conditions; it must express the proper meaning, and it must express it in such a way as to secure the rhyme. The best poems are probably those in which the poet's effort to find a rhyme is unconscious, and in which both thoughts have from the beginning exercised a mutual influence in the selection of their verbal expressions, which can then be made to rhyme by a means of slight remodification.

In some cases change of expression serves the purposes of dream condensation more directly, in making possible the invention of a verbal construction which is ambiguous and therefore suited to the expression of more than one dream-thought. The whole range of word-play is thus put at the service of the dream activity. The part played by words y in the formation of dreams ought not to surprise us. A word being a point of junction for a number of conceptions, it possesses, so to speak, a predestined ambiguity, and neuroses (obsessions, phobias) take advantage of the conveniences which words offer for the purposes of condensation and disguise quite as readily as the dream.[18] That dream conception also profits by this displacement of expression is easily demonstrated. It is naturally confusing if an ambiguous word is put in the place of two ambiguous ones; and the employment of a figurative expression instead of the sober

300

everyday one thwarts our understanding, especially since the dream never tells us whether the elements which it shows are to be interpreted literally or figuratively, or whether they refer to the dream material directly or only through the agency of interpolated forms of speech.[19] Several examples of representations in the dream which are held together only by ambiguity have already been cited ("her mouth opens without difficulty," in the dream of Irma's injection; "I cannot go yet," in the last dream reported, p. 340), &c. I shall now cite a dream in the analysis of which the figurative expression of abstract thought plays a greater part. The difference between such dream interpretation and interpretation by symbolism may again be sharply distinguished; in the symbolic interpretation of dreams the key to the symbolism is arbitrarily chosen by the interpreter, while in our own cases of verbal disguise all these keys are universally known and are taken from established customs of speech. If the correct notion occurs at the right opportunity, it is possible to solve dreams of this sort completely or in part, independently of any statements made by the dreamer.

A lady, a friend of mine, dreams: *She is in the opera-house. It is a Wagnerian performance which has lasted till 7.45 in the morning. In the parquette and parterre there are tables, around which people dine and drink. Her cousin and his young wife, who have just returned from their honeymoon, sit next to her at one of these tables, and next to them sits one of the aristocracy. Concerning the latter the idea is that the young wife has brought him back with her from the wedding journey. It is quite above board, just as if she were bringing back a hat from her trip. In the midst of the parquette there is a high tower, on the top of which is a platform surrounded by an iron grating. There, high up, stands the conductor with the features of Hans Richter; he is continually running around behind the grating, perspiring awfully, and from this position conducting the orchestra, which is arranged around the base of the tower. She herself sits in a box with a lady friend (known to me). Her youngest sister tries to hand her from the parquette a big piece of coal with the idea that she did not know that it would last so long and that she must by this time be terribly cold. (It was a little as if the boxes had to be heated during the long performance.)*

The dream is senseless enough, though the situation is well developed too—the tower in the midst of the parquette from which the conductor leads the orchestra; but, above all, the coal which her sister hands her! I purposely asked for no analysis of this dream. With the knowledge I have of the personal relations of the dreamer, I was able to interpret parts of it independently. I knew that she had entertained warm feelings for a musician whose career had been prematurely blasted by insanity. I therefore decided to take the tower in the parquette verbally. It was apparent, then, that the man whom she wished to see in the place of Hans Richter *towered* above all the other members of the orchestra. This tower must, therefore, be designated as a composite picture formed by an apposition; with its pedestal it represents the greatness of the man, but with its gratings on top, behind which he runs around like a prisoner or an animal in a cage (an allusion to the name of the unfortunate man), it represents his later fate. "Lunatic-tower" is perhaps the word in which both thoughts might have met.

Now that we have discovered the dream's method of representation, we may try with the same key to open the second apparent absurdity,—that of the coal which her sister hands her. "Coal" must mean "secret love."

> *"No coal,* no *fire* so hotly glows
> As the *secret love* which no one knows."

She and her friend remain seated while her younger sister, who still has opportunities to marry, hands her up the coal "because she did not know it would last so long." What would last so long is not told in the dream. In relating it we would supply "the performance"; but in the dream we must take the sentence as it is, declare it ambiguous, and add "until she marries." The interpretation "secret love" is then confirmed by the mention of the cousin who sits with his wife in the parquette, and by the open love-affair attributed to the latter. The contrasts between secret and open love, between her fire and the coldness of the young wife, dominate the dream. Moreover, here again there is a person "in high position" as a middle term between the aristocrat and the musician entitled to high hopes.

By means of the above discussion we have at last brought to light a third factor, whose part in the transformation of the dream thoughts into the dream content is not to be considered trivial; *it is the regard for presentability (German: Darstellbarkeit) in the peculiar psychic material which the dream makes use of,*—that is fitness for representation, for the most part by means of visual images. Among the various subordinate ideas associated with the essential dream thoughts, that one will be preferred which permits of a visual representation, and the dream-activity does not hesitate promptly to recast the inflexible thought into another verbal form, even if it is the more unusual one, as long as this form makes dramatisation possible, and thus puts an end to the psychological distress caused by cramped thinking. This pouring of the thought content into another mould may at the same time be put at the service of the condensation work, and may establish relations with another thought which would otherwise not be present. This other thought itself may perhaps have previously changed its original expression for the purpose of meeting these relations half-way.

In view of the part played by puns, quotations, songs, and proverbs in the intellectual life of educated persons, it would be entirely in accordance with our expectation to find disguises of this sort used with extraordinary frequency. For a few kinds of material a universally applicable dream symbolism has been established on a basis of generally known allusions and equivalents. A good part of this symbolism, moreover, is possessed by the dream in common with the psychoneuroses, and with legends and popular customs.

Indeed, if we look more closely, we must recognise that in employing this method of substitution the dream is generally doing nothing original. For the attainment of its purpose, which in this case is the possibility of dramatisation without interference from the censor, it simply follows the paths which it finds already marked out in unconscious thought, and gives preference to those transformations of the suppressed material which may become conscious also in the form of wit and allusion, and with which all the fancies of neurotics are filled. Here all at once we come to understand Scherner's method of dream interpretation, the essential truth of which I have defended elsewhere. The occupation of one's fancy with one's own body is by no means peculiar to, or characteristic of the dream alone. My analyses have shown me that this is a regular occurrence in the unconscious thought of neurotics, and goes back to sexual curiosity, the object of which for the adolescent youth or maiden is found in the genitals of the opposite sex, or even of the same sex. But, as Scherner and Volkelt very appropriately declare, the house is not the only group of ideas which is used for the symbolisation of the body—either in the dream or in the unconscious fancies of the neurosis. I know some patients, to be sure, who have steadily adhered to an architectural symbolism for the body and the genitals (sexual interest certainly extends far beyond the region of the external genital organs), to whom posts and pillars signify legs (as in the "Song of Songs"), to whom every gate suggests a bodily opening ("hole"), and every water-main a urinary apparatus, and the like. But the group of associations belonging to plant life and to the kitchen is just as eagerly chosen to conceal sexual images; in the first case the usage of speech, the result of phantastic comparisons dating from the most ancient times, has made abundant preparation (the "vineyard" of the Lord, the "seeds," the "garden" of the girl in the "Song of Songs"). The ugliest as well as the most intimate details of sexual life may be dreamed about in apparently harmless allusions to culinary operations, and the symptoms of hysteria become practically unintelligible if we forget that sexual symbolism can conceal itself behind the most commonplace and most inconspicuous matters, as its best hiding-place. The fact that some neurotic children cannot look at blood and raw meat, that they vomit at the sight of eggs and noodles, and that the dread of snakes, which is natural to mankind, is monstrously exaggerated in neurotics, all of this has a definite sexual meaning. Wherever the neurosis employs a disguise of this sort, it treads the paths once trodden by the whole of humanity in the early ages of civilisation—paths of whose existence customs of speech, superstitions, and morals still give testimony to this day.

I here insert the promised flower dream of a lady patient, in which I have italicised everything which is to be sexually interpreted. This beautiful dream seemed to lose its entire charm for the dreamer after it had been interpreted.

(*a*) Preliminary dream: *She goes to the two maids in the kitchen and scolds them for taking so long to prepare "a little bite of food." She also sees a great many coarse dishes standing in the kitchen inverted so that the water may drip off them, and heaped up in a pile.* Later addition: *The two maids go to fetch water, and must, as it were, step into a river which reaches up into the house or into the yard.*[20]

(*b*) Main dream:[21] *She is descending from a high place*[22] *over balustrades that are curiously fashioned or fences which are united into big squares and consist of a conglomeration of little squares.*[23] *It is really not intended for climbing upon; she is worried about finding a place for her foot, and she is glad her dress doesn't get caught anywhere, and that she remains so respectable while she is going.*[24] *She is also carrying a large bough in her hand,*[25] *really a bough of a tree, which is thickly studded with red blossoms; it has many branches, and spreads out.*[26] *With this is connected the idea of cherry blossoms, but they look like full-bloom camelias, which of course do not grow on trees. While she is descending, she first has one, then suddenly two, and later again only one.*[27] *When she arrives at the bottom of the lower blossoms they have already fallen off to a considerable extent. Now that she is at the bottom, she sees a porter who is combing—as she would like to express it—just such a tree—that is, who is* *plucking thick bunches of hair from it, which hang from it like moss. Other workmen have chopped off such boughs in a garden, and have thrown them upon the street, where they lie about, so that many people take some of them. But she asks whether that is right, whether anybody may take one.*[28] In the garden there stands a young man (having a personality with which she is acquainted, not a member of her family) *up to whom she goes in order to ask him how it is possible to transplant such boughs into her own garden.*[29] He embraces her, whereat she resists and asks him what he means, whether it is permissible to embrace her in such a manner. He says that there is no wrong in it, that it is permitted.*[30] He then declares himself willing to go with her into the other garden, in order to show her the transplanting, and he says something to her which she does not correctly understand: "Besides this three metres—(later on she says: square metres) *or three fathoms of ground are lacking." It seems as though the man were trying to ask her something in return for his affability, as though he had the intention of indemnifying himself in her garden, as though he wanted to evade some law or other, to derive some advantage from it without causing her an injury. She does not know whether or not he really shows her anything.*[31]

I must mention still another series of associations which often serves the purpose of concealing sexual meaning both in dreams and in the neurosis,—I refer to the change of residence series. To change one's residence is readily replaced by "to remove," an ambiguous expression which may have reference to clothing. If the dream also contains a "lift" (elevator), one may think of the verb "to lift," hence of lifting up the clothing.

I have naturally an abundance of such material, but a report of it would carry us too far into the discussion of neurotic conditions. Everything leads to the same conclusion, that no special symbolising activity of the mind in the formation of dreams need be assumed; that, on the contrary, the dream makes use of such symbolisations as are to be found ready-made in unconscious thought, because these better satisfy the requirements of dream formation, on account of their dramatic fitness, and particularly on account of their exemption from the censor.

(E)   Examples—Arithmetic Speeches in the Dream

Before I proceed to assign to its proper place the fourth of the factors which control the formation of the dream, I shall cite several examples from my collection of dreams for the purpose partly of illustrating the co-operation of the three factors with which we are acquainted, and partly of supplying proof for assertions which have been made without demonstration or of drawing irrefutable inferences from them. For it has been very difficult for me in the foregoing account of the dream activity to demonstrate my conclusions by means of examples. Examples for the individual thesis are convincing only when considered in connection with a dream interpretation; when they are torn from their context they lose their significance, and, furthermore, a dream interpretation, though not at all profound, soon becomes so extensive that it obscures the thread of the discussion which it is intended to illustrate. This technical motive may excuse me for now mixing together all sorts of things which have nothing in common but their relation to the text of the foregoing chapter.

We shall first consider a few examples of very peculiar or unusual methods of representation in the dream. The dream of a lady is as follows: *A servant girl is standing on a ladder as though to clean the unndows, and has with her a chimpanzee and a gorilla cat* (later corrected—angora cat). *She throws the animals at the dreamer; the chimpanzee cuddles up to her, and this is disgusting to her.* This dream has accomplished its purpose by the simplest possible means, namely by taking a mere mode of speech literally and representing it according to the meaning of its words. "Ape," like the names of animals in general, is an epithet of opprobrium, and the situation of the dream means nothing but "to hurl invectives." This same collection will soon furnish us with further examples of the use of this simple artifice.

Another dream proceeds in a very similar manner: *A woman with a child that has a conspicuously deformed cranium; the dreamer has heard that the child got into this condition owing to its position in its mother's womb. The doctor says that the cranium might be given a better shape by means of compression, but that would harm the brain. She thinks that because it is a boy it won't suffer so much from deformity.* This dream contains a plastic-representation of the concept: *"Childish impressions,"* which the dreamer has heard of in the course of explanations concerning the treatment.

In the following example, the dream activity enters upon a different path. The dream contains a recollection of an excursion to the Hilmteich, near Graz: *There is a terrible storm outside; a miserable hotel—the water is dripping from the walls, and the beds are damp.* (The latter part of the content is less directly expressed than I give it.) The dream signifies "superfluous." The abstract idea occurring in the dream thoughts is first made equivocal by a certain straining of language; it has, perhaps, been replaced by "overflowing" or by "fluid" and "super-fluid (-fluous)" and has then been given representation by an accumulation of like impressions. Water within, water without, water in the beds in the form of dampness—everything fluid and "super" fluid. That, for the purposes of the dream representation, the spelling is much less regarded than the sound of words ought not surprise us when we remember that rhyme exercises similar privileges.

The fact that language has at its disposal a great number of words which were originally intended in a picturesque and concrete sense but are at present used in a faded abstract sense has in other cases made it very easy for the dream to represent its thoughts. The dream need only restore to these words their full significance, or follow the evolution of their meaning a little way back. For example, a man dreams that his friend, who is struggling to get out of a very tight place, calls upon him to help him. The analysis shows that the tight place is a hole, and that the dream uses symbolically his very words to his friend, "Be careful, or you'll get yourself into a hole."[32] Another dreamer climbs upon a mountain from which he sees a very extraordinary broad view. He identifies himself with his brother who is editing a "review" which deals with relations to the Farthest East.

It would be a separate undertaking to collect such methods of representation and to arrange them according to the principles upon which they are based. Some of the representations are quite witty. They give the impression that they would have never been divined if the dreamer himself had not reported them.

1. A man dreams that he is asked for a name, which, however, he cannot recall. He himself explains that this means: *It does not occur to me in the dream.*

2. A female patient relates a dream in which all the persons concerned were especially big. "That means," she adds, "that it must deal with an episode of my early childhood, for at that time all grown up people naturally seemed to me immensely big."

     The transference into childhood is also expressed differently in other dreams by translating time into space. One sees the persons and scenes in question as if at a great distance, at the end of a long road, or as if looked at through the wrong end of the opera-glass.

3. A man, who in waking life shows an inclination to abstract and indefinite expressions, but who is otherwise endowed with wit

enough, dreams in a certain connection that he is at a railroad station while a train is coming in. But then the station platform approaches the train, which stands still; hence an absurd inversion of the real state of affairs. This detail is again nothing but an index to remind one that something else in the dream should be turned about. The analysis of the same dream brings back the recollection of a picture-book in which men are represented standing on their heads and walking on their hands.

4. The same dreamer on another occasion relates a short dream which almost recalls the technique of a rebus. His uncle gives him a kiss in an automobile. He immediately adds the interpretation, which I should never have found: it means *Autoerotism*. This might have been made as a joke in the waking state.

The dream work often succeeds in representing very awkward material, such as proper names, by means of the forced utilisation of very far-fetched references. In one of my dreams the elder Bruecke *has given me a task. I compound a preparation, and skim something from it which looks like crumpled tinfoil.* (More of this later on.) The notion corresponding to this, which was not easy to find, is "stanniol," and now I know that I have in mind the name of the author Stannius, which was borne by a treatise on the nervous system of fishes, which I regarded with awe in my youthful years. The first scientific task which my teacher gave me was actually concerned with the nervous system of a fish—the *Ammocoetes*. Obviously the latter name could never have been used in a picture puzzle.

I shall not omit here to insert a dream having a curious content, which is also remarkable as a child's dream, and which is very easily explained by the analysis. A lady relates: "I can remember that when I was a child I repeatedly dreamed, that the *dear Lord had a pointed paper hat on his head*. They used to make me wear such a hat at table very often, so that I might not be able to look at the plates of the other children and see how much they had received of a particular dish. Since I have learned that God is omniscient, the dream signifies that I know everything in spite of the hat which I am made to wear."

Wherein the dream work consists, and how it manages its material, the dream thoughts, can be shown in a very instructive manner from the numbers and calculations which occur in dreams. Moreover, numbers in dreams are regarded as of especial significance by superstition. I shall therefore give a few more examples of this kind from my own collection.

I

The following is taken from the dream of a lady shortly before the close of her treatment:

She wants to pay for something or other; her daughter takes 3 florins and 65 kreuzer from her pocket-book; but the mother says: "What are you doing? It only costs 21 kreuzer." This bit of dream was immediately intelligible to me without further explanation from my knowledge of the dreamer's circumstances. The lady was a foreigner who had provided for her daughter in an educational institution in Vienna, and who could continue my treatment as long as her daughter stayed in the city. In three weeks the daughter's school year was to end, and with that the treatment also stopped. On the day before the dream the principal of the institute had urged her to make up her mind to allow her child to remain with her for another year. She had then obviously worked out this suggestion to the conclusion that in this case she would be able to continue the treatment for one year more. Now, this is what the dream refers to, for a year is equal to 365 days; the three weeks that remain before the close of the school year and of the treatment are equivalent to 21 days (though the hours of treatment are not as many as that). The numerals, which in the dream thoughts referred to time, are given money values in the dream, not without also giving expression to a deeper meaning for "time is money." 365 kreuzer, to be sure, are 3 *florins and* 65 *kreuzer*. The smallness of the sums which appear in the dream is a self-evident wish-fulfilment; the wish has reduced the cost of both the treatment and the year's instruction at the institution.

II

The numerals in another dream involve more complicated relations. A young lady, who, however, has already been married a number of years, learns that an acquaintance of hers of about her own age, Elsie L., has just become engaged. There upon she dreams: *She is sitting in the theatre with her husband, and one side of the orchestra is quite unoccupied. Her husband tells her that Elsie L. and her husband had also wanted to go, but that they had been able to get nothing but poor seats, three for 1 florin and 50 kreuzer, and of course they could not take those. She thinks that they didn't lose much either.*

Where do the 1 *florin and* 50 *kreuzer* come from? From an occurrence of the previous day which is really indifferent. The dreamer's sister-in-law had received 150 florins as a present from her husband, and had quickly got rid of them by buying some jewelry. Let us note that 150 florins is 100 times more than 1 florin and 50 kreuzer. Whence the 3 which stands before the theatre seats? There is only one association for this, namely, that the bride is that many months—three—younger than herself. Information concerning the significance of the feature that one side of the orchestra remains empty leads to the solution of the dream. This feature is an undisguised allusion to a little occurrence which has given her husband good cause for teasing her. She had decided to go to the theatre during the week, and had been careful to get tickets a few days before, for which she had to pay the pre-emption charge. When they got to the theatre they found that one side of the house was almost empty; she certainly did not need *to be in such a hurry.*

I shall now substitute the dream thoughts for the dream: "It surely was nonsense to marry so early; there was *no need for my being in such a hurry.* From the case of Elsie L., I see that I should have got a husband just the same—and one who is a *hundred times* better (husband, sweetheart, treasure)—if I had only *waited* (antithesis to the haste of her sister-in-law). I could have bought *three* such men for the money (the dowry!). Our attention is drawn to the fact that the numerals in this dream have changed their meanings and relations to a much greater extent than in the one previously considered. The transforming and disfiguring activity of the dream has in this case been greater, a fact which we interpret as meaning that these dream thoughts had to overcome a particularly great amount of inner psychic resistance up to the point of their representation. We must also not overlook the circumstance that the dream contains an absurd element, namely, that *two* persons take *three* seats. We digress to the interpretation of the absurdity of dreams when we remark that this absurd detail of the dream content is intended to represent the most strongly emphasized detail of the dream thoughts: "It was *nonsense* to marry so early." The figure 3 belonging to a quite subordinate relation of the two compared persons (three months' difference in age) has thus been skilfully used to produce the nonsense demanded by the dream. The reduction of the actual 150 florins to 1 florin and 50 kreuzer corresponds to her disdain of her husband in the suppressed thoughts of the dreamer.

III

Another example displays the arithmetical powers of the dream, which have brought it into such disrepute. A man dreams: *He is sitting at B——'s* (a

310

family of his earlier acquaintance) *and says, "It was nonsense for you not to give me Amy in marriage." Thereupon he asks the girl, "How old are you?" Answer: "I was born in 1882." "Ah, then you are 28 years old."*

Since the dream occurs in the year 1898, this is obviously poor arithmetic, and the inability of the dreamer to calculate may be compared to that of the paralytic, if there is no other way of explaining it. My patient was one of those persons who are always thinking about every woman they see. The person who followed him in my office, regularly for several months, was a young lady, whom he used to meet, about whom he used to ask frequently, and to whom he was very anxious to be polite. This was the lady whose age he estimated at 28 years. So much for explaining the result of the apparent calculation. But 1882 was the year in which he had married. He had been unable to refrain from engaging in conversation with the two females whom he met at my house—two girls, by no means youthful, who alternately opened the door for him, and as he did not find them very responsive, he had given himself the explanation that they probably considered him an elderly "settled" gentleman.

IV

For another number dream with its interpretation,—a dream distinguished by its obvious determination, or rather over-determination, I am indebted to B. Dattner:

My host, a policeman in the municipal service, dreamed that he was standing at his post in the street, which was a wish-realisation. The inspector then came over to him, having on his gorget the numbers 22 and 62 or 26—at all events there were many two's on it. Division of the number 2262 in the reproduction of the dream at once points to the fact that the components have separate meanings. It occurs to him that the day before, while on duty, they were discussing the duration of their time of service. The occasion for this was furnished by an inspector who had been pensioned at 62 years. The dreamer had only completed 22 years of service, and still needed 2 years and 2 months to make him eligible for a 90 per cent. pension. The dream first shows him the fulfilment of a long wished for wish, the rank of inspector. The superior with 2262 on his collar is himself; he takes care to do his duty on the street, which is another preferred wish; he has served his 2 years and 2 months, and can now be retired from the service with full pension, like the 62-year-old inspector.

If we keep in mind these examples and similar ones (to follow), we may say: Dream activity does not calculate at all, whether correctly or incorrectly; it joins together in the form of a calculation numerals which occur in the dream thoughts, and which may serve as allusions to material which is incapable of being represented. It thus utilises numerals as material for the expression of its purposes in the same manner as it does names and speeches known as word presentations.

For the dream activity cannot compose a new speech. No matter how many speeches and answers may occur in dreams, which may be sensible or absurd in themselves, analysis always shows in such cases that the dream has only taken from the dream thoughts fragments of speeches which have been delivered or heard, and dealt with them in a most arbitrary manner. It has not only torn them from their context and mutilated them, taken up one piece and rejected another, but it has also joined them together in a new way, so that the speech which seems coherent in the dream falls into three or four sections in the course of analysis. In this new utilisation of the words, the dream has often put aside the meaning which they had in the dream thoughts, and has derived an entirely new meaning from them.[33] Upon closer inspection the more distinct and compact constituents of the dream speech may be distinguished from others which serve as connectives and have probably been supplied, just as we supply omitted letters and syllables in reading. The dream speech thus has the structure of breccia stones, in which larger pieces of different material are held together by a solidified cohesive mass.

In a very strict sense this description is correct, to be sure, only for those speeches in the dream which have something of the sensational character of a speech, and which are described as "speeches." The others which have not, as it were, been felt as though heard or spoken (which have no accompanying acoustic or motor emphasis in the dream) are simply thoughts such as occur in our waking thought activity, and are transferred without change into many dreams. Our reading, also, seems to furnish an abundant and not easily traceable source of material for speeches, this material being of an indifferent nature. Everything, however, which appears conspicuously in the dream as a speech can be referred to real speeches which have been made or heard by the dreamer himself.

We have already found examples for the explanation of such dream speeches in the analysis of dreams cited for other purposes. Here is one example in place of many, all of which lead to the same conclusion.

A large courtyard in which corpses are cremated. The dreamer says: "I'm going away from here, I can't look at this." (Not a distinct speech.) *Then he meets two butcher boys and asks: "Well, did it taste good?" One of them answers: "No, it wasn't good." As though it had been human flesh.*

The harmless occasion for this dream is as follows: After taking supper with his wife, the dreamer pays a visit to his worthy but by no means appetising neighbour. The hospitable old lady is just at her evening meal, and *urges* him (instead of this word a composite sexually-significant word is jocosely used among men) to taste of it. He declines, saying that he has no appetite. *"Go on,* you can stand some more," or something of the kind. The dreamer is thus forced to taste and praise what is offered. "But that's good!" After he is alone again with his wife, he scolds about the neighbour's importunity and about the quality of the food he has tasted. "I can't stand the sight of it," a phrase not appearing even in the dream as an actual speech, is a thought which has reference to the physical charms of the lady who invites him, and which would be translated as meaning that he does not want to look at her.

The analysis of another dream which I cite at this point for the sake of the very distinct speech that forms its nucleus, but which I shall explain only when we come to consider emotions in the dream—will be more instructive. I dream very distinctly: *I have gone to Bruecke's laboratory at night, and upon hearing a soft knocking at the door, I open it to (the deceased) Professor Fleischl, who enters in the company of several strangers, and after saying a few words sits down at his table.* Then follows a second dream: *My friend Fl. has come to Vienna in July without attracting much attention; I meet him on the street while he is in conversation with my (deceased) friend P., and I go somewhere or other with these two, and they sit down opposite each other as though at a little table, while I sit at the narrow end of the table facing them. Fl. tells about his sister and says: "In three-quarters of an hour she was dead," and then something like: "That is the threshold." As P. does not understand him, Fl. turns to me, and asks me how much I have told of his affairs. Whereupon, seized by strange emotions, I want to tell Fl. that P. (can't possibly know anything because he) is not alive. But, noticing the mistake myself, I say: "Non vixit." Then I look at P. searchingly, and under my gaze he becomes pale and blurred, his eyes a morbid blue—and at last he dissolves. I rejoice greatly at this; I now understand that Ernest Fleischel, too, was only an apparition, a revenant, and I find that it is quite possible for such a person to exist only as long as one wants him to, and that he can be made to disappear by the wish of another person.*

This beautiful dream unites so many of the characteristics of the dream content which are problematic—the criticism made in the dream itself in that I myself notice my mistake in having said "Non vixit" instead of "Non vivit"; the unconstrained intercourse with dead persons, whom the dream itself declares to be dead; the absurdity of the inference and the intense satisfaction which the inference gives me—that "by my life" I should like to give a complete solution of these problems. But in reality I am incapable of doing this—namely, the thing I do in the dream—of sacrificing such dear persons to my ambition. With every revelation of the true meaning of the dream, with which I am well acquainted, I should have been put to shame. Hence I am content with selecting a few of the elements of the dream, for interpretation, some here, and others later on another page.

The scene in which I annihilate P. by a glance forms the centre of the dream. His eyes become strange and weirdly blue, and then he dissolves. This scene is an unmistakable copy of one really experienced. I was a demonstrator at the physiological institute, and began my service in the early hours, and *Bruecke* learned that I had been late several times in getting to the school laboratory. So one morning he came promptly for the opening of the class and waited for me. What he said to me was brief and to the point; but the words did not matter at all. What overwhelmed me was the terrible blue eyes through which he looked at me and before which I melted away—as P. does in the dream, for P. has changed rôles with him much to my relief. Anyone who remembers the eyes of the great master, which were wonderfully beautiful until old age, and who has ever seen him in anger, can easily imagine the emotions of the young transgressor on that occasion.

But for a long time I was unable to account for the "Non Vixit," with which I execute sentence in the dream, until I remembered that these two words possessed such great distinctness in the dream, not because they were heard or spoken, but because they were seen. Then I knew at once where they came from. On the pedestal of the statue of Emperor Joseph in the Hofburg at Vienna, may be read the following beautiful words:

Saluti patriae *vixit*
non diu sed totus.

I had culled from this inscription something which suited the one inimical train of thought in the dream thoughts and which now intended to mean: "That fellow has nothing to say, he is not living at all." And I now recalled that the dream was dreamed a few days after the unveiling of the memorial to *Fleischl* in the arcades of the university, upon which occasion I had again seen *Bruecke's* statue and must have thought with regret (in the unconscious) how my highly gifted friend P. with his great devotion to science had forfeited his just claim to a statue in these halls by his premature death. So I set up this memorial to him in the dream; the first name of my friend P. is Joseph.[34]

According to the rules of dream interpretation, I should still not be justified in replacing *non vivit,* which I need, by *non vixit,* which is placed at my disposal by the recollection of the Joseph monument. Something now calls my attention to the fact that in the dream scene, two trains of thought concerning my friend P. meet, one hostile, the other friendly—of which the former is superficial, the latter veiled, and both are given representation in the same words: *non vixit.* Because my friend P. has deserved well of science, I erect a statue to him; but because he has been guilty of an evil wish (which is expressed at the end of the dream) I destroy him. I have here constructed a sentence of peculiar resonance, and I must have been influenced by some model. But where can I find similar antithesis, such a parallel between two opposite attitudes towards the same person, both claiming to be entirely valid, and yet both trying not to encroach upon each other? Such a parallel is to be found in a single place, where, however, a deep impression is made upon the reader—in Brutus' speech of justification in Shakespeare's *Julius Caesar:* "As Caesar loved me, I weep for him; as he was fortunate, I rejoice at it; as he was valiant, I honour him; but, as he was ambitious I slew him." Is not this which I have discovered, the same sentence structure and thought contrast as in the dream thought? I thus play Brutus in the dream. If I could only find in the dream thoughts, one further trace of confirmation for this astonishing collateral connection! I think the following might be such: My friend comes to Vienna in *July.* This detail finds no support whatever in reality. To my knowledge my friend has never been in Vienna during the month of *July.* But the month of *July* is named after *Julius Caesar,* and might therefore very well furnish the required allusion to the intermediary thought that I am playing the part of Brutus.[35]

Strangely enough I once actually played the part of Brutus. I presented the scene between Brutus and Caesar from Schiller's poems to an audience of children when I was a boy of fourteen years. I did this with my nephew, who was a year older than I, and who had come to us from England— also a *revenant*—for in him I recognised the playmate of my first childish years. Until the end of my third year we had been inseparable, had loved each other and scuffled with each other, and, as I have already intimated, this childish relation has constantly determined my later feelings in my intercourse with person? of my own age. My nephew John has since found many incarnations, which have revivified first one aspect, then another, of this character which is so ineradicably fixed in my unconscious memory. Occasionally he must have treated me very badly, and I must have shown courage before my tyrant, for in later years I have often been told of the short speech with which I vindicated myself when my father—his grandfather— called me to account: "I hit him because he hit me." This childish scene must be the one which causes *non vivit* to branch off into *non vixit,* for in the language of later childhood striking is called *wichsen* (German, *wichsen*—to smear with shoe-polish, to tan, *i.e.,* to flog); the dream activity does not hesitate to take advantage of such connections. My hostility towards my friend P., which has so little foundation in reality—he was far superior to me, and might therefore have been a new edition of the playmate of my childhood—can certainly be traced to my complicated relations with John during our infancy. I shall, however, return to this dream later.

(F) Absurd Dreams—Intellectual Performances in the Dream

In our interpretation of dreams thus far we have come upon the element of *absurdity* in the dream-content so often that we must no longer postpone an investigation of its cause and significance. We remember, of course, that the absurdity of dreams has furnished the opponents of dream investigation with their chief argument for considering the dream nothing but the meaningless product of a reduced and fragmentary activity of the mind.

I begin with specimens in which the absurdity of the dream-content is only apparent and immediately disappears when the dream is more thoroughly examined. There are a few dreams which—accidentally one is at first inclined to think—are concerned with the dead father of the dreamer.

I

Here is the dream of a patient who had lost his father six years before:

A terrible accident has occurred to his father. He was riding in the night train when a derailment took place, the seats came together, and his head was crushed from side to side. The dreamer sees him lying on the bed with a wound over his left eyebrow, which runs off vertically. The dreamer is surprised that his father has had a misfortune (since he is dead already, as the dreamer adds in telling his dream). His father's eyes are so clear.

According to the standards prevailing in dream criticism, this dream-content would have to be explained in the following manner: At first, when the dreamer is picturing his father's misfortune, he has forgotten that his father has already been in his grave for years; in the further course of the dream this memory comes to life, and causes him to be surprised at his own dream even while he is still dreaming. Analysis, however, teaches us that it is entirely useless to attempt such explanations. The dreamer had given an artist an order for a bust of his father, which he had inspected two days before the dream. This is the thing which seems to him to have met with an *accident*. The sculptor has never seen the father, and is working from photographs which have been given him. On the very day before the dream the pious son had sent an old servant of the family to the studio in order to see whether he would pass the same judgment upon the marble head, namely, that it had turned out too *narrow from side to side,* from temple to temple. Now follows the mass of recollections which has contributed to the formation of this dream. The dreamer's father had a habit, whenever he was harassed by business cares or family difficulties, of pressing his temples with both hands, as though he were trying to compress his head, which seemed to grow too large for him. When our dreamer was four years old he was present when the accidental discharge of a pistol blackened his father's eyes (*his eyes are so clear*). While alive his father had had a deep wrinkle at the place where the dream shows the injury, whenever he was thoughtful or sad. The fact that in the dream this wrinkle is replaced by a wound points to the second occasion of the dream. The dreamer had taken a photograph of his little daughter; the plate had fallen from his hand, and when picked up showed a crack that ran like a vertical furrow across the forehead and reached as far as the orbital curve. He could not then get the better of his superstitious forebodings, for, on the day before his mother's death, a photographic plate with her likeness had cracked as he was handling it.

Thus the absurdity of the dream is only the result of an inaccuracy of verbal expression, which does not take the trouble to distinguish the bust and the photograph from the original. We are all accustomed to say of a picture, "Don't you think father is good?" Of course the appearance of absurdity in this dream might easily have been avoided. If it were permissible to pass judgment after a single experience, one might be tempted to say that this semblance of absurdity is admitted or desired.

II

Here is another very similar example from my own dreams (I lost my father in the year 1896):

After his death my father has been politically active among the Magyars, and has united them into a political body; to accompany which I see a little indistinct picture: *a crowd of people as in the Reichstag; a person who is standing on one or two benches, others round about him. I remember that he looked very like Garibaldi on his death-bed, and I am glad that this promise has really come true.*

This is certainly absurd enough. It was dreamed at the time that the Hungarians got into a lawless condition, through Parliamentary obstruction, and passed through the crisis from which Koloman Szell delivered them. The trivial circumstance that the scene beheld in the dream consists of such little pictures is not without significance for the explanation of this element. The usual visual representation of our thoughts results in pictures which impress us as being life-size; my dream picture, however, is the reproduction of a wood-cut inserted in the text of an illustrated history of Austria, representing Maria Theresa in the Reichstag of Pressburg—the famous scene of "Moriamur pro rege nostro."[36] Like Maria Theresa, my father, in the dream, stands surrounded by the multitude; but he is standing on one or two benches, and thus like a judge on the bench. (He has *united* them—here the intermediary is the phrase, "We shall need no *judge.*") Those of us who stood around the death-bed of my father actually noticed that he looked much like Garibaldi. He had a *post-mortem* rise of temperature, his cheeks shone redder and redder . . . involuntarily we continue: "And behind him lay in phantom radiance that which subdues us all—the common thing."

This elevation of our thoughts prepares us for having to deal with this very "common thing." The *post-mortem* feature of the rise in temperature corresponds to the words, "after his death" in the dream content. The most agonising of his sufferings had been a complete paralysis of the intestines (*obstruction*), which set in during the last weeks. All sorts of disrespectful thoughts are connected with this. A man of my own age who had lost his father while he was still at the Gymnasium, upon which occasion I was profoundly moved and tendered him my friendship, once told me, with derision, about the distress of a lady relative whose father had died on the street and had been brought home, where it turned out upon undressing the corpse, that at the moment of death, or *post-mortem,* an evacuation of the bowels had taken place. The daughter of the dead man was profoundly unhappy at having this ugly detail stain her memory of her father. We have now penetrated to the wish that is embodied in this dream. *To stand before one's children pure and great after one's death,* who would not wish that? What has become of the absurdity of the dream? The appearance of it has been caused only by the fact that a perfectly permissible mode of speech—in the case of which we are accustomed to ignore the absurdity that happens to exist between its parts—has been faithfully represented in the dream. Here, too, we are unable to deny that the semblance of absurdity is one which is desired and has been purposely brought about.[37]

III

In the example which I now cite I can detect the dream activity in the act of purposely manufacturing an absurdity for which there is no occasion at all in the subject-matter. It is taken from the dream that I had as a result of meeting Count Thun before my vacation trip. *"I am riding in a one-horse carriage, and give orders to drive to a railway station. 'Of course I cannot ride with you on the railway line itself,' I say, after the driver made an objection as though I had tired him out; at the same time it seems as though I had already driven with him for a distance which one usually rides on the train."* For this confused and senseless story the analysis gives the following explanation: During the day I had hired a one-horse carriage which was to take me to a remote street in Durnbach. The driver, however, did not know the way, and kept on driving in the manner of those good people until I noticed the fact and showed him the way, not sparing him a few mocking remarks withal. From this driver a train of thought led to the aristocratic personage whom I was destined to meet later. For the present I shall only remark that what strikes us middle-class plebeians about the aristocracy is that they like to put themselves in the driver's seat. Does not Count Thun guide the Austrian car of state?

The next sentence in the dream, however, refers to my brother, whom I identify with the driver of the one-horse carriage. I had this year refused to take the trip through Italy with him ("of course I cannot ride with you on the railway line itself"), and this refusal was a sort of punishment for his wonted complaint that I usually tired him out on this trip (which gets into the dream unchanged) by making him take hurried trips and see too many nice things in one day. That evening my brother had accompanied me to the railroad station, but shortly before getting there had jumped out, at the state railway division of the Western Station, in order to take a train to Purkersdorf. I remarked to him that he could stay with me a little longer, inasmuch as he did not go to Purkersdorf by the state railway but by the Western Railway. This is how it happens that in the dream I rode in the wagon a distance *which one usually rides on the train*. In reality, however, it was just the opposite; I told my brother: The distance which you ride on the state railway you could ride in my company on the Western Railway. The whole confusion of the dream is therefore produced by my inserting in the dream the word "wagon" instead of "state railway," which, to be sure, does good service in bringing together the driver and my brother. I then find in the dream some nonsense which seems hardly straightened out by my explanation, and which almost forms a contradiction to my earlier speech ("Of course I cannot ride with you on the railway line itself"). But as I have no occasion whatever for confounding the state railway with the one-horse carriage, I must have intentionally formed the whole puzzling story in the dream in this way.

But with what intention? We shall now learn what the absurdity in the dream signifies, and the motives which admitted it or created it. The solution of the mystery in the case in question is as follows: In the dream I needed something absurd and incomprehensible in connection with "riding" (Fahren) because in the dream thoughts I had a certain judgment which required representation. On an evening at the house of the hospitable and clever lady who appears in another scene of the same dream as the "hostess," I heard two riddles which I could not solve. As they were known to the other members of the party, I presented a somewhat ludicrous figure in my unsuccessful attempts to find a solution. They were two equivoques turning on the words "Nachkommen" (to come after—offspring) and "vorfahren" (to ride in advance—forefathers, ancestry). They read as follows:

The coachman does it
At the master's behest;
Everyone has it;
In the grave does it rest.
(Ancestry.)

It was confusing to find half of the second riddle identical with the first.

The coachman does it
At the master's behest;
Not everyone has it,
In the cradle does it rest.
(Offspring.)

As I had seen Count Thun ride in advance (vorfahren), so high and mighty, and had merged into the Figaro-mood which finds the merit of aristocratic gentlemen in the fact that they have taken the trouble to be born (Nachkommen—to become offspring), the two riddles became intermediary thoughts for the dream-work. As aristocrats can be readily confounded with coachmen, and as coachmen were in our country formerly called brothers-in-law, the work of condensation could employ my brother in the same representation. But the dream thought at work in the background was as follows: *It is nonsense to be proud of one's ancestry.* (*Vorfahren.*) *I would rather be myself an ancestor.* (*Vorfahr.*) For the sake of this judgment, "it is nonsense," we have the nonsense in the dream. We can now also solve the last riddle in this obscure passage of the dream, namely, that I have already driven before (vorher gefahren, vorgefahren) with the coachman.

Thus the dream is made absurd if there occurs as one of the elements in the dream thoughts the judgment *"That is nonsense,"* and in general if disdain and criticism are the motives for one of the trains of unconscious thought. Hence absurdity becomes one of the means by which the dream activity expresses contradiction, as it does by reversing a relation in the material between the dream thoughts and dream content, and by utilising sensations of motor impediment. But absurdity in the dream is not simply to be translated by "no"; it is rather intended to reproduce the disposition of the dream thoughts, this being to show mockery and ridicule along with the contradiction. It is only for this purpose that the dream activity produces anything ridiculous. Here again it transforms *a part of the latent content into a manifest form.*[38]

As a matter of fact we have already met with a convincing example of the significance of an absurd dream. The dream-interpreted without analysis, of the Wagnerian performance lasting until 7.45 in the morning, in which the orchestra is conducted from a tower, &c. (see p. 344) is apparently trying to say: It is a *crazy* world and an *insane* society. He who deserves a thing doesn't get it, and he who doesn't care for anything has it—and in this she means to compare her fate with that of her cousin. The fact that dreams concerning a dead father were the first to furnish us with examples of absurdity in dreams is by no means an accident. The conditions necessary for the creations of absurd dreams are here grouped together in a typical manner. The authority belonging to the father has at an early age aroused the criticism of the child, and the strict demands he has made have caused the child to pay particularly close attention to every weakness of the father for its own extenuation; but the piety with which the father's personality is surrounded in our thoughts, especially after his death, increases the censorship which prevents the expressions of this criticism from becoming conscious.

IV

The following is another absurd dream about a dead father:

I receive a notice from the common council of my native city concerning the costs of a confinement in the hospital in the year 1851, which was necessitated by an attack from which I suffered. I make sport of the matter, for, in the first place, I was not yet alive in the year 1851, and, in the second place, my father, to whom the notice might refer, is already dead. I go to him in the adjoining room, where he is lying on a bed, and tell him about it. To my astonishment he recalls that in that year—1851—he was once drunk and had to be locked up or confined. It was when he was working for the house of T—. "Then you drank, too?" I ask. "You married soon after?" I figure that I was born in 1856, which appears to me as though immediately following.

In view of the preceding discussion, we shall translate the insistence with which this dream exhibits its absurdities as the sure sign of a particularly embittered and passionate controversy in the dream thoughts. With all the more astonishment, however, we note that in this dream the controversy is waged openly, and the father designated as the person against whom the satire is directed. This openness seems to contradict our assumption

322

of a censor as operative in the dream activity. We may say in explanation, however, that here the father is only an interposed person, while the conflict is carried on with another one, who makes his appearance in the dream by means of a single allusion. While the dream usually treats of revolt against other persons, behind which the father is concealed, the reverse is true here; the father serves as the man of straw to represent others, and hence the dream dares thus openly to concern itself with a person who is usually hallowed, because there is present the certain knowledge that he is not in reality intended. We learn of this condition of affairs by considering the occasion of the dream. Now, it occurred after I had heard that an older colleague, whose judgment is considered infallible, had expressed disapproval and astonishment at the fact that one of my patients was then continuing psychoanalytical work with me for the fifth year. The introductory sentences of the dream point with transparent disguise to the fact that this colleague had for a time taken over the duties which my father could no longer perform (*expenses, fees at the hospital*); and when our friendly relations came to be broken I was thrown into the same conflict of feelings which arises in the case of misunderstanding between father and son in view of the part played by the father and his earlier functions. The dream thoughts now bitterly resent the reproach that *I am not making better progress,* which extends itself from the treatment of this patient to other things. Does this colleague know anyone who can get on faster? Does he not know that conditions of this sort are usually incurable and last for life? What are four or five years in comparison to a whole life, especially when life has been made so much easier for the patient during the treatment?

The impression of absurdity in this dream is brought about largely by the fact that sentences from different divisions of the dream thoughts are strung together without any reconciling transition. Thus the sentence, *I go to him in the adjoining room,* &c., leaves the subject dealt with in the preceding sentences, and faithfully reproduces the circumstances under which I told my father about my marriage engagement. Thus the dream is trying to remind me of the noble disinterestedness which the old man showed at that time, and to put it in contrast with the conduct of another, a new person. I now perceive that the dream is allowed to make sport of my father for the reason that in the dream thought he is held up as an example to another man, in full recognition of his merit. It is in the nature of every censorship that it permits the telling of untruth about forbidden things rather than truth. The next sentence, in which my father remembers having *once been drunk,* and having been *locked up for it,* also contains nothing which is actually true of my father. The person whom he covers is here a no less important one than the great Meynert, in whose footsteps I followed with such great veneration, and whose attitude towards me was changed into undisguised hostility after a short period of indulgence. The dream recalls to me his own statement that in his youth he was addicted to the *chloroform* habit, and that for this he had to enter a sanatorium. It recalls also a second experience with him shortly before his death. I carried on an embittered literary controversy with him concerning hysteria in the male, the existence of which he denied, and when I visited him in his last illness and asked him how he felt, he dwelt upon the details of his condition and concluded with the words: "You know, I have always been one of the prettiest cases of masculine hysteria." Thus, to my satisfaction, and *to my astonishment,* he admitted what he had so long and so stubbornly opposed. But the fact that in this scene I can use my father to cover Meynert is based not upon the analogy which has been found to exist between the two persons, but upon the slight, but quite adequate, representation of a conditional sentence occurring in the dream thoughts, which in full would read as follows: "Of course if I were of the second generation, the son of a professor or of a court-councillor, I should have *progressed more rapidly.*" In the dream I now make a court-councillor and a professor of my father. The most obvious and most annoying absurdity of the dream lies in the treatment of the date 1851, which seems to me to be hardly distinguishable from 1856, as though *a difference of five years would signify nothing whatever.* But it is just this idea of the dream thoughts which requires expression. *Four or five years*—that is the length of time which I enjoyed the support of the colleague mentioned at the outset; but it is also the time during which I kept my bride waiting before I married her; and, through a coincidence that is eagerly taken advantage of by the dream thoughts, it is also the time during which I am now keeping one of my best patients waiting for the completion of his cure. *"What are five years?"* ask the dream thoughts. *"That is no time at all for me— that doesn't come into consideration.* I have time enough ahead of me, and just as what you didn't want to believe came true at last, so I shall accomplish this also." Besides the number 51, when separated from the number of the century, is determined in still another manner and in an opposite sense; for which reason it occurs in the dream again. Fifty-one is an age at which a man seems particularly exposed to danger, at which I have seen many of my colleagues suddenly die, and among them one who had been appointed to a professorship a few days before, after he had been waiting a long time.

V

Another absurd dream which plays with figures, runs as follows:

One of my acquaintances, Mr. M., has been attacked in an essay by no less a person than Goethe, with justifiable vehemence, we all think. Mr. M. has, of course, been crushed by this attack. He complains of it bitterly at a dinner party; but he says that his veneration for Goethe has not suffered from this personal experience. I try to find some explanation of the chronological relations, which seem improbable to me. Goethe died in 1832; since his attack upon M. must of course have taken place earlier, Mr. M. was at the time a very young man. It seems plausible to me that he was 18 years old. But I do not know exactly what year it is at present, and so the whole calculation lapses into obscurity. The attack, moreover, is contained in Goethe's well-known essay entitled "Nature."

We shall soon find means to justify the nonsense of this dream. Mr. M., with whom I became acquainted *at a dinner-party,* had recently requested me to examine his brother, who showed signs of *paralytic insanity.* The conjecture was right; the painful thing about this visit was that the patient exposed Ms brother by alluding to his youthful pranks when there was no occasion in the conversation for his doing so. I had asked the patient to tell me the year of his birth, and had got him to make several small calculations in order to bring out the weakness of his memory—all of which tests he passed fairly well. I see now that I am acting like a paralytic in the dream (*I do not know exactly what year it is at present*). Other subject-matter in the dream is drawn from another recent source. The editor of a medical journal, a friend of mine, had accepted for his paper a very unfavourable, a *"crushing,"* criticism of the last book of my friend Fl. of Berlin, the author of which was a very *youthful* reviewer, who was not very competent to pass judgment. I thought I had a right to interfere, and called the editor to account; he keenly regretted the acceptance of the criticism, but would not promise redress. Thereupon I broke off relations with the journal, and in my letter of resignation expressed the hope that *our personal relations would not suffer from the incident.* The third source of this dream is an account given by a female patient—it was fresh in my memory at the time—of the mental disease of her brother who had fallen into a frenzy, crying "Nature, Nature." The physicians in attendance thought that the cry was derived from a reading of Goethe's beautiful *essay,* and that it pointed to overwork in the patient in the study of natural philosophy. I thought rather of the sexual sense in which even less cultured people with us use the word "Nature," and the fact that the unfortunate man later mutilated his genitals seemed to show that I was not far wrong. Eighteen years was the age of this patient at the time when the attack of frenzy occurred.

If I add further that the book of my friend so severely criticised ("It is a question whether the author is crazy or we are" had been the opinion of another critic) treats of the *temporal relations of life* and refers the duration of Goethe's life to the multiple of a number significant from the point of view of biology, it will readily be admitted that I am putting myself in the place of my friend in the dream. (*I try to find some explanation of the chronological relations.*) But I behave like a paralytic, and the dream revels in absurdity. This means, then, as the dream thoughts say ironically. "Of course he is the fool, the lunatic, and you are the man of genius who knows better. Perhaps, however, it is the other way around?" Now, this other way around is explicitly represented in the dream, in that Goethe has attacked the young man, which is absurd, while it is perfectly possible even to-day for a young fellow to attack the immortal Goethe. and in that I figure from the year of Goethe's death, while I caused the paralytic to calculate from the year of his birth.

But I have already promised to show that every dream is the result of egotistical motives. Accordingly, I must account for the fact that in this dream I make my friend's cause my own and put myself in his place. My rational conviction in waking thought is not adequate to do this. Now, the story of the eighteen-year-old patient and of the various interpretations of his cry, "Nature," alludes to my having brought myself into opposition to most physicians by claiming sexual etiology for the psychoneuroses. I may say to myself: "The same kind of criticism your friend met with you will meet with too, and have already met with to some extent," and now I may replace the "he" in the dream thoughts by "we." "Yes, you are right; we two are the fools." That *mea res agitur,* is clearly shown by the mention of the short, incomparably beautiful essay of Goethe, for it was a public reading of this essay which induced me to study the natural science while I was still undecided in the graduating class of the Gymnasium.

VI

I am also bound to show of another dream in which my ego does not occur that it is egotistic. On page 243 I mentioned a short dream in which Professor M. says: "My son, the myopic . . . "; and I stated that this was only a preliminary dream to another one, in which I play a part. Here is the main dream, omitted above, which challenges us to explain its absurd and unintelligible word-formation.

On account of some happenings or other in the city of Rome it is necessary for the children to flee, and this they do. The scene is then laid before a gate, a two-winged gate in antique style (the Porta Romana in Siena, as I know while I am still dreaming). I am sitting on the edge of a well, and am very sad; I almost weep. A feminine person—nurse, nun—brings out the two boys and hands them over to their father, who is not myself. The elder of the two is distinctly my eldest son, and I do not see the face of the other; the woman who brings the boy asks him for a parting kiss. She is distinguished by a red nose. The boy denies her the kiss, but says to her, extending his hand to her in parting, "Auf Geseres," and to both of us (or to one of us) "Auf Ungeseres." I have the idea that the latter indicates an advantage.

This dream is built upon a tangle of thoughts induced by a play I saw at the theatre, called *Das neue Ghetto* ("The New Ghetto.") The Jewish question, anxiety about the future of my children who cannot be given a native country of their own, anxiety about bringing them up so that they may have the right of native citizens—all these features may easily be recognised in the accompanying dream thoughts.

"We sat by the waters of Babylon and wept." Siena, like Rome, is famous for its beautiful fountains. In the dream I must find a substitute of some kind for Rome (*cf.* p. 179) in localities which are known to me. Near the Porta Romana of Siena we saw a large, brightly illuminated building, which we found to be the Manicomio, the insane asylum. Shortly before the dream I had heard that a co-religionist had been forced to resign a position at a state asylum which he had secured with great effort.

Our interest is aroused by the speech: *"Auf Geseres"*—where we might expect, from the situation maintained throughout the dream, *"Auf Wiedersehen"* (*Au revoir*)—and by its quite meaningless opposite, *"Auf Ungeseres."*

According to information I have received from Hebrew scholars, *Geseres* is a genuine Hebrew word derived from the verb *goiser,* and may best be rendered by "ordained sufferings, fated disaster." From its use in the Jewish jargon one might think it signified "wailing and lamentation." *Ungeseres* is a coinage of my own and first attracts my attention; but for the present it baffles me. The little observation at the end of the dream, that *Ungeseres* indicates an advantage over *Geseres* opens the way to the associations and to an explanation. The same relation holds good with caviare; the unsalted kind[39] is more highly prized than the salted. Caviare to the general, "noble passions"; herein lies concealed a joking allusion to a member of my household, of whom I hope—for she is younger than I—that she will watch over the future of my children; this, too, agrees with the fact that another member of my household, our worthy nurse, is clearly indicated in the nurse (or nun) of the dream. But a connecting link is wanting between the pair, *salted* and *unsalted,* and *Geseres—ungeseres.* This is to be found in *soured and unsoured.* In their flight or exodus out of Egypt, the children of Israel did not have time to allow their bread to be leavened, and in memory of the event to this day they eat unsoured bread at Easter time. Here I can also find room for the sudden notion which came to me in this part of the analysis. I remembered how we promenaded about the city of Breslau, which was strange to us, at the end of the Easter holidays, my friend from Berlin and I. A little girl asked me to tell her the way to a certain street; I had to tell her I did not know it, whereupon I remarked to my friend, "I hope that later on in life the little one will show more perspicacity in selecting the persons by whom she allows herself to be guided." Shortly afterwards a sign caught my eye: "Dr, *Herod,* office hours. . . . " I said to myself: "I hope this colleague does not happen to be a children's specialist." Meanwhile my friend had been developing his views on the biological significance of *bilateral* symmetry, and had begun a sentence as follows: "If we had but one eye in the middle of our foreheads like *Cyclops.* . . . " This leads us to the speech of the professor in the preliminary dream: "My son, the *myopic.*" And now I have been led to the chief source for *Geseres.* Many years ago, when this son of Professor M., who is to-day an independent thinker, was still sitting on his school-bench, he contracted a disease of the eye, which the doctor declared gave cause for anxiety. He was of the opinion that as long as it remained in one eye it would not matter; if, however, it should extend to the other eye, it would be serious. The disease healed in the one eye without leaving any bad effects; shortly afterwards, however, its symptoms actually appeared in the other eye. The terrified mother of the boy immediately summoned the physician to the seclusion of her country resort. But he took another view of the matter. *"What sort of 'Geseres' is this you are making?"* he said to his mother with impatience. "If one side got well, the other side will get well too." And so it turned out.

And now as to the connection between this and myself and those dear to me. The school-bench upon which the son of Professor M. learned his first lessons has become the property of my eldest son—it was given to his mother—into whose lips I put the words of parting in the dream. One of the wishes that can be attached to this transference may now easily be guessed. This school-bench is intended by its construction to guard the child from becoming shortsighted and one-sided. Hence, myopia (and behind the Cyclops) and the discussion about *bilateralism*. The concern about one-sidedness is of two-fold signification; along with the bodily one-sidedness, that of intellectual development may be referred to. Does it not seem as though the scene in the dream, with all its madness, were putting its negative on just this anxiety? After the child has said his word of parting *on the one side,* he calls out its opposite on the *other side,* as though in order to establish an equilibrium. *He is acting, as it were, in obedience to bilateral symmetry!*

Thus the dream frequently has the profoundest meaning in places where it seems most absurd. In all ages those who had something to say and were unable to say it without danger to themselves gladly put on the cap and bells. The listener for whom the forbidden saying was intended was more likely to tolerate it if he was able to laugh at it, and to flatter himself with the comment that what he disliked was obviously something absurd. The dream proceeds in reality just as the prince does in the play who must counterfeit the fool, and hence the same thing may be said of the dream which Hamlet says of himself, substituting an unintelligible witticism for the real conditions: "I am but mad north-north-west; when the wind is southerly I know a hawk from a handsaw."[40]

Thus my solution of the problem of the absurdity of dreams is that the dream thoughts are never absurd—at least not those belonging to the dreams of sane persons—and that the dream activity produces absurd dreams and dreams with individual absurd elements if criticism, ridicule, and derision in the dream thoughts are to be represented by it in its manner of expression. My next concern is to show that the dream activity is primarily brought about by the co-operation of the three factors which have been mentioned—and of a fourth one which remains to be cited—that it accomplishes nothing short of a transposition of the dream thoughts, observing the three conditions which are prescribed for it, and that the question whether the mind operates in the dream with all its faculties, or only with a portion of them, is deprived of its cogency and is inapplicable to the actual circumstances. But since there are plenty of dreams in which judgments are passed, criticisms made, and facts recognised, in which astonishment at some single element of the dream appears, and arguments and explanations are attempted, I must meet the objections which may be inferred from these occurrences by the citation of selected examples.

My answer is as follows: *Everything in the dream which occurs as an apparent exercise of the critical faculty is to be regarded, not as an intellectual accomplishment of the dream activity, but as belonging to the material of the dream thoughts, and it has found its way from them as a finished structure to the manifest dream content.* I may go even further than this. Even the judgments which are passed upon the dream as it is remembered after awakening and the feelings which are aroused by the reproduction of the dream, belong in good part to the latent dream content, and must be fitted into their place in the interpretation of the dream.

I

A striking example of this I have already given. A female patient does not wish to relate her dream because it is too vague. She has seen a person in the dream, and does not know whether it is her husband or her father. Then follows a second dream fragment in which there occurs a "manure-can," which gives rise to the following reminiscence. As a young housewife, she once jokingly declared in the presence of a young relative who frequented the house that her next care would be to procure a new manure-can. The next morning one was sent to her, but it was filled with lilies of the valley. This part of the dream served to represent the saying, "Not grown on your own manure."[41] When we complete the analysis we find that in the dream thoughts it is a matter of the after-effects of a story heard in youth, to the effect that a girl had given birth to a child *concerning whom it was not clear who was the real father*. The dream representation here goes over into the waking thought, and allows one element of the dream thoughts to be represented by a judgment expressed in the waking state upon the whole dream.

II

A similar case: One of my patients has a dream which seems interesting to him, for he says to himself immediately after awakening: *"I must tell that to the doctor."* The dream is analysed, and shows the most distinct allusion to an affair in which he had become involved during the treatment, and of which he had decided *to tell me nothing.'* [42]

III

Here is a third example from my own experience:

I go to the hospital with P. through a region in which houses and gardens occur. With this comes the idea that I have already seen this region in dreams several times. I do not know my way very well; P. shows me a way which leads through a corner to a restaurant (a room, not a garden); here I ask for Mrs. Doni, and I hear that she is living in the background in a little room with three children. I go there, and while on the way I meet an indistinct person with my two little girls, whom I take with me after I have stood with them for a while. A kind of reproach against my wife for having left them there.

Upon awakening I feel great *satisfaction,* the cause for this being the fact that I am now going to learn from the analysis what is meant by the idea *"I have already dreamed of that."* [43] But the analysis of the dream teaches me nothing on the subject; it only shows me that the satisfaction belongs to the latent dream content, and not to my judgment upon the dream. It is satisfaction over the fact that I have had children by my marriage. P. is a person in whose company I walked the path of life for a certain space, but who has since far outdistanced me socially and materially— whose marriage, however, has remained childless. The two occasions for the dream furnishing the proof of this may be found by means of complete analysis. On the previous day I had read in the paper the obituary notice of a certain Mrs. Dona A—y (out of which I make Doni), who had died in childbirth; I was told by my wife that the dead woman had been nursed by the same midwife she herself had had at the birth of our two youngest boys. The name Dona had caught my attention, for I had recently found it for the first time in an English novel. The other occasion for the dream may be found in the date on which it was dreamed; it was on the night before the birthday of my eldest boy, who, it seems, is poetically gifted.

IV

The same satisfaction remained with me after awakening from the absurd dream that my father, after his death, had played a political part among the Magyars, and it is motivated by a continuance of the feeling which accompanied the last sentence of the dream: *"I remember that on his deathbed he looked so much like Garibaldi, and I am glad that it has really come true.* (*Here belongs a forgotten continuation.*) I can now supply from the analysis what belongs in this gap of the dream. It is the mention of my second boy, to whom I have given the first name of a great historical personage, who attracted me powerfully during my boyhood, especially during my stay in England. I had to wait for a year after making up my mind to use this name in case the expected child should be a son, and I greeted him with it *in high satisfaction* as soon as he was born. It is easy to see how the father's lust for greatness is transferred in his thoughts to his children; it will readily be believed that this is one of the ways in which the suppression of this lust which becomes necessary in life is brought about. The little fellow won a place in the text of this dream by virtue of the fact that the same accident—quite pardonable in a child or a dying person—of soiling his clothes had happened to him. With this may be compared the allusion *"Stuhlrichter"* (judge on the stool-bench, *i.e.* presiding judge) and the wish of the dream: To stand before one's children great and pure.

V. I am now called upon to find expressions of judgment which remain in the dream itself, and are not retained in or transferred to our waking thoughts, and I shall consider it a great relief if I may find examples in dreams, which have already been cited for other purposes. The dream about Goethe's attacking Mr. M. seems to contain a considerable number of acts of judgment. *I try to find some explanation of the chronological relations, which seem improbable to me.* Does not this look like a critical impulse directed against the nonsensical idea that Goethe should have made a literary attack upon a young man of my acquaintance? *"It seems plausible to me* that he was 18 years old." That sounds quite like the result of a dull-witted calculation; and *"I do not know exactly what year it is"* would be an example of uncertainty or doubt in the dream.

But I know from analysis that these acts of judgment, which seem to have been performed in the dream for the first time, admit of a different construction in the light of which they become indispensable for interpreting the dream, and at the same time every absurdity is avoided. With the sentence, *"I try to find some explanation of the chronological relations,"* I put myself in the place of my friend who is actually trying to explain the chronological relations of life. The sentence then loses its significance as a judgment that objects to the nonsense of the previous sentences. The interposition, *"which seems improbable to me,"* belongs to the subsequent *"it seems plausible to me."* In about the same words I had answered the lady who told me the story of her brother's illness: *"It seems improbable to me* that the cry of 'Nature, Nature,' had anything to do with Goethe; *it appears much more plausible* that it had the sexual significance which is known to you." To be sure, a judgment has been passed here, not, however, in the dream but in reality, on an occasion which is remembered and utilised by the dream thoughts. The dream content appropriates this judgment like any other fragment of the dream thoughts.

The numeral 18, with which the judgment in the dream is meaninglessly connected, still preserves a trace of the context from which the real judgment was torn. Finally, *"I am not certain what year it is"* is intended for nothing else than to carry out my identification with the paralytic, in the examination of whom this point of confirmation had actually been established.

In the solution of these apparent acts of judgment, in the dream, it may be well to call attention to the rule of interpretation which says that the coherence which is fabricated hi the dream between its constituent parts is to be disregarded as specious and unessential, and that every dream element must be taken by itself and traced to its source. The dream is a conglomeration, which is to be broken up into its elements for the purposes of investigation. But other circumstances call our attention to the fact that a psychic force is expressed in dreams which establishes this apparent coherence—that is to say, which subjects the material that is obtained by the dream activity to a *secondary elaboration.* We are here confronted with manifestations of this force, upon which we shall later fix our attention as being the fourth of the factors which take part in the formation of the dream.

VI

I select other examples of critical activity in the dreams which have already been cited. In the absurd dream about the communication from the common council I ask the question: *"You married shortly after? I figure that I was born in* 1856, *which appears to me as though following immediately.* This quite takes the form of an *inference.* My father married shortly after his attack in the year 1851; I am the oldest son, born in 1856; this agrees perfectly. We know that this inference has been interpolated by the wish-fulfilment, and that the sentence which dominates the dream thoughts is to the following effect: 4 *or 5 years, that is no time at all, that need not enter the calculation.*But every part of this chain of inferences is to be determined from the dream thoughts in a different manner, both as to its content and as to its form. It is the patient—about whose endurance my colleague complains—who intends to marry immediately after the close of the treatment. The manner in which I deal with my father in the dream recalls an *inquest* or *examination,* and with that the person of a university instructor who was in the habit of taking a complete list of credentials at the enrolment of his class: "You were born when?" In 1856. "Patre?" Then the applicant gave the first name of his father with a Latin ending, and we students assumed that the Aulic Councillor drew *inferences* from the first name of the father which the name of the enrolled student would not always have supplied. According to this, the *drawing of inferences* in the dream would be merely a repetition of the *drawing of inferences* which appears as part of the subject-matter in the dream thoughts. From this we learn something new. If an inference occurs in the dream content, it invariably comes from the dream thoughts; it may be contained in these as a bit of remembered material, or it may serve as a logical connective in a series of dream thoughts. In any case an inference in the dream represents an inference in the dream thoughts.[44]

The analysis of this dream should be continued here. With the inquest of the Professor there is connected the recollection of an index (published in Latin during my time) of the university students; also of my course of studies. The *five years* provided for the study of medicine were as usual not enough for me. I worked along unconcernedly in the succeeding years; in the circle of my acquaintances I was considered a loafer, and there was doubt as to whether I would "get through." Then all at once I decided to take my examinations; and I got "through," *in spite of the postponement.* This is a new confirmation of the dream thoughts, which I defiantly hold up to my critics: "Even though you are unwilling to believe it, because I take my time, I shall reach a conclusion (German *Schluss,* meaning either end or conclusion, *inference*). It has often happened that way."

In its introductory portion this dream contains several sentences which cannot well be denied the character of an argumentation. And this argumentation is not at all absurd; it might just as well belong to waking thought. *In the dream I make sport of the communication of the Common Council, for in the first place I was not yet in the world in 1851, and in the second place, my father, to whom it might refer, is already dead.* Both are not only correct in themselves, but coincide completely with the arguments that I should use in case I should receive a communication of the sort mentioned. We know from our previous analysis that this dream has sprung from deeply embittered and scornful dream thoughts; if we may assume further that the motive for censorship is a very strong one, we shall understand that the dream activity has every reason to create a *flawless refutation of a baseless insinuation* according to the model contained in the dream thoughts. But analysis shows that in this case the dream activity has not had the task of making a free copy, but it has been required to use subject-matter from the dream thoughts for its purpose. It is as if in an algebraic equation there occurred plus and minus signs, signs of powers and of roots, besides the figures, and as if someone, in copying this equation without understanding it, should take over into his copy the signs of operation as well as the figures, and fail to distinguish between the two kinds. The two arguments may be traced to the following material. It is painful for me to think that many of the assumptions upon which I base my solution of psychoneuroses, as soon as they have become known, will arouse scepticism and ridicule. Thus I must maintain that impressions from the second year of life, or even from the first, leave a lasting trace upon the temperament of persons who later become diseased, and that these impressions—greatly distorted it is true, and exaggerated by memory—are capable of furnishing the original and fundamental basis of hysterical symptoms. Patients to whom I explain this in its proper place are in the habit of making a parody upon the explanation by declaring themselves willing to look for reminiscences of the period *when they were not yet alive.*It would quite accord with my expectation, if enlightenment on the subject of the unsuspected part played by the father in the earliest sexual impulses of feminine patients should get a similar reception. (*Cf.* the discussion on p. 234.) And, nevertheless, both positions are correct according to my well-founded conviction. In confirmation I recall certain examples in which the death of the father happened when the child was very young, and later events, otherwise inexplicable, proved that the child had unconsciously preserved recollections of the persons who had so early gone out of its life. I know that both of my assertions are based upon *inferences* the validity of which will be attacked. If the subject-matter of these very inferences which I fear will be contested is used by the dream activity for setting up *incontestable inferences,* this is a performance of the wish-fulfilment.

VII

In a dream which I have hitherto only touched upon, astonishment at the subject to be broached is distinctly expressed at the outset.

"The elder Bruecke must have given me some task or other; strangely enough it relates to the preparation of my own lower body, pelvis and legs, which I see before me as though in the dissecting room, but without feeling my lack of body and without a trace of horror. Louise N. is standing near, and doing her work next to me. The pelvis is eviscerated; now the upper, now the lower view of the same is seen, and the two views mingle. Thick fleshy red lumps (which even in the dream make me think of haemorrhoids) are to be seen. Also something had to be carefully picked out, which lay over these and which looked like crumpled tin-foil.[45] *Then I was again in possession of my legs and made a journey through the city, but took a wagon (owing to my fatigue). To my astonishment the wagon drove into a house door, which opened and allowed it to pass into a passage that was snapped off at the end, and finally led further on into the open.*[46] *At last I wandered through changing landscapes with an Alpine guide, who carried my things. He carried me for some way, out of consideration for my tired legs. The ground was muddy, and we went along the edge; people sat on the ground, a girl among them. like Indians or Gypsies. Previously I had moved myself along on the slippery ground, with constant astonishment that I was so well able to do it after the preparation. At last we came to a small wooden house which ended in an open window. Here the guide set me down, and laid two wooden boards which stood in readiness on the window sill, in order that in this way the chasm might be bridged which had to be crossed in order to get to the window. Now, I grew really frightened about my legs. Instead of the expected crossing, I saw two grown-up men lying upon wooden benches which were on the walls of the hut, and something like two sleeping children next to them. It seems as though not the boards but the children were intended to make possible the crossing. I awakened with frightened thoughts.*

Anyone who has formed a proper idea of the abundance of dream condensation will easily be able to imagine how great a number of pages the detailed analysis of this dream must fill. Luckily for the context, I shall take from it merely the one example of astonishment, in the dream, which makes its appearance in the parenthetical remark, *"strangely enough."* Let us take up the occasion of the dream. It is a visit of this lady, Louise N., who assists at the work in the dream. She says: "Lend me something to read." I offer her *She,* by Rider Haggard. "A *strange* book, but full of hidden sense," I try to explain to her; "the eternal feminine, the immortality of our emotions—" Here she interrupts me: "I know that book already. Haven't you something of your own?" "No, my own immortal works are still unwritten." "Well, when are you going to publish your so-called latest revelations which you promised us would be good reading?" she asks somewhat sarcastically. I now perceive that she is a mouthpiece for someone else, and I become silent. I think of the effort it costs me to publish even my work on the Dream, in which I have to surrender so much of my own intimate character. "The best that you know you can't tell to the children." The preparation of *my own body,* which I am ordered to make in the dream, is thus the *self-analysis* necessitated in the communication of my dreams. The elder Bruecke very properly finds a place here; in these first years of my scientific work it happened that I neglected a discovery, until his energetic commands forced me to publish it. But the other trains of thought which start from my conversation with Louise N. go too deep to become conscious; they are side-tracked by way of the related material which has been awakened in me by the mention of Rider Haggard's *She.* The comment "strangely enough" goes with this book, and with another by the same author, *The Heart of the World,* and numerous elements of the dream are taken from these two fantastic novels. The muddy ground over which the dreamer is carried, the chasm which must be crossed by means of the boards that have been brought along, come from *She;* the Indians, the girl, and the wooden house, from the *Heart of the World.* In both novels a woman is the leader, both treat of dangerous wanderings; *She* has to do with an adventurous journey to the undiscovered country, a place almost untrodden by foot of man. According to a note which I find in my record of the dream, the fatigue in my legs was a real sensation of those days. Doubtless in correspondence with this came a tired frame of mind and the doubting question: "How much further will my legs carry me?" The adventure in *She* ends with the woman leader's meeting her death in the mysterious fire at the centre of the earth, instead of attaining immortality for herself and others. A fear of this sort has unmistakably arisen in the dream thoughts. The "wooden house," also, is surely the coffin— that is, the grave. But the dream activity has performed its masterpiece in representing this most unwished-for of all thoughts by means of a wish-fulfilment. I have already once been in a grave, but it was an empty Etruscan grave near Orvieto—a narrow chamber with two stone benches on the walls, upon which the skeletons of two grown-up persons had been laid. The ulterior of the wooden house in the dream looks exactly like this, except that wood has been substituted for stone. The dream seems to say:

"If you must Bo soon lie in your grave, let it be this Etruscan grave," and by means of this interpolation it transforms the saddest expectation into one that is really to be desired. As we shall learn, it is, unfortunately, only the idea accompanying an emotion which the dream can change into its opposite, not usually the emotion itself. Thus I awake with "frightened thoughts," even after the dream has been forced to represent my idea— that perhaps the children will attain what has been denied to the father—a fresh allusion to the strange novel in which the identity of a person is preserved through a series of generations covering two thousand years.

VIII

In the context of another dream there is a similar expression of astonishment at what is experienced in the dream. This, however, is connected with a striking and skilfully contrived attempt at explanation which might well be called a stroke of genius—so that I should have to analyse the whole dream merely for the sake of it, even if the dream did not possess two other features of interest. I am travelling during the night between the eighteenth and the nineteenth of July on the Southern Railway, and in my sleep I hear someone call out: *"Hollthurn, 10 minutes." I immediately think of Holothurian—of a museum of natural history—that here is a place where brave men have vainly resisted the domination of their overlord. Yes, the counter reformation in Austria! As though it were a place in Styria or the Tyrol. Now I distinctly see a little museum in which the remains or the possessions of these men are preserved. I wish to get off, but I hesitate to do so. Women with fruit are standing on the platform; they crouch on the floor, and in that position hold out their baskets in an inviting manner. I hesitate, in doubt whether we still have time, but we are still standing. I am suddenly in another compartment in which the leather and the seats are so narrow that one's back directly touches the back rest.*[47] *I am surprised at this, but I may have changed cars while asleep. Several people, among them an English brother and sister; a row of books distinctly on a shelf on the wall. I see* The Wealth of Nations, *then* Matter and Motion (*by Maxwell*)—*the books are thick and bound in brown linen. The man asks his sister for a book by Schiller, and whether she has forgotten it. These are books which first seem mine, then seem to belong to the brother and sister. At this point I wish to join in the conversation in order to confirm and support what is being said—.* I awaken sweating all over my body, because all the windows are shut. The train stops at Marburg.

While writing down the dream, a part of it occurs to me which my memory wished to omit. *I say to the brother and sister about a certain work: "It is from . . . " but I correct myself: "It is by . . . " The man remarks to his sister: "He said it correctly."*

The dream begins with the name of a station, which probably must have partially awakened me. For this name, which was Marburg, I substituted Hollthurn. The fact that I heard Marburg when it was first called, or perhaps when it was called a second time, is proved by the mention in the dream of Schiller, who was born in Marburg, though not in the one in Styria.[48] Now this time, although I was travelling first-class, it was under very disagreeable circumstances. The train was overcrowded; I had met a gentleman and lady in my compartment who seemed persons of quality, but who did not have the good breeding or who did not think it worth while to conceal their displeasure at my intrusion. My polite salutation was not answered, and although the man and the woman sat next each other (with their backs in the direction in which we were riding), the woman made haste to pre-empt the place opposite her and next the window with her umbrella; the door was immediately closed and demonstrative remarks about the opening of windows were exchanged. Probably I was quickly recognised as a person hungry for fresh air. It was a hot night, and the air in the compartment, thus shut on all sides, was almost suffocating. My experience as a traveller leads me to believe that such inconsiderate, obtrusive conduct marks people who have only partly paid for their tickets, or not at all. When the conductor came, and I presented my dearly bought ticket, the lady called out ungraciously, and as though threateningly: "My husband has a pass." She was a stately figure with sour features, in age not far from the time set for the decay of feminine beauty; the man did not get a chance to say anything at all, and sat there motionless. I tried to sleep. In the dream I take terrible revenge on my disagreeable travelling companions; no one would suspect what insults and humiliations are concealed behind the disjointed fragments of the first half of the dream. After this desire has been satisfied, the second wish, to exchange my compartment for another, makes itself evident. The dream makes changes of scene so often, and without raising the least objection to such changes, that it would not have been in the least remarkable if I had immediately replaced my travelling companions by more pleasant ones for my recollection. But this was one of the cases where something or other objected to the change of scene and considered explanation of the change necessary. How did I suddenly get into another compartment? I surely could not remember having changed cars. So there was only one explanation: *I must have left the carriage while asleep,* a rare occurrence, examples for which, however, are furnished by the experience of the neuropathologist. We know of persons who undertake railroad journeys in a crepuscular state without betraying their abnormal condition by any sign, until some station on the journey they completely recover consciousness, and are then surprised at the gap in their memory. Thus, while I am still dreaming, I declare my own case to be such a one of *"Automatisme ambulatoire."*

Analysis permits another solution. The attempt at explanation, which so astounds me if I am to attribute it to the dream activity, is not original, but is copied from the neurosis of one of my patients. I have already spoken on another page of a highly cultured and, in conduct, kind-hearted man, who began, shortly after the death of his parents, to accuse himself of murderous inclinations, and who suffered because of the precautionary measures he had to take to insure himself against these inclinations. At first walking along the street was made painful for him by the compulsion impelling him to demand an accounting of all the persons he met as to whither they had vanished; if one of them suddenly withdrew from his pursuing glance, there remained a painful feeling and a thought of the possibility that he might have put the man out of the way. This compulsive idea concealed, among other things, a Cain-fancy, for "all men are brothers." Owing to the impossibility of accomplishing his task, he gave up taking walks and spent his life imprisoned within his four walls. But news of murderous acts which have been committed outside constantly reached his room through the papers, and his conscience in the form of a doubt kept accusing him of being the murderer. The certainty of not having left his dwelling for weeks protected him against these accusations for a time, until one day there dawned upon him the possibility that he might have left *his house while in an unconscious condition,* and might thus have committed the murder without knowing anything about it. From that time on he locked his house door, and handed the key over to his old housekeeper, and strictly forbade her to give it into his hands even if he demanded it.

This, then, is the origin of the attempted explanation, that I may have changed carriages while in an unconscious condition—it has been transferred from the material of the dream thoughts to the dream in a finished state, and is obviously intended to identify me with the person of that patient. My memory of him was awakened by an easy association.

I had made my last night journey with this man a few weeks before. He was cured, and was escorting me into the country, to his relatives who were summoning me; as we had a compartment to ourselves, we left all the windows open through the night, and, as long as I had remained awake, we had a delightful conversation. I knew that hostile impulses towards his father from the time of his childhood, in connection with sexual material, had been at the root of his illness. By identifying myself with him, I wanted to make an analogous confession to myself. The second scene of the dream really resolves itself into a wanton fancy to the effect that my two elderly travelling companions had acted so uncivilly towards me for the reason that my arrival prevented them from exchanging love-tokens during the night as they had intended. This fancy, however, goes back to an early childhood scene in which, probably impelled by sexual inquisitiveness, I intruded upon the bedroom of my parents, and was driven from it by my father's emphatic command.

I consider it superfluous to multiply further examples. All of them would confirm what we have learned from those which have been already cited, namely, that an act of judgment in the dream is nothing but the repetition of a prototype which it has in the dream thoughts. In most cases it is an inappropriate repetition introduced in an unfitting connection; occasionally, however, as in our last example, it is so artfully disposed that it may give the impression of being an independent thought activity in the dream. At this point we might turn our attention to that psychic activity which indeed does not seem to co-operate regularly in the formation of dreams, but whose effort it is, wherever it does co-operate, to fuse together those dream elements that are incongruent on account of their origins in an uncontradictory and intelligible manner. We consider it best, however, first to take up the expressions of emotion which appear in the dream, and to compare them with the emotions which analysis reveals to us in the dream thoughts.

(G) The Affects in the Dream

A profound remark of Stricker's has called our attention to the fact that the expressions of emotion in the dream do not permit of being disposed of in the slighting manner in which we are accustomed to shake off the dream itself, after we have awakened. "If I am afraid of robbers in the dream, the robbers, to be sure, are imaginary, but the fear of them is real," and the same is true if I am glad in the dream. According to the testimony of our feelings, the emotion experienced in the dream is in no way less valid than one of like intensity experienced in waking life, and the dream makes its claim to be taken up as a part of our real mental experiences, more energetically on account of its emotional content than on account of its ideal content. We do not succeed in accomplishing this separation in waking life, because we do not know how to estimate an emotion psychically except in connection with a presentation content. If in kind or in intensity an affect and an idea are incongruous, our waking judgment becomes confused.

The fact that in dreams the presentation content does not entail the affective influence which we should expect as necessary in waking thought has always caused astonishment. Strümpell was of the opinion that ideas in the dream are stripped of their psychic values. But neither does the dream lack opposite instances, where the expression of intense affect appears in a content, which seems to offer no occasion for its development. I am in a horrible, dangerous, or disgusting situation in the dream, but I feel nothing of fear or aversion; on the other hand, I am sometimes terrified at harmless things and glad at childish ones.

This enigma of the dream disappears more suddenly and more completely than perhaps any other of the dream problems, if we pass from the manifest to the latent content. We shall then no longer be concerned to explain it, for it will no longer exist. Analysis teaches us *that presentation contents have undergone displacements and substitutions, while affects have remained unchanged.* No wonder, then, that the presentation content which has been altered by dream disfigurement no longer fits the affect that has remained intact; but there is no cause for wonder either after analysis has put the correct content in its former place.

In a psychic complex which has been subjected to the influence of the resisting censor the affects are the unyielding constituent, which alone is capable of guiding us to a correct supplementation. This state of affairs is revealed in psychoneuroses even more distinctly than in the dream. Here the affect is always in the right, at least as far as its quality goes; its intensity may even be increased by means of a displacement of neurotic attention. If a hysteric is surprised that he is so very afraid of a trifle, or if the patient with compulsive ideas is astonished that he develops such painful self-reproach out of a nonentity, both of them err in that they regard the presentation content—the trifle or the nonentity—as the essential thing, and they defend themselves in vain because they make this presentation content the starting point in their thought. Psychoanalysis, however, shows them the right way by recognising that, on the contrary, the affect is justified, and by searching for the presentation which belongs to it and which has been suppressed by means of replacement. The assumption is here made that the development of affect and the presentation content do not constitute such an indissoluble organic union as we are accustomed to think, but that the two parts may be, so to speak, soldered together in such a way that they may be detached from one another by means of analysis. Dream interpretation shows that this is actually the case.

I give first an example in which analysis explains the apparent absence of affect in a presentation content which ought to force a development of emotion.

I

The dreamer sees three lions in a desert, one of which is laughing, but she is not afraid of them. Then, however, she must have fled from them, for she is trying to climb a tree, but she finds that her cousin, who is a teacher of French, is already up in the tree, &c.

The analysis gives us the following material for this dream: A sentence in the dreamer's English lesson had become the indifferent occasion for it: "The lion's greatest beauty is his mane." Her father wore a beard which surrounded his face like a mane. The name of her English teacher was Miss *Lyons*. An acquaintance of hers had sent her the ballads of *Loewe* (German, Loewe—lion). These, then, are the three lions; why should she have been afraid of them? She has read a story in which a negro who has incited his fellows to revolt is hunted with bloodhounds and climbs a tree to save himself. Then follow fragments in wanton mood, like the following. Directions for catching lions from *Die Fliegende Blaetter*: "Take a desert and strain it; the lions will remain." Also a very amusing, but not very proper anecdote about an official who is asked why he does not take greater pains to win the favour of his superior officer, and who answers that he has been trying to insinuate himself, but that the man ahead of him *is already up*. The whole matter becomes intelligible as soon as one learns that on the day of the dream the lady had received a visit from her husband's superior. He was very polite to her, kissed her hand, and *she was not afraid of him at all*, although he is a "big bug" (German—*Grosses Tier* = "big animal") and plays the part of a "social lion" in the capital of her country. This lion is, therefore, like the lion in the *Midsummer Night's Dream,* who unmasks as Snug, the joiner, and of such stuff are all dream lions made when one is not afraid.

II

As my second example, I cite the dream of the girl who saw her sister's little son lying dead in a coffin, but who, I may now add, felt no pain or sorrow thereat. We know from analysis why not. The dream only concealed her wish to see the man she loved again; the affect must be attuned to the wish, and not to its concealment. There was no occasion for sorrow at all.

In a number of dreams the emotion at least remains connected with that presentation content which has replaced the one really belonging to it. In others the breaking up of the complex is carried further. The affect seems to be entirely separated from the idea belonging to it, and finds a place somewhere else in the dream where it fits into the new arrangement of the dream elements. This is similar to what we have learned of acts of judgment of the dream. If there is a significant inference in the dream thoughts, the dream also contains one; but in the dream the inference may be shifted to entirely different material. Not infrequently this shifting takes place according to the principle of antithesis.

I illustrate the latter possibility by the following dream, which I have subjected to the most exhaustive analysis.

III

A castle by the sea; afterwards it lies not directly on the sea, but on a narrow canal that leads to the sea. A certain Mr. P. is the governor of it. I stand with him in a large salon with three windows, in front of which rise the projections of a wall, like battlements of a fort. I belong to the garrison, perhaps as a volunteer marine officer. We fear the arrival of hostile warships, for we are in a state of war. Mr. P. has the intention of leaving; he gives me instructions as to what must be done in case the dreaded event happens. His sick wife is in the threatened castle with her children. As soon as the bombardment begins the large hall should be cleared. He breathes heavily, and tries to get away; I hold him back, and ask him in what way I should send him news in case of need. He says something else, and then all at once falls over dead. I have probably taxed him unnecessarily with my questions. After his death, which makes no further impression upon me, I think whether the widow is to remain in the castle, whether I should give notice of the death to the commander-in-chief, and whether I should take over the direction of the castle as the next in command. I now stand at the window, and muster the ships as they pass by; they

344

are merchantmen that dart past upon the dark water, several of them with more than one smokestack, others with bulging decks (that are quite similar to the railway stations in the preliminary dream which has not been told). *Then my brother stands next to me, and both of us look out of the window on to the canal. At the sight of a ship we are frightened, and call out: "Here comes the warship."' It turns out, however, that it is only the same ships which I have already known that are returning. Now comes a little ship, strangely cut off, so that it ends in the middle of its breadth; curious things like cups or salt-cellars are seen on the deck. We call as though with one voice: "That is the breakfast-ship."*

The rapid motion of the ships, the deep blue of the water, the brown smoke of the funnels, all this together makes a highly tense, sombre impression.

The localities in this dream are put together from several journeys to the Adriatic Sea (Miramare, Duino, Venice. Aquileja). A short but enjoyable Easter trip to Aquileja with my brother, a few weeks before the dream, was still fresh in my memory. Besides, the naval war between America and Spain, and the worry connected with it about my relatives living in America, play a part. Manifestations of emotion appear at two places in this dream. In one place an emotion that would be expected is lacking—it is expressly emphasized that the death of the governor makes no impression upon me; at another point, where I see the warships I am *frightened,* and experience all the sensations of fright while I sleep. The distribution of affects in this well-constructed dream has been made in such a way that every obvious contradiction is avoided. For there is no reason why I should be frightened at the governor's death, and it is fitting that as the commander of the castle I should be alarmed by the sight of the warship. Now analysis shows that Mr. P. is nothing but a substitute for my own Ego (in the dream I am his substitute). I am the governor who suddenly dies. The dream thoughts deal with the future of those dear to me after my premature death. No other disagreeable thought is to be found among the dream thoughts. The fright which is attached to the sight of the warship must be transferred from it to this disagreeable thought. Inversely, the analysis shows that the region of the dream thoughts from which the warship comes is filled with most joyous reminiscences. It was at Venice a year before, one charmingly beautiful day, that we stood at the windows of our room on the Riva Schiavoni and looked upon the blue lagoon, in which more activity could be seen that day than usually. English ships were being expected, they were to be festively received; and suddenly my wife called out, happy as a child: *"There come the*

English warships!" In the dream I am frightened at the very same words; we see again that speeches in the dream originate from speeches in life. I shall soon show that even the element "English" in this speech has not been lost for the dream activity. I thus convert joy into fright on the way from the dream thoughts to the dream content, and I need only intimate that by means of this very transformation I give expression to a part of the latent dream content. The example shows, however, that the dream activity is at liberty to detach the occasion for an affect from its context in the dream thoughts, and to insert it at any other place it chooses in the dream content.

I seize the opportunity which is incidentally offered, of subjecting to closer analysis the "breakfast ship," whose appearance in the dream so nonsensically concludes a situation that has been rationally adhered to. If I take a closer view of this object in the dream, I am now struck by the fact that it was black, and that on account of its being cut off at its greatest breadth it closely resembled, at the end where it was cut off, an object which had aroused our interest in the museums of the Etruscan cities. This object was a rectangular cup of black clay with two handles, upon which stood things like coffee cups, or tea cups, very similar to our modern *breakfast* table service. Upon inquiring, we learned that this was the toilet set of an Etruscan lady, with little boxes for rouge and powder; and we said jokingly to each other that it would not be a bad idea to take a thing like that home to the lady of the house. The dream object, therefore, signifies *"black toilet"* (German, *toilette*—dress)—mourning—and has direct reference to a death. The other end of the dream object reminds us of the "boat" (German, *Nachen*), from the root [Greek], as a philological friend has told me, upon which corpses were laid in prehistoric times and were left to be buried by the sea. With this circumstance is connected the reason for the return of the ships in the dream.

"Quietly the old man on his rescued boat drifts into the harbour."

It is the return voyage after the ship *wreck* (German, schiff *bruch;* ship-*breaking, i.e.* shipwreck), the breakfast-ship looks as though it were *broken* off in the middle. But whence comes the name "breakfast"-ship? Here is where the "English" comes in, which we have left over from the warships. *Breakfast—a breaking of the fast.* Breaking again belongs to ship-*wreck* (Schiff *bruch*), and *fasting* is connected with the mourning dress.

The only thing about this breakfast-ship, which has been newly created by the dream, is its name. The thing has existed in reality, and recalls to me the merriest hours of my last journey. As we distrusted the fare in Aquileja, we took some food with us from Goerz, and bought a bottle of excellent Istrian wine in Aquileja, and while the little mail-steamer slowly travelled through the Canal delle Mee and into the lonely stretch of lagoon towards Grado, we took our breakfast on deck—we were the only passengers—and it tasted to us as few breakfasts have ever tasted. This, then, was the *"breakfast-ship,"* and it is behind this very recollection of great enjoyment that the dream hides the saddest thoughts about an unknown and ominous future.

The detachment of emotions from the groups of ideas which have been responsible for their development is the most striking thing that happens to them in the course of dream formation, but it is neither the only nor even the most essential change which they undergo on the way from the dream thoughts to the manifest dream. If the affects in the dream thoughts are compared with those in the dream, it at once becomes clear that wherever there is an emotion in the dream, this is also to be found in the dream thoughts; the converse, however, is not true. In general, the dream is less rich in affects than the psychic material from which it is elaborated. As soon as I have reconstructed the dream thoughts I see that the most intense psychic impulses are regularly striving in them for self-assertion, usually in conflict with others that are sharply opposed to them. If I turn back to the dream, I often find it colourless and without any of the more intense strains of feeling. Not only the content, but also the affective tone of my thoughts has been brought by the dream activity to the level of the indifferent. I might say that a *suppression of the affects* has taken place. Take, for example, the dream of the botanical monograph. It answers to a passionate plea for my freedom to act as I am acting and to arrange my life as seems right to me and to me alone. The dream which results from it sounds indifferent; I have written a monograph; it is lying before me; it is fitted with coloured plates, and dried plants are to be found with each copy. It is like the peacefulness of a battlefield; there is no trace left of the tumult of battle.

It may also turn out differently—vivid affective expressions may make their appearance in the dream; but we shall first dwell upon the unquestionable fact that many dreams appear indifferent, while it is never possible to go deeply into the dream thoughts without deep emotion.

A complete theoretical explanation of this suppression of emotions in the course of the dream activity cannot be given here; it would require a most careful investigation of the theory of the emotions and of the mechanism of suppression. I shall find a place here for two thoughts only. I am forced—on other grounds—to conceive the development of affects as a centrifugal process directed towards the interior of the body, analogous to the processes of motor and secretory innervation. Just as in the sleeping condition the omission of motor impulses towards the outside world seems to be suspended, so a centrifugal excitement of emotions through unconscious thought may be made more difficult during sleep. Thus the affective impulses aroused during the discharge of the dream thoughts would themselves be weak excitements, and therefore those getting into the dream would not be stronger. According to this line of argument the "suppression of the affects" would not be a result of the dream activity at all, but a result of the sleeping condition. This may be so. but this cannot possibly be all. We must also remember that all the more complex dreams have shown themselves to be a compromised result from the conflict of psychic forces. On the one hand, the thoughts that constitute the wish must fight the opposition of a censorship; on the other hand, we have often seen how, even in unconscious thinking, each train of thought is harnessed to its contradictory opposite. Since all of these trams of thought are capable of emotion, we shall hardly make a mistake, broadly speaking. if we regard the suppression of emotion as the result of the restraint which the contrasts impose upon one another and which the censor imposes upon the tendencies which it has suppressed. *The restraint of affects would accordingly be the second result of the dream censor as the disfigurement of the dream was the first.*

I shall insert an example of a dream in which the indifferent affective tone of the dream content may be explained by a contrast in the dream thoughts. I have the following short dream to relate, which every reader will read with disgust:

IV

A bit of rising ground, and on it something like a toilet in the open; a very long bench, at the end of which is a large toilet aperture. All of the back edge is thickly covered with little heaps of excrement of all sizes and degrees of freshness. A shrub behind the bench. I urinate upon the bench; a long stream of urine rinses everything clean, the patches of excrement easily come off and fall into the opening. It seems as though something remained at the end nevertheless.

Why did I experience no disgust in this dream?

Because, as the analysis shows, the most pleasant and satisfying thoughts have co-operated in the formation of this dream. Upon analysing it I immediately think of the Augean stables cleansed by Hercules. I am this Hercules. The rising ground and the shrub belong to Aussee, where my children are now staying. I have discovered the infantile etiology of the neuroses and have thus guarded my own children from becoming ill. The bench (omitting the aperture, of course) is the faithful copy of a piece of furniture which an affectionate female patient has made me a present of. This recalls how my patients honour me. Even the museum of human excrement is susceptible of less disagreeable interpretation. However much I am disgusted with it, it is a souvenir of the beautiful land of Italy, where in little cities, as everyone knows, water-closets are not equipped in any other way. The stream of urine that washes everything clean is an unmistakable allusion to greatness. It is in this manner that Gulliver extinguishes the great fire in Lilliput; to be sure, he thereby incurs the displeasure of the tiniest of queens. In this way, too, Gargantua, the superman in Master Rabelais, takes vengeance upon the Parisians, straddling Notre Dame and training his stream of urine upon the city. Only yesterday I was turning over the leaves of Garnier's illustrations of Rabelais before I went to bed. And, strangely enough, this is another proof that I am the superman! The platform of Notre Dame was my favourite nook in Paris; every free afternoon I was accustomed to go up into the towers of the church and climb about among the monsters and devil-masks there. The circumstances that all the excrement vanishes so rapidly before the stream correspond to the motto: *Afflavit et dissipati sunt,* which I shall some day make the title of a chapter on the therapeutics of hysteria.

And now as to the occasion giving rise to the dream. It had been a hot afternoon in summer; in the evening I had given a lecture on the relation between hysteria and the perversions, and everything which I had to say displeased me thoroughly, appeared to me stripped of all value. I was tired, found no trace of pleasure in my difficult task, and longed to get away from this rummaging in human filth, to see my children and then the beauties of Italy. In this mood I went from the auditorium to a café to find some modest refreshment in the open air, for my appetite had left me. But one of my audience went with me; he begged for permission to sit with me while I drank my coffee and gulped down my roll, and began to say flattering things to me. He told me how much he had learned from me, and that he now looked at everything through different eyes, that I had cleansed the Augean stables, *i.e.* the theory of the neuroses, of its errors and prejudices—in short, that I was a very great man. My mood was ill-suited to his song of praise; I struggled with disgust, and went home earlier in order to extricate myself. Before I went to sleep I turned over the leaves of Rabelais, and read a short story by C. F. Meyer entitled *Die Leiden eines Knaben* (The Hardships of a Boy).

The dream had been drawn from these materials, and the novel by Meyer added the recollection of childish scenes (*cf.* the dream about Count Thun, last scene). The mood of the day, characterised by disgust and annoyance, is continued in the dream in the sense that it is permitted to furnish nearly the entire material for the dream content. But during the night the opposite mood of vigorous and even exaggerated self-assertion was awakened, and dissipated the earlier mood. The dream had to take such a form as to accommodate the expression of self-depreciation and exaggerated self-assertion in the same material. This compromise formation resulted in an ambiguous dream content, but likewise in an indifferent strain of feeling owing to the restraint of the contrasts upon each other.

According to the theory of wish-fulfilment this dream could not have happened had not the suppressed, but at the same time pleasurable, train of thought concerning personal aggrandisement been coupled with the opposing thoughts of disgust. For disagreeable things are not intended to be represented by the dream; painful thoughts that have occurred during the day can force their way into the dream only if they lend a cloak to the wish-fulfilment. The dream activity can dispose of the affects in the dream thoughts in still another way, besides admitting them or reducing them to zero. *It can change them into their opposite.* We have already become acquainted with the rule of interpretation that every element of the dream may be interpreted by its opposite, as well as by itself. One can never tell at the outset whether to set down the one or the other; only the connection can decide this point. A suspicion of this state of affairs has evidently got into popular consciousness; dream books very often proceed according to the principle of contraries in their interpretation. Such transformation into opposites is made possible by the intimate concatenation of associations, which in our thoughts finds the idea of a thing in that of its opposite.

Like every other displacement this serves the purposes of the censor, but it is also often the work of the wish-fulfilment, for wish-fulfilment consists precisely in this substitution of an unwelcome thing by its opposite. The emotions of the dream thoughts may appear in the dream transformed into their opposites just as well as the ideas, and it is probable that this inversion of emotions is usually brought about by the dream censor. The *suppression* and *inversion of affects* are useful in social life, as the current analogy for the dream censor has shown us—above all, for purposes of dissimulation. If I converse with a person to whom I must show consideration while I am saying unpleasant things to him, it is almost more important that I should conceal the expression of my emotion from him, than that I modify the wording of my thoughts. If I speak to him in polite words, but accompany them by looks or gestures of hatred and disdain, the effect which I produce upon this person is not very different from what it would have been if I had recklessly thrown my contempt into his face. Above all, then, the censor bids me suppress my emotions, and if I am master of the art of dissimulation, I can hypocritically show the opposite emotion—smiling where I should like to be angry, and pretending affection where I should like to destroy.

We already know of an excellent example of such an inversion of emotion for the purposes of the dream censor. In the dream about my uncle's beard I feel great affection for my friend R., at the same time that, and because, the dream thoughts berate him as a simpleton. We have drawn our first proof for the existence of the censor from this example of the inversion of emotions. Nor is it necessary here to assume that the dream activity creates a counter emotion of this kind out of nothing; it usually finds it lying ready in the material of the dream thoughts, and intensifies it solely with the psychic force of the resisting impulse until a point is reached where the emotion can be won over for the formation of the dream. In the dream of my uncle, just mentioned, the affectionate counter emotion has probably originated from an infantile source (as the continuation of the dream would suggest), for the relation between uncle and nephew has become the source of all my friendships and hatreds, owing to the peculiar nature of my childish experiences (*cf.* analysis on p. 361).

There is a class of dreams deserving the designation "hypocritical," which puts the theory of wish-fulfilment to a severe test. My attention was called to them when Mrs. Dr. M. Hilferding brought up for discussion in the Vienna Psychoanalytic Society the dream reported by Rosegger, which is reprinted below.

In *Waldheimat,* vol. xi., Rosegger writes as follows in his story, *Fremd gemacht,* p. 303:

"I have usually enjoyed healthful sleep, but I have lost the rest of many a night. With my modest existence as a student and literary man, I have for long years dragged along with me the shadow of a veritable tailor's life, like a ghost from which I could not become separated. I cannot say that I have occupied myself so often and so vividly with thoughts of my past during the day. An assailer of heaven and earth arising from the skin of the Philistine has other things to think about. Nor did I, as a dashing young fellow, think about my nocturnal dreams; only later, when I got into the habit of thinking about everything or when the Philistine within me again asserted itself, it struck me that whenever I dreamed I was always the journeyman tailor, and was always working in my master's shop for long hours without any remuneration. As I sat there and sewed and pressed I was quite aware that I no longer belonged there, and that as a burgess of a town I had other things to attend to; but I was for ever having vacations, and going out into the country, and it was then that I sat near my boss and assisted him. I often felt badly, and regretted the loss of time which I might spend for better and more useful purposes. If something did not come up to the measure and cut exactly, I had to submit to a reproach from the boss. Often, as I sat with my back bent in the dingy shop, I decided to give notice that I was going to quit. On one occasion I actually did so, but the boss took no notice of it, and the next time I was again sitting near him and sewing.

"How happy I was when I woke up after such weary hours! And I then resolved that, if this dream came intruding again, I would throw it off with energy and would cry aloud: 'It is only a delusion, I am in bed, and I want to sleep.' . . . And the next night I would be sitting in the tailor shop again.

"Thus years passed with dismal regularity. While the boss and I were working at Alpelhofer's, at the house of the peasant where I began my apprenticeship, it happened that he was particularly dissatisfied with my work. 'I should like to know where in the world your thoughts are?' cried he, and looked at me gloomily, I thought the most sensible thing for me to do would be to get up and explain to the boss that I was with him only as a favour, and then leave. But I did not do this. I submitted, however, when the boss engaged an apprentice, and ordered me to make room for him on the bench. I moved into the corner, and kept on sewing. On the same day another tailor was engaged; he was bigoted, as he was a Czech who had worked for us nineteen years before, and then had fallen into the lake on his way home from the public-house. When he tried to sit down there was no room for him. I looked at the boss inquiringly, and he said to me, 'You have no talent for the tailoring business; you may go; you are free.' My fright on that occasion was so overpowering that I awoke.

"The morning gray glimmered through the clear window of my beloved home. Objects of art surrounded me; in the tasteful bookcase stood the eternal Homer, the gigantic Dante, the incomparable Shakespeare, the glorious Goethe—all shining and immortal. From the adjoining room resounded the clear little voices of the children, who were waking and prattling with their mother. I felt as if I had found again that idyllically sweet, that peaceful, poetical, and spiritual life which I have so often and so deeply conceived as the contemplative fortune of mankind. And still I was vexed that I had not given my boss notice first, instead of allowing him to discharge me.

"And how remarkable it is; after the night when the boss 'discharged me' I enjoyed rest; I no longer dreamed of my tailoring—of this experience which lay in the remote past, which in its simplicity was really happy, and which, nevertheless, threw a long shadow over the later years of my life."

I

In this dream, the series of the poet who, in his younger years, has been a journeyman tailor, it is hard to recognise the domination of the wish-fulfilment. All the delightful things occurred during the waking state, while the dream seemed to drag along the ghostlike shadow of an unhappy existence which had been long forgotten. My own dreams of a similar nature have put me in a position to give some explanation for such dreams. As a young doctor I for a long time worked in the chemical institute without being able to accomplish anything in that exacting science, and I therefore never think in my waking state about this unfruitful episode in my life, of which I am really ashamed. On the other hand, it has become a recurring dream with me that I am working in the laboratory, making analyses, and having experiences there, &c.; like the examination dreams, these dreams are disagreeable, and they are never very distinct. During the analysis of one of these dreams my attention was directed to the word "analysis," which gave me the key to an understanding of these dreams. For I had since become an "analyst." I make analyses which are highly praised—to be sure, psychoanalyses. I then understood that when I grew proud of these analyses of the waking state, and wanted to boast how much I had accomplished thereby, the dream would hold up to me at night those other unsuccessful analyses of which I had no reason to be proud; they are the punitive dreams of the upstart, like those of the tailor who became a celebrated poet. But how is it possible for the dream to place itself at the service of self-criticism in its conflict with parvenu-pride, and to take as its content a rational warning instead of the fulfilment of a prohibitive wish? I have already mentioned that the answer to this question entails many difficulties. We may conclude that the foundation of the dream was at first formed by a phantasy of overweening ambition, but that only its suppression and its abashment reached the dream content in its stead. One should remember that there are masochistic tendencies in the psychic life to which such an inversion might be attributed. But a more thorough investigation of the individual dreams allows the recognition of still another element. In an indistinct subordinate portion of one of my laboratory dreams, I was just at the age which placed me in the most gloomy and most unsuccessful year of my professional career; I still had no position and no means of support, when I suddenly found that I had the choice of many women whom I could marry! I was, therefore, young again, and, what is more, she was young again—the woman who has shared with me all these hard years. In this way one of the wishes which constantly frets the heart of the ageing man was revealed as the unconscious dream inciter. The struggle raging in the other psychic strata between vanity and self-criticism has certainly determined the dream content, but the more deeply-rooted wish of youth has alone made it possible as a dream. One may say to himself even in the waking state: To

354

be sure it is very nice now, and times were once very hard; but it was nice, too, even then, you were still so young.

In considering dreams reported by a poet one may often assume that he has excluded from the report those details which he perceived as disturbing and which he considered unessential. His dreams, then, give us a riddle which could be readily solved if we had an exact reproduction of the dream content.

O. Rank has called my attention to the fact that in Grimm's fairy tale of the valiant little tailor, or "Seven at one Stroke," a very similar dream of an upstart is related. The tailor, who became the hero and married the king's daughter, dreamed one night while with the princess, his wife, about his trade; the latter, becoming suspicious, ordered armed guards for the following night, who should listen to what was spoken in the dream, and who should do away with the dreamer. But the little tailor was warned, and knew enough to correct his dream.

The complex of processes—of suspension, subtraction, and inversion—through which the affects of the dream thoughts finally become those of the dream, may well be observed in the suitable synthesis of completely analysed dreams. I shall here treat a few cases of emotional excitement in the dream which furnish examples of some of the cases discussed.

In the dream about the odd task which the elder Bruecke gives me to perform—of preparing my own pelvis—the *appropriate horror is absent in the dream itself.* Now this is a wish-fulfilment in various senses. Preparation signifies self-analysis, which I accomplish, as it were, by publishing my book on dreams, and which has been so disagreeable to me that I have already postponed printing the finished manuscript for more than a year. The wish is now actuated that I may disregard this feeling of opposition. and for that reason I feel no horror (*Grauen,* which also means to grow grey) in the dream. I should also like to escape the horror—in the other (German) sense—of growing grey; for I am already growing grey fast. and the grey in my hair warns me withal to hold back no longer. For we know that at the end of the dream the thought secures expression in that I should have to leave my children to get to the goal of their difficult journey.

In the two dreams that shift the expression of satisfaction to the moments immediately after awakening, this satisfaction is in the one case motivated by the expectation that I am now going to learn what is meant by "I have already dreamed of it," and refers in reality to the birth of my first child, and in the other case it is motivated by the conviction that "that which has been announced by a sign" is now going to happen, and the latter satisfaction is the same which I felt at the arrival of my second son. Here the same emotions that dominated in the dream thoughts have remained in the dream, but the process is probably not so simple as this in every dream. If the two analyses are examined a little, it will be seen that this satisfaction which does not succumb to the censor receives an addition from a source which must fear the censor; and the emotion drawn from this source would certainly arouse opposition if it did not cloak itself in a similar emotion of satisfaction that is willingly admitted, if it did not, as it were, sneak in behind the other. Unfortunately, I am unable to show this in the case of the actual dream specimen, but an example from another province will make my meaning intelligible. I construct the following case: Let there be a person near me whom I hate so that a strong feeling arises in me that I should be glad if something were to happen to him. But the moral part of my nature does not yield to this sentiment; I do not dare to express this ill-wish, and when something happens to him which he does not deserve, I suppress my satisfaction at it, and force myself to expressions and thoughts of regret. Everyone will have found himself in such a position. But now let it happen that the hated person draws upon himself a well-deserved misfortune by some fault; now I may give free rein to my satisfaction that he has been visited by a just punishment, and I express opinion in the matter which coincides with that of many other people who are impartial. But I can see that my satisfaction turns out to be more intense than that of the others, for it has received an addition from another source—from my hatred, which has hitherto been prevented by the inner censor from releasing an emotion, but which is no longer prevented from doing so under the altered circumstances. This case is generally typical of society, where persons who have aroused antipathy or are adherents of an unpopular minority incur guilt. Their punishment does not correspond to their transgression but to their transgression *plus* the ill-will directed against them that has hitherto been ineffective. Those who execute the punishment doubtless commit an injustice, but they are prevented from becoming aware of it by the satisfaction arising from the release within themselves of a suppression of long standing. In such cases the emotion is justified according to its quality, but not according to its quantity; and the self-criticism that has been appeased as to the one point is only too ready to neglect examination of the second point. Once you have opened the doors, more people get through than you originally intended to admit.

The striking feature of the neurotic character, that incitements capable of producing emotion bring about a result that is qualitatively justified but is quantitatively excessive, is to be explained in this manner, in so far as it admits of a psychological explanation at all. The excess is due to sources of emotion which have remained unconscious and have hitherto been suppressed, which can establish in the associations a connection with the actual incitement, and which can thus find release for its emotions through the vent which the unobjectionable and admitted source of emotion opens. Our attention is thus called to the fact that we may not consider the relation of mutual restraint as obtaining exclusively between the suppressed and the suppressing psychic judgment. The cases in which the two judgments bring about a pathological emotion by co-operation and mutual strengthening deserve just as much attention. The reader is requested to apply these hints regarding the psychic mechanism for the purpose of understanding the expressions of emotion in the dream. A satisfaction which makes its appearance in the dream, and which may readily be found at its proper place in the dream thoughts, may not always be fully explained by means of this reference. As a rule it will be necessary to search for a second source in the dream thoughts, upon which the pressure of the censor is exerted, and which under the pressure would have resulted not in satisfaction, but in the opposite emotion—which, however, is enabled by the presence of the first source to free its satisfaction affect from suppression and to reinforce the satisfaction springing from the other source. Hence emotions in the dream appear as though formed by the confluence of several tributaries, and as though over-determined in reference to the material of the dream thoughts; *sources of affect which can furnish the same affect join each other in the dream activity in order to produce it.*[49]

Some insight into these tangled relations is gained from analysis of the admirable dream in which "Non vixit" constitutes the central point (*cf.* p. 358). The expressions of emotion in this dream, which are of different qualities, are forced together at two points in the manifest content. Hostile and painful feelings (in the dream itself we have the phrase, "seized by strange emotions") overlap at the point where I destroy my antagonistic friend with the two words. At the end of the dream I am greatly pleased, and am quite ready to believe in a possibility which I recognise as absurd when I am awake, namely, that there are revenants who can be put out of the way by a mere wish.

I have not yet mentioned the occasion for this dream. It is an essential one, and goes a long way towards explaining it. I had received the news from my friend in Berlin (whom I have designated as F.) that he is about to undergo an operation and that relatives of his living in Vienna would give me information about his condition. The first few messages after the operation were not reassuring, and caused me anxiety. I should have liked best to go to him myself, but at that time I was affected with a painful disease which made every movement a torture for me. I learn from the dream thoughts that I feared for the life of my dear friend. I knew that his only sister, with whom I had not been acquainted, had died early after the shortest possible illness. (In the dream F. *tells about his sister, and says: "In three-quarters of an hour she was dead."*) I must have imagined that his own constitution was not much stronger, and that I should soon be travelling, in spite of my health, in answer to far worse news—and that I should arrive too *late,* for which I should reproach myself for ever.[50] This reproach about arriving too late has become the central point of the dream, but has been represented in a scene in which the honoured teacher of my student years—Bruecke—reproaches me for the same thing with a terrible look from his blue eyes. The cause of this deviation from the scene will soon be clear; the dream cannot reproduce the scene itself in the manner in which it occurred to me. To be sure, it leaves the blue eyes to the other man, but it gives me the part of the annihilator, an inversion which is obviously the result of the wish-fulfilment. My concern for the life of my friend, my self-reproach for not having gone to him, my shame (he had repeatedly come to me in Vienna), my desire to consider myself excused on account of my illness—all of this makes up a tempest of feeling which is distinctly felt in sleep, and which raged in every part of the dream thoughts.

But there was another thing about the occasion for the dream which had quite the opposite effect. With the unfavourable news during the first days of the operation, I also received the injunction to speak to no one about the whole affair, which hurt my feelings, for it betrayed an unnecessary distrust of my discretion. I knew, of course, that this request did not proceed from my friend, but that it was due to clumsiness or excessive timidity on the part of the messenger. but the concealed reproach made me feel very badly because it was not altogether unjustified. Only reproaches which "have something in them" have power to irritate, as everyone knows. For long before, in the case of two persons who were friendly to each other and who were willing to honour me with their friendship, I had quite needlessly tattled what the one had said about the other; to be sure this incident had nothing to do with the affairs of my friend F. Nor have I forgotten the reproaches which I had to listen to at that time. One of the two friends between whom I was the trouble-maker was Professor Fleischl; the other one I may name Joseph, a name which was also borne by my friend and antagonist P., who appears in the dream.

Two dream elements, first *inconspicuously,* and secondly the question of *Fl. as to how much of his affairs I have mentioned to P.,* give evidence of the reproach that I am incapable of keeping anything to myself. But it is the admixture of these recollections which transposes the reproach for arriving too late from the present to the time when I was living in Bruecke's laboratory; and by replacing the second person in the annihilation scene of the dream by a Joseph I succeed in representing not only the first reproach that I arrive too late, but also a second reproach, which is more rigorously suppressed, that I keep no secrets. The condensing and replacing activity of this dream, as well as the motives for it, are now obvious.

My anger at the injunction not to give anything away, originally quite insignificant, receives confirmation from sources that flow far below the surface, and so become a swollen stream of hostile feelings towards persons who are in reality dear to me. The source which furnishes the confirmation is to be found in childhood. I have already said that my friendships as well as my enmities with persons of my own age go back to my childish relations with my nephew, who was a year older than I. In these he had the upper hand, and I early learned how to defend myself; we lived together inseparably, loved each other, and at the same time, as statements of older persons testify, scuffled with and accused each other. In a certain sense all my friends are incarnations of this first figure, "which early appeared to my blurred sight"; they are all *revenants.* My nephew himself returned in the years of adolescence, and then we acted Caesar and Brutus. An intimate friend and a hated enemy have always been indispensable requirements for my emotional life; I have always been able to create them anew, and not infrequently my childish ideal has been so closely approached that friend and enemy coincided in the same person, not simultaneously, of course, nor in repeated alterations, as had been the case in my first childhood years.

I do not here wish to trace the manner in which a recent occasion for emotion may reach back to one in childhood—through connections like these I have just described—in order to find a substitute for itself, in this earlier occasion for the sake of increased emotional effect. Such an investigation would belong to the psychology of the unconscious, and would find its place in a psychological explanation of neuroses. Let us assume for the purposes of dream interpretation that a childhood recollection makes its appearance or is formed by the fancy, say to the following effect: Two children get into a fight on account of some object—just what we shall leave undecided, although memory or an allusion of memory has a very definite one in mind—and each one claims that he got to it first, and that he, therefore, has first right to it. They come to blows, for might makes right; and, according to the intimation of the dream, I must have known that I was in the wrong (*noticing the error myself*), but this time I remain the stronger and take possession of the battlefield; the defeated combatant hurries to my father, his grandfather, and accuses me, and I defend myself with the words which I know from my father: *"I hit him because he hit me."* Thus this recollection, or more probably fancy, which forces itself upon my attention in the course of the analysis—from my present knowledge I myself do not know how—becomes an intermediary of the dream thoughts that collects the emotional excitements obtaining in the dream thoughts, as the bowl of a fountain collects the streams of water flowing into it. From this point the dream thoughts flow along the following paths: "It serves you quite right if you had to vacate your place for me; why did you try to force me out of my place? I don't need you; I'll soon find someone else to play with," &c. Then the ways are opened through which these thoughts again follow into the representation of the dream. For such an "ôte-toi que je m'y mette" I once had to reproach my deceased friend Joseph. He had been next to me in the line of promotion in Bruecke's laboratory, but advancement there was very slow. Neither of the two assistants budged from bis place, and youth became impatient. My friend, who knew that his time of life was limited, and who was bound by no tie to his superior, was a man seriously ill; the wish for his removal permitted an objectionable interpretation—he might be moved by something besides promotion. Several years before, the same wish for freedom had naturally been more intense in my own case;

wherever in the world there are gradations of rank and advancement, the doors are opened for wishes needing suppression. Shakespeare's Prince Hal cannot get rid of the temptation to see how the crown fits even at the bed of his sick father. But, as may easily be understood, the dream punishes this ruthless wish not upon me but upon him.[51]

"As he was ambitious, I slew him." As he could not wait for the other man to make way for him, he himself has been put out of the way. I harbour these thoughts immediately after attending the unveiling of the statue to the other man at the university. A part of the satisfaction which I feel in the dream may therefore be interpreted: Just punishment; it served you right.

At the funeral of this friend a young man made the following remark, which seemed out of place: "The preacher talked as though the world couldn't exist without this one human being." The displeasure of the sincere man, whose sorrow has been marred by the exaggeration, begins to arise in him. But with this speech are connected the dream thoughts: "No one is really irreplaceable; how many men have I already escorted to the grave, but I am still living, I have survived them all, I claim the field." Such a thought at the moment when I fear that when I travel to see him I shall find my friend no longer among the living, permits only of the further development that I am glad I am surviving someone, that it is not I who have died, but he—that I occupy the field as I once did in the fancied scene in childhood. This satisfaction, coming from sources in childhood, at the fact that I claim the field, covers the larger part of the emotion which appears in the dream. I am glad that I am the survivor—I express this sentiment with the naïve egotism of the husband who says to his wife: "If one of us dies, I shall move to Paris." It is such a matter of course for my expectation that I am not to be the one.

It cannot be denied that great self-control is necessary to interpret one's dreams and to report them. It is necessary for you to reveal yourself as the one scoundrel among all the noble souls with whom you share the breath of life. Thus, I consider it quite natural that *revenants* exist only as long as they are wanted, and that they can be obviated by a wish. This is the thing for which my friend Joseph has been punished. But the *revenants* are the successive incarnations of the friend of my childhood; I am also satisfied at the fact that I have replaced this person for myself again and again, and a substitute will doubtless soon be found even for the friend whom I am about to lose. No one is irreplaceable.

But what has the dream censor been doing meanwhile? Why does it not raise the most emphatic objection to a train of thought characterised by such brutal selfishness, and change the satisfaction that adheres to it into profound repugnance? I think it is because other unobjectionable trains of thought likewise result in satisfaction and cover the emotion coming from forbidden infantile sources with their own. In another stratum of thought I said to myself at that festive unveiling: "I have lost so many dear friends, some through death, some through the dissolution of friendship—is it not beautiful that I have found substitutes for them, that I have gained one who means more to me than the others could, whom I shall from now on always retain, at the age when it is not easy to form new friendships?" The satisfaction that I have found this substitute for lost friends can be taken over into the dream without interference, but behind it there sneaks in the inimical satisfaction from the infantile source. Childish affection undoubtedly assists in strengthening the justifiable affection of to-day; but childish hatred has also found its way into the representation.

But besides this there is distinct reference in the dream to another chain of thoughts, which may manifest itself in the form of satisfaction. My friend had shortly before had a little daughter born, after long waiting. I knew how much he had grieved for the sister whom he lost at an early age, and I wrote to him that he would transfer to this child the love he had felt for her. This little girl would at last make him forget his irreparable loss.

Thus this chain also connects with the intermediary thoughts of the latent dream content, from which the ways spread out in opposite directions: No one is irreplaceable. You see, nothing but *revenants;* all that one has lost comes back. And now the bonds of association between the contradictory elements of the dream thoughts are more tightly drawn by the accidental circumstance that the little daughter of my friend bears the same name as the girl playmate of my own youth, who was just my own age and the sister of my oldest friend and antagonist. I have heard the name "Pauline" with *satisfaction,* and in order to allude to this coincidence I have replaced one Joseph in the dream by another Joseph, and have not overlooked the similarity in sound between the names Fleischl and F. From this point a train of thought runs to the naming of my own children. I insisted that the names should not be chosen according to the fashion of the day but should be determined by regard for the memory of beloved persons. The children's names make them *"revenants."* And, finally, is not the having of children the only access to immortality for us all?

I shall add only a few remarks about the emotions of the dream from another point of view. An emotional inclination—what we call a mood—may occur in the mind of a sleeping person as its dominating element, and may induce a corresponding mood in the dream. This mood may be the result of the experiences and thoughts of the day, or it may be of somatic origin; in either case it will be accompanied by the chains of thought that correspond to it. The fact that in the one case this presentation content conditions the emotional inclination primarily, and that in the other case it is brought about secondarily by a disposition of feeling of somatic origin remains without influence upon the formation of the dream. This formation is always subject to the restriction that it can represent only a wish-fulfilment, and that it may put its psychic motive force at the service only of the wish. The mood that is actually present will receive the same treatment as the sensation which actually comes to the surface during sleep (*cf.* p. 205), which is either neglected or reinterpreted so as to signify a wish-fulfilment. Disagreeable moods during sleep become a motive force of the dream by actuating energetic wishes, which the dream must fulfil. The material to which they are attached is worked over until it finally becomes suitable for the expression of the fulfilled wish. The more intense and the more dominating the element of the disagreeable mood in the dream thought, the more surely will the wish-impulses that have been most rigorously suppressed take advantage of the opportunity to secure representation, for they find that the difficult part of the work necessary in securing representation has already been accomplished in that the repugnance is already actually in existence, which they would otherwise have had to produce by their own effort. With this discussion we again touch upon the problem of anxiety dreams, which we may regard as bounding the province of the dream activity.

(H) Secondary Elaboration

We may at last proceed to an exposition of the fourth of the factors which take part in the formation of the dream.

If we continue the examination of the dream content, in the manner already outlined—that is, by testing striking occurrences as to their origin in the dream thoughts—we encounter elements which can be explained only by making an entirely new assumption. I have in mind cases where one shows astonishment, anger, or resistance in a dream, and that, too, against a party of the dream content itself. Most of these exercises of the critical faculty in dreams are not directed against the dream content, but prove to be portions of dream material which have been taken over and suitably made use of, as I have shown by fitting examples. Some things of this sort, however, cannot be disposed of in such a way; their correlative cannot be found in the dream material. What, for instance, is meant by the criticism not infrequent in dreams: "Well, it's only a dream"? This is a genuine criticism of the dream such as I might make if I were awake. Not at all infrequently it is the forerunner to waking; still oftener it is preceded by a painful feeling, which subsides when the certainty of the dream state has been established. The thought: "But it's only a dream," occurring during the dream, has the same object which is meant to be conveyed on the stage through the mouth of the beautiful Helen von Offenbach; it wants to minimise what has just occurred and secure indulgence for what is to follow. Its purpose is to reassure and, so to speak, put to sleep a certain instance which at the given moment has every reason to be active and to forbid the continuation of the dream—or the scene. It is pleasanter to go on sleeping and to tolerate the dream, "because it's only a dream anyway." I imagine that the disparaging criticism. "But it's only a dream," enters into the dream at the moment when the censor, which has never been quite asleep, feels that it has been surprised by the already admitted dream. It is too late to suppress the dream, and the instance therefore carries with it that note of fear or of painful feeling which presents itself in the dream. It is an expression of the *esprit d'escalier* on the part of the psychic censor.

In this example we have faultless proof that not everything which the dream contains comes from the dream thoughts, but that a psychic function which cannot be differentiated from our waking thoughts may make contributions to the dream content. The question now is, does this occur only in altogether exceptional cases, or does the psychic instance which is usually active only as censor take a regular part in the formation of dreams?

One must decide unhesitatingly for the latter view. It is indisputable that the censoring instance, whose influence we have so far recognised only in limitations and omissions in the dream content, is also responsible for interpolations and amplifications in this content. Often these interpolations are easily recognised; they are reported irresolutely, prefaced by an "as if," they are not in themselves particularly vivid, and are regularly inserted at points where they may serve to connect two portions of the dream content or improve the sequence between two sections of the dream. They manifest less ability to stick in the memory than genuine products of the dream material; if the dream is subject to forgetting, they are the first to fall away, and I am strongly inclined to believe that our frequent complaint that we have dreamed so much, that we have forgotten most of this and have remembered only fragments of it, rests on the immediate falling away of just these cementing thoughts. In a complete analysis these interpolations are often betrayed by the fact that no material is to be found for them in the dream thoughts. But after careful examination I must designate this case as a rare one; usually interpolated thoughts can be traced to an element in the dream thoughts, which, however, can claim a place in the dream neither on account of its own merit nor on account of over-determination. The psychic function in dream formation, which we are now considering, aspires to the original creations only in the most extreme cases; whenever possible, it makes use of anything available it can find in the dream material.

The thing which distinguishes and reveals this part of the dream activity is its tendency. This function proceeds in a manner similar to that which the poet spitefully attributes to the philosopher; with its scraps and rags, it stops up the breaches in the structure of the dream. The result of its effort is that the dream loses the appearance of absurdity and incoherence, and approaches the pattern of an intelligible experience. But the effort is not always crowned with complete success. Thus dreams occur which may seem faultlessly logical and correct upon superficial examination; they start from a possible situation, continue it by means of consistent changes, and end up—although this is very rare—with a not unnatural conclusion. These dreams have been subjected to the most thorough elaboration at the hands of a psychic function similar to our waking thought; they seem to have a meaning, but this meaning is very far removed from the real signification of the dream. If they are analysed, one is convinced that the secondary elaboration has distorted the material very freely, and has preserved its proper relations as little as possible. These are the dreams which have, so to speak, already been interpreted before we subject them to waking interpretation. In other dreams this purposeful elaboration has been successful only to a certain point; up to this point consistency seems to be dominant, then the dream becomes nonsensical or confused, and perhaps finally it lifts itself for a second time in its course to an appearance of rationality. In still other dreams the elaboration has failed completely; we find ourselves helpless in the presence of a senseless mass of fragmentary contents.

I do not wish to deny to this fourth dream-moulding power, which will soon seem to us a familiar one—it is in reality the only one among the four dream-moulders with which we are familiar,—I do not wish to deny this fourth factor the capability of creatively furnishing the dream with new contributions. But surely its influence, like that of the others, manifests itself preponderatingly in the preferring and choosing of already created psychic material in the dream thoughts. Now there is a case where it is spared the work, for the most part, of building, as it were, a façade to the dream, by the fact that such a structure, waiting to be used, is already to be found complete in the material of the dream thoughts. The element of the dream thoughts which I have in mind, I am in the habit of designating as a "phantasy"; perhaps I shall avoid misunderstanding if I immediately adduce the day dream of waking life as an analogy.[52] The part played by this element in our psychic life has not yet been fully recognised and investigated by the psychiatrists; in this study M. Benedikt has, it seems to me, made a highly promising beginning. The significance of the day dream has not yet escaped the unerring insight of poets; the description of the day dreams of one of his subordinate characters which A. Daudet gives us in *Nabab* is universally known. A study of the psychoneuroses discloses the astonishing fact that these phantasies or day dreams are the immediate predecessors of hysterical symptoms—at least of a great many of them; hysterical symptoms directly depend not upon the memories themselves, but upon phantasies built on the basis of memories. The frequent occurrence of conscious day phantasies brings these formations within the scope of our knowledge; but just as there are such conscious phantasies, so there are a great many unconscious ones, which must remain unconscious on account of their content and on account of their origin from repressed material. A more thorough examination into the character of these day phantasies shows with what good reason the same name has been given to these formations as to the products of our nocturnal thought,—dreams. They possess an essential part of their properties in common with nocturnal dreams; an examination of them would really have afforded the shortest and best approach to an understanding of night dreams.

Like dreams, they are fulfilments of wishes; like dreams a good part of them are based upon the impressions of childish experiences; like dreams their creations enjoy a certain amount of indulgence from the censor. If we trace their formation, we see how the wish motive, which is active in their production, has taken the material of which they are built, mixed it together, rearranged it, and composed it into a new unit. They bear the same relation to the childish memories, to which they go back, as some of the quaint palaces of Rome bear to the ancient ruins, whose freestones and pillars have furnished the material for the structure built in modern form.

In the "secondary elaboration" of the dream content which we have ascribed to our fourth dream-making factor, we again find the same activity which in the creation of day dreams is allowed to manifest itself unhampered by other influences. We may say without further preliminary that this fourth factor of ours seeks to form something *like a day dream* from the material at hand. Where, however, such a day dream has already been formed in connection with the dream thought, this factor of the dream-work will preferably get control of it, and strive to introduce it into the dream content. There are dreams which consist merely of the repetition of such a day fancy, a fancy which has perhaps remained unconscious—as, for instance, the dream of the boy that he is riding with the heroes of the Trojan war in a war chariot. In my dream "Autodidasker," at least the second part of the dream is the faithful repetition of a day phantasy—harmless in itself—about my dealings with Professor N. The fact that the phantasy thus provided more often forms only one part of the dream, or that only one part of the phantasy that makes its way to the dream content, has its origin in the complexity of the conditions which the dream must satisfy at its genesis. On the whole, the phantasy is treated like any other component of the latent material; still it is often recognisable in the dream as a whole. In my dreams parts often occur which are emphasized by an impression different from that of the rest. They seem to me to be in a state of flux, to be more coherent and at the same time more transient than other pieces of the same dream. I know that these are unconscious phantasies which get into the dream by virtue of their association, but I have never succeeded in registering such a phantasy. For the rest these phantasies, like all other component parts of the dream thoughts, are jumbled together and condensed, one covered up by another, and the like; but there are all degrees, from the case where they may constitute the dream content or at least the dream façade unchanged to the opposite case, where they are represented in the dream content by only one of their elements or by a remote allusion to such an element. The extent to which the phantasies are able to withstand the demands of the censor and the tendency to condensation are, of course, also decisive of their fate among the dream thoughts.

In my choice of examples for dream analysis I have, wherever possible, avoided those dreams in which unconscious fancies play a somewhat important part, because the introduction of this psychic element would have necessitated extensive discussion of the psychology of unconscious thought. But I cannot entirely omit the "phantasy" even in this matter of examples, because it often gets fully into the dream and still more often distinctly pervades it. I may mention one more dream, which seems to be composed of two distinct and opposed phantasies, overlapping each other at certain places, of which the first is superficial, while the second becomes, as it were, the interpreter of the first.[53]

The dream—it is the only one for which I have no careful notes—is about to this effect: The dreamer—an unmarried young man—is sitting in an inn, which is seen correctly; several persons come to get him, among them someone who wants to arrest him. He says to his "table companions, "I will pay later, I am coming back." But they call to him, laughing scornfully: "We know all about that; that's what everybody says." One guest calls after him: "There goes another one." He is then led to a narrow hall, where he finds a woman with a child in her arms. One of his escorts says: "That is Mr. Müller." A commissioner or some other official is running through a bundle of tickets or papers repeating Müller, Müller, Müller. At last the commissioner asks him a question, which he answers with "Yes." He then takes a look at the woman, and notices that she has grown a large beard.

The two component parts are here easily separated. What is superficial is the *phantasy of being arrested;* it seems to be newly created by the dream-work. But behind it appears the *phantasy of marriage,* and this material, on the contrary, has undergone but slight change at the hands of the dream activity. The features which are common to both phantasies come into distinct prominence as in a Galton's composite photograph. The promise of the bachelor to come back to his place at the club table, the scepticism of the drinking companions, sophisticated in their many experiences, the calling after: "There goes (marries) another one,"—all these features can easily be capable of the other interpretation. Likewise the affirmative answer given to the official. Running through the bundle of papers with the repetition of the name, corresponds to a subordinate but well-recognised feature of the marriage ceremonies—the reading aloud of the congratulatory telegrams which have arrived irregularly, and which, of course, are all addressed to the same name. In the matter of the bride's personal appearance in this dream, the marriage phantasy has even got the better of the arrest phantasy which conceals it. The fact that this bride finally displays a beard, I can explain from an inquiry—I had no chance to make an analysis. The dreamer had on the previous day crossed the street with a friend who was just as hostile to marriage as himself, and had called his friend's attention to a beautiful brunette who was coming towards them. The friend had remarked: "Yes, if only these women wouldn't get beards, as they grow older, like their fathers."

Of course there is no lack of elements in this dream, on which the dream disfigurement has done more thorough work. Thus the speech: "I will pay later," may have reference to the conduct of the father-in-law in the matter of dowry—which is uncertain. Obviously all kinds of scruples are preventing the dreamer from surrendering himself with pleasure to the phantasy of marrying. One of these apprehensions—lest one's freedom be lost when one marries—has embodied itself in the transformation to a scene of arrest.

Let us return to the thesis that the dream activity likes to make use of a phantasy which is finished and at hand, instead of creating one afresh from the material of the dream thoughts; we shall perhaps solve one of the most interesting riddles of the dream if we keep this fact in mind. I have on page 29 related the dream of Maury, who is struck on the back of the neck with a stick, and who awakes in the possession of a long dream—a complete romance from the time of the French Revolution. Since the dream is represented as coherent and as explicable by reference to the disturbing stimulus alone, about the occurrence of which stimulus the sleeper could suspect nothing, only one assumption seems to be left, namely, that the whole richly elaborated dream must have been composed and must have taken place in the short space of time between the falling of the stick on Maury's cervical vertebra and the awakening induced by the blow. We should not feel justified in ascribing such rapidity to the waking mental activity, and so are inclined to credit the dream activity with a remarkable acceleration of thought as one of its characteristics.

Against this inference, which rapidly becomes popular, more recent authors (Le Lorrain, Egger, and others) have made emphatic objection. They partly doubt the correctness with which the dream was reported by Maury, and partly try to show that the rapidity of our waking mental capacity is quite as great as that which we may concede without reservation to the dream activity. The discussion raises fundamental questions, the settlement of which I do not think concerns me closely. But I must admit that the argument, for instance, of Egger has not impressed me as convincing against the guillotine dream of Maury. I would suggest the following explanation of this dream: Would it be very improbable that the dream of Maury exhibits a phantasy which had been preserved in his memory in a finished state for years, and which was awakened—I should rather say alluded to—at the moment when he became aware of the disturbing stimulus? The difficulty of composing such a long story with all its details in the exceedingly short space of time which is here at the disposal of the dreamer then disappears; the story is already composed. If the stick had struck Maury's neck when he was awake there would perhaps have been time for the thought: "Why, that's like being guillotined." But as he is struck by the stick while asleep, the dream activity quickly finds occasion in the incoming stimulus to construct a wish-fulfilment, as though it thought (this is to be taken entirely figuratively): "Here is a good opportunity to realise the wish phantasy which I formed at such and such a time while I was reading." That this dream romance is just such a one as a youth would be likely to fashion under the influence of powerful impressions does not seem questionable to me. Who would not have been carried away—especially a Frenchman and a student of the history of civilisation—by descriptions of the Reign of Terror, in which the aristocracy, men and women, the flower of the nation, showed that it was possible to die with a light heart, and preserved their quick wit and refinement of life until the fatal summons? How tempting to fancy one's self in the midst of all this as one of the young men who parts from his lady with a kiss of the hand to climb fearlessly upon the scaffold! Or perhaps ambition is the ruling motive of the phantasy—the ambition to put one's self in the place of one of those powerful individuals who merely, by the force of their thinking and their fiery eloquence, rule the city in which the heart of mankind is beating so convulsively, who are impelled by conviction to send thousands of human beings to their death, and who pave the way for the transformation of Europe; who, meanwhile, are not sure of their own heads, and may one day lay them under the knife of the guillotine, perhaps in the rôle of one of the Girondists or of the hero Danton? The feature, "accompanied by an innumerable multitude," which is preserved in the memory, seems to show that Maury's phantasy is an ambitious one of this sort.

But this phantasy, which has for a long time been ready, need not be experienced again in sleep; it suffices if it is, so to speak, "touched off." What I mean is this: If a few notes are struck and someone says, as in *Don Juan:* "That is from *Figaro's Wedding* by Mozart," memories suddenly surge up within me, none of which I can in the next moment recall to consciousness. The characteristic phrase serves as an entrance station from which a complete whole is simultaneously put in motion. It need not be different in the case of unconscious thought. The psychic station which opens the way to the whole guillotine phantasy is set in motion by the waking stimulus. This phantasy, however, is not passed in review during sleep, but only afterwards in waking memory. Upon awakening one remembers the details of the phantasy, which in the dream was regarded as a whole. There is, withal, no means of making sure that one really has remembered anything which has been dreamed. The same explanation, namely, that one is dealing with finished phantasies which have been set in motion as wholes by the waking stimulus, may be applied to still other dreams which proceed from a waking stimulus—for instance to the battle dream of Napoleon at the explosion of the bomb. I do not mean to assert that all waking dreams admit of this explanation, or that the problem of the accelerated discharge of ideas in dreams is to be altogether solved in this manner.

We must not neglect the relation of this secondary elaboration of the dream content to the other factors in the dream activity. Might the procedure be as follows: the dream-creating factors, the impulse to condense, the necessity of evading the censor, and the regard for dramatic fitness in the psychic resources of the dream—these first of all create a provisional dream content, and this is then subsequently modified until it satisfies the exactions of a second instance? This is hardly probable. It is necessary rather to assume that the demands of this instance are from the very beginning lodged in one of the conditions which the dream must satisfy, and that this condition, just like those of condensation, of censorship, and of dramatic fitness, simultaneously affect the whole mass of material in the dream thoughts in an inductive and selective manner.

But of the four conditions necessary for the dream formation, the one last recognised is the one whose exactions appear to be least binding upon the dream. That this psychic function, which undertakes the so-called secondary elaboration of the dream content is identical with the work of our waking thought may be inferred with great probability from the following consideration:—Our waking (foreconscious) thought behaves towards a given object of perception just exactly as the function in question behaves towards the dream content. It is natural for our waking thought to bring about order in the material of perception, to construct relationships, and to make it subject to the requirements of an intelligible coherence. Indeed, we go too far in doing this; the tricks of prestidigitators deceive us by taking advantage of this intellectual habit. In our effort to put together the sensory impressions which are offered to us in a comprehensible manner, we often commit the most bizarre errors and even distort the truth of the material we have before us. Proofs for this are too generally familiar to need more extended consideration here. We fail to see errors in a printed page because our imagination pictures the proper words. The editor of a widely-read French paper is said to have risked the wager that he could print the words "from in front" or "from behind" in every sentence of a long article without any of his readers noticing it. He won the wager. A curious example of incorrect associations years ago caught my attention in a newspaper. After the session of the French chamber, at which Dupuy quelled a panic caused by the explosion of a bomb thrown into the hall by an anarchist by saying calmly, "La séance continue," the visitors in the gallery were asked to testify as to their impression of the attempted assassination. Among them were two provincials. One of these told that immediately after the conclusion of a speech he had heard a detonation, but had thought that it was the custom in parliament to fire a shot whenever a speaker had finished. The other, who had apparently already heard several speakers, had got the same idea, with the variation, however, that he supposed this shooting to be a sign of appreciation following an especially successful speech.

Thus the psychic instance which approaches the dream content with the demand that it must be intelligible, which subjects it to preliminary interpretation, and in doing so brings about a complete misunderstanding of it, is no other than our normal thought. In our interpretation the rule will be in every case to disregard the apparent coherence of the dream as being of suspicious origin, and, whether the elements are clear or confused, to follow the same regressive path to the dream material.

We now learn upon what the scale of quality in dreams from confusion to clearness—mentioned above, page 334—essentially depends. Those parts of the dream with which the secondary elaboration has been able to accomplish something seem to us clear; those where the power of this activity has failed seem confused. Since the confused parts of the dream are often also those which are less vividly imprinted, we may conclude that the secondary dream-work is also responsible for a contribution to the plastic intensity of the individual dream structures.

If I were to seek an object of comparison for the definitive formation of the dream as it manifests itself under the influence of normal thinking, none better offers itself than those mysterious inscriptions with which *Die Fliegende Blaetter* has so long amused its readers. The reader is supposed to find a Latin inscription concealed in a given sentence which, for the sake of contrast, is in dialect and as scurrilous as possible in significance. For this purpose the letters are taken from their groupings in syllables and are newly arranged. Now and then a genuine Latin word results, at other places we think that we have abbreviations of such words before us, and at still other places in the inscription we allow ourselves to be carried along over the senselessness of the disjointed letters by the semblance of disintegrated portions or by breaks in the inscription. If we do not wish to respond to the jest we must give up looking for an inscription, must take the letters as we see them, and must compose them into words of our mother tongue, unmindful of the arrangement which is offered.

I shall now undertake a résumé of this extended discussion of the dream activity. We were confronted by the question whether the mind exerts all its capabilities to the fullest development in dream formation, or only a fragment of its capabilities, and these restricted in their activity. Our investigation leads us to reject such a formulation of the question entirely as inadequate to our circumstances. But if we are to remain on the same ground when we answer as that on which the question is urged upon us, we must acquiesce in two conceptions which are apparently opposed and mutually exclusive. The psychic activity in dream formation resolves itself into two functions—the provision of the dream thoughts and the transformation of these into the dream content. The dream thoughts are entirely correct, and are formed with all the psychic expenditure of which we are capable; they belong to our thoughts which have not become conscious, from which our thoughts which have become conscious also result by means of a certain transposition. Much as there may be about them which is worth knowing and mysterious, these problems have no particular relation to the dream, and have no claim to be treated in connection with dream problems. On the other hand, there is that second portion of the activity which changes the unconscious thoughts into the dream content, an activity peculiar to dream life and characteristic of it. Now, this peculiar dream-work is much further removed from the model of waking thought than even the most decided depredators of psychic activity in dream formation have thought. It is not, one might say, more negligent, more incorrect, more easily forgotten, more incomplete than waking thought; it is something qualitatively altogether different from waking thought, and therefore not in any way comparable to it. It does not in general think, calculate, or judge at all, but limits itself to transforming. It can be exhaustively described if the conditions which must be satisfied at its creation are kept in mind. This product, the dream, must at any cost be withdrawn from the censor, and for this purpose the dream activity makes use of the *displacement of psychic intensities* up to the transvaluation of all psychic values; thoughts must exclusively or predominatingly be reproduced in the material of visual and acoustic traces of memory, and this requirement secures for the dream-work the *regard for presentability,* which meets the requirement by furnishing new displacements. Greater intensities are (probably) to be provided than are each night at the disposal of the dream thoughts, and this purpose is served by the prolific *condensation* which is undertaken with the component parts of the dream thoughts. Little attention is paid to the logical relations of the thought material; they ultimately find a veiled representation in the *formal* peculiarities of the dream. The affects of the dream thoughts undergo lesser changes than their presentation content. As a rule they are suppressed; where they are preserved they are freed from the presentations and put together according to their similarity. Only one part of the dream-work—the revision varying in amount, made by the partially roused conscious thought—at all agrees with the conception which the authors have tried to extend to the entire activity of dream formation.

Notes:

1. In estimating this description of the author one may recall the significance of stairway dreams, referred to on p. 266.

2. The fantastic nature of the situation relating to the nurse of the dreamer is shown by the objectively ascertained circumstance that the nurse in this case was his mother. Furthermore, I may call attention to the regret of the young man in the anecdote (p. 189), that he had not taken better advantage of his opportunity with the nurse as probably the source of the present dream.

3. This is the real inciter of the dream.

4. By way of supplement. Such books are poison to a young girl. She herself in youth had drawn much information from forbidden books.

5. A further train of thought leads to *Penthesilsia* by the same author: cruelty towards her lover.

6. Given by translator as author's example could not be translated.

7. The same analysis and synthesis of syllables—a veritable chemistry of syllables—serves us for many a jest in waking life. "What is the cheapest method of obtaining silver? You go to a field where silver-berries are growing and pick them; then the berries are eliminated and the silver remains in a free state." The first person who read and criticised this book made the objection to me—which other readers will probably repeat—"that the dreamer often appears too witty." That is true, as long as it applies to the dreamer; it involves a condemnation only when its application is extended to the interpreter of the dream. In waking reality I can make very little claim to the predicate "witty"; if my dreams appear witty, this is not the fault of my individuality, but of the peculiar psychological conditions under which the dream is fabricated, and is intimately connected with the theory of wit and the comical. The dream becomes witty because the shortest and most direct way to the expression of its thoughts is barred for it: the dream is under constraint. My readers may convince themselves that the dreams of my patients give the impression of being witty (attempting to be witty), in the same decree and in a greater than my own. Nevertheless this reproach impelled me to compare the technique of wit with the dream activity, which I have done in a book published in 1905, on *Wit and its Relation to the Unconscious*. (Author.)

8. Lasker died of progressive paralysis, that is of the consequences of an infection caught from a woman (lues); Lasalle, as is well known, was killed in a duel on account of a lady.

9. In the case of a young man who was suffering from obsessions, but whose intellectual functions were intact and highly developed, I recently found the only exception to this rule. The speeches which occurred in his dreams did not originate in speeches which he had heard or had made himself, but corresponded to the undisfigured wording of his obsessive thoughts, which only came to his consciousness in a changed state while he was awake.

10. Psychic intensity, value, and emphasis due to the interest of an idea are, of course, to be kept distinct from sensational intensity, and from intensity of that which is conceived.

11. Since I consider this reference of dream disfigurement to the censor as the essence of my dream theory, I here insert the latter portion of a story "Traumen wie Wachen" from *Phantasien eines Realisten,* by Lynkeus, Vienna, (second edition, 1900), in which I find this chief feature of my theory reproduced:—

"Concerning a man who possesses the remarkable quality of never dreaming nonsense. . . . "

"Your marvellous characteristic of dreaming as you wake is based upon your virtues, upon your goodness, your justice, and your love for truth; it is the moral clearness of your nature which makes everything about you intelligible."

"But if you think the matter over carefully," replied the other, "I almost believe that all people are created as I am, and that no human being ever dreams nonsense! A dream which is so distinctly remembered that it can be reproduced, which is therefore no dream of delirium, always has a meaning; why, it cannot be otherwise! For that which is in contradiction with itself can never be grouped together as a whole. The fact that time and space are often thoroughly shaken up detracts nothing from the real meaning of the dream, because neither of them has had any significance whatever for its essential contents. We often do the same thing in waking life; think of the fairy-tale, of many daring and profound phantastic creations, about which only an ignorant person would say: 'That is nonsense! For it is impossible.'"

"If it were only always possible to interpret dreams correctly, as you have just done with mine!" said the friend.

"That is certainly not an easy task, but the dreamer himself ought always to succeed in doing it with a little concentration of attention. . . . You ask why it is generally impossible? Your dreams seem to conceal something secret, something unchaste of a peculiar and higher nature, a certain mystery in your nature which cannot easily be revealed by thought; and it is for that reason that your dreaming seems so often to be without meaning, or even to be a contradiction. But in the profoundest sense this is by no means the case; indeed it cannot be true at all, for it is always the same person, whether he is asleep or awake."

12. I have since given the complete analysis and synthesis of two dreams in the *Bruchstueck einer Hysterieanalyse,* 1905.

13. From a work of K. Abel, *Der Gegensinn der Urworte,* 1884 (see my review of it in the Bleuler-Freud *Jahrbuch,* II., 1910), I learned with surprise a fact which is confirmed by other philologists, that the oldest languages behaved in this regard quite like the dream. They originally had only one word for both extremes in a series of qualities or activities (strong—weak, old—young, far—near, to tie—to separate), and formed separate designations for the two extremes only secondarily through slight modifications of the common primitive word. Abel demonstrated these relationships with rare exceptions in the old Egyptian, and he was able to show distinct remnants of the same development in the Semitic and Indo-Germanic languages.

14. If I do not know behind which of the persons which occur in the dream I am to look for my ego, I observe the following rule: That person in the dream who is subject to an emotion which I experience while asleep, is the one that conceals my ego.

15. The hysterical attack sometimes uses the same device—the inversion of time-relations—for the purpose of concealing its meaning from the spectator. The attack of a hysterical girl, for example, consists in enacting a little romance, which she has unconsciously fancied in connection with an encounter in the street car. A man, attracted by the beauty of her foot, addresses her while she is reading, whereupon she goes with him and experiences a stormy love scene. Her attack begins with the representation of this scene in writhing movements of the body (accompanied by motions of the lips to signify kissing, entwining of the arms for embraces), whereupon she hurries into another room, sits down in a chair, lifts her skirt in order to show her foot, acts as though she were about to read a book, and speaks to me (answers me).

16. Accompanying hysterical symptoms: Failure to menstruate and profound depression, which was the chief ailment of the patient.

17. A reference to a childhood experience is after complete analysis shown to exist by the following intermediaries: "The Moor has done his duty, the Moor *may go.*" And then follows the waggish question: "How old is the Moor when he has done his duty? One year. Then he may go." (It is said that I came into the world with so much black curly hair that my young mother declared me to be a Moor.) The circumstance that I do not find my hat is an experience of the day which has been turned to account with various significations. Our servant, who is a genius at stowing away things, had hidden the hat. A suppression of sad thoughts about death is also concealed behind the conclusion of the dream: "I have not nearly done my duty yet; I may not go yet." Birth and death, as in the dream that occurred shortly before about Goethe and the paralytic (p. 372).

18. *Cf. Der Witz und seine Beziehung zum Unbewussten,* 2nd edit. 1912, and "word-bridges," in the solutions of neurotic symptoms.

19. In general it is doubtful in the interpretation of every element of the dream whether it—
    (*a*) is to be regarded as having a negative or a positive sense (relation of opposition);
    (*b*) is to be interpreted historically (as a reminiscence);
    (*c*) is symbolic; or whether
    (*d*) its valuation is to be based upon the sound of its verbal expression.

    In spite of this manifold signification, it may be said that the representation of the dream activity does not impose upon the translator any greater difficulties than the ancient writers of hieroglyphics imposed upon their readers.

20. For the interpretation of this preliminary dream, which is to be regarded as "casual," see p. 321.

21. Her career.

22. High birth, the wish contrast to the preliminary dream.

23. A composite image, which unites two localities, the so-called garret (German *Boden*—floor, garret) of her father's house, in which she played with her brother, the object of her later fancies, and the garden of a malicious uncle, who used to tease her.

24. Wish contrast to an actual memory of her uncle's garden, to the effect that she used to expose herself while she was asleep.

25. Just as the angel bears a lily stem in the Annunciation.

26. For the explanation of this composite image, see p. 325; innocence, menstruation, Camille.

27. Referring to the plurality of the persons who serve the purpose of her fancy.

28. Whether it is permitted to "pull one off," *i.e.* to masturbate.

29. The bough has long since been used to represent the male genital, and besides that it contains a very distinct allusion to the family name of the dreamer.

30. Refers to matrimonial precautions, as does that which follows.

31. An analogous "biographical" dream was reported on p. 263, as the third of the examples of dream symbolism; a second example is the one fully reported by Rank under the title "Traum dor sich selbst deutet"; for another one which must be read in the "opposite direction," see Stekel, p. 486.

32. Given by translator as author's example could not be translated.

33. The neurosis also proceeds in the same manner. I know a patient who involuntarily—contrary to her own wishes—hears (hallucinatory) songs or fragments of songs without being able to understand their meaning to her psychic life. She is surely not a paranoiac. Analysis showed that she wrongly utilised the text of these songs by means of a certain license. "Oh thou blissful one, Oh thou happy one," is the beginning of a Christmas song. By not continuing it to the word "Christmas time" she makes a bridal song out of it, &c. The same mechanism of disfigurement may take place also without hallucinations as a mere mental occurrence.

34. As a contribution to the over-determination: My excuse for coming late was that after working late at night I had in the morning to make the long journey from Kaiser Josef Street to Waehringer Street.

35. In addition Caesar—Kaiser.

36. I have forgotten in what author I found a dream mentioned that was overrun with unusually small figures, the source of which turned out to be one of the engravings of Jacques Callot, which the dreamer had looked at during the day. These engravings contained an enormous number of very small figures; a series of them treats of the horrors of the Thirty Years' War.

37. The frequency with which in the dream dead persons appear as living, act, and deal with us, has called forth undue astonishment and given rise to strange explanations, from which our ignorance of the dream becomes strikingly evident. And yet the explanation for these dreams lies very close at hand. How often we have occasion to think: *"If father were still alive, what would he say to it?"* The dream can express this *if* in no other way than by present time in a definite situation. Thus, for instance, a young man, whose grandfather has left him a great inheritance, dreams that his grandfather is alive and demands an accounting of him, upon an occasion when the young man had been reproached for making too great an expenditure of money. What we consider a resistance to the dream—the objection made by our better knowledge, that after all the man is already dead—is in reality a consolation, because the dead person did not have this or that experience, or satisfaction at the knowledge that he has nothing more to say.

Another form of absurdity found in dreams of deceased relatives does not express folly and absurdity, but serves to represent the most extreme rejection; as the representation of a repressed thought which one would gladly have appear as something least thought of. Dreams of this kind are only solvable if one recalls that the dream makes no distinction between things desired and realities. Thus, for example, a man who nursed his father during his sickness, and who felt his death very keenly, sometime afterward dreamed the following senseless dream: *The father was again living, and conversed with him as usual, but* (the remarkable thing about it) *he had nevertheless died, though he did not know it.* This dream can be understood if after "he had nevertheless died," one inserts *in consequence of the dreamer's wish, and* if after "but he did not know it" one adds *that the dreamer has entertained this wish.* While nursing his father, the son often wishes his father's death; *i.e.* he entertained the really compassionate desire that death finally put an end to his suffering. While mourning after his death, this very wish of compassion became an unconscious reproach, as if it had really contributed to shorten the life of the sick man. Through the awakening of early infantile feelings against the father, it became possible to express this reproach as a dream; and it was just because of the world-wide contrast between the dream inciter and day thought that this dream had to come out so absurdly (*cf.* with this, "Formulierungen über die zwei Prizipien des seelischen Geschehens, *Jahrbuch,* Bleuler-Freud, III, 1, 1911).

38. Here the dream activity parodies the thought which it designates as ridiculous, in that it creates something ridiculous in relation to it. Heine does something similar when he tries to mock the had rhymes of the King of Bavaria. He does it in still worse rhymes:

> "Herr Ludwig ist ein grosser Poet
> Und singt er, so stuerzt Apollo
> Vor ihm auf die Knie und bittet und flelit,
> 'Halt ein, ich werde sonst toll oh!'"

39. Note the resemblance of *Geseres* and *Ungeseres* to the German words for salted and unsalted—*gesalzen* and *ungesalzen;* also to the German words for soured and unsoured—*gesauert* and *ungesauert.* (Translator.)

40. This dream also furnishes a good example for the general thesis that dreams of the same night, even though they be separated in memory, spring from the same thought material. The dream situation in which I am rescuing my children from the city of Rome, moreover, is disfigured by a reference to an episode belonging to my childhood. The meaning is that I envy certain relatives who years ago had occasion to transplant their children to another soil.

41. This German expression is equivalent to our saying "You are not responsible for that," or "That has not been acquired through your own efforts." (Translator.)

42. The injunction or purpose contained in the dream, "I must tell that to the doctor," which occurs in dreams that are dreamed in the course of psychoanalytical treatment, regularly corresponds to a great resistance to the confession involved in the dream, and is not infrequently followed by forgetting of the dream.

43. A subject about which an extensive discussion has taken place in the volumes of the *Revue Philosophique*—(Paramnesia in the Dream).

44. These results correct in several respects my earlier statements concerning the representation of logical relations (p. 318). The latter described the general conditions of dream activity, but they did not take into consideration its finest and most careful performances.

45. Stanniol, allusion to *Stannius,* the nervous system of fishes; *cf.* p. 352.

46. The place in the corridor of my apartment house where the baby carriages of the other tenants stand; it is also otherwise several times over-determined.

47. This description is not intelligible even to myself, but I follow the principle of reproducing the dream in those words which occur to me while I am writing it down. The wording itself is a part of the dream representation.

48. Schiller was not born in one of the Marburgs, but in Marbach, as every graduate of a Gymnasium knows, and as I also knew. This again is one of those errors (*cf.* p. 388) which are included as substitutes for an intended deception at another place—an explanation of which I have attempted in the *Psychopathologie des Alltagslebens*).

49. As analogy to this, I have since explained the extraordinary effect of pleasure produced by "tendency" wit.

50. It is this fancy from the unconscious dream thoughts which peremptorily demands *non vivit* instead of *non vixit*. "You have come too late, he is no longer alive." The fact that the manifest situation also tends towards "non vivit" has been mentioned on page 362.

51. It is striking that the name Joseph plays such a large part in my dreams (see the dream about my uncle). I can hide my ego in the dream behind persons of this name with particular ease, for Joseph was the name of the *dream interpreter* in the Bible.

52. Rêve, petit roman—day-dream, story.

53. I have analysed a good example of a dream of this kind having its origin in the stratification of several phantasies, in the *Bruchstück einer Hysteria Analyse*, 1905. Moreover I undervalued the significance of such phantasies for dream formation, as long as I was working chiefly with my own dreams, which were based rarely upon day dreams, most frequently upon discussions and mental conflicts. With other persons it is often much easier to prove the *full analogy between the nocturnal dream and the day dream*. It is often possible in an hysterical patient to replace an attack by a dream; it is then obvious that the phantasy of day dreams is the first step for both psychic formations.

7

The Psychology
of the Dream Activities

A mong the dreams which I have heard from others there is one which at this point is especially worthy of our attention. It was told to me by a female patient who in turn had heard it in a lecture on dreams. Its original source is unknown to me. This dream evidently made a deep impression upon the lady, as she went so far as to imitate it, *i.e.* to repeat the elements of this dream in a dream of her own in order to express by this transference her agreement with it in a certain point.

The essential facts of this illustrative dream are as follows: For days and nights a father had watched at the sick-bed of his child. After the child died, he retired to rest in an adjoining room, leaving the door ajar, however, so as to enable him to look from his room into the other, where the corpse lay surrounded by burning candles. An old man, who was left as a watch, sat near the corpse murmuring prayers. After sleeping a few hours the father dreamed that *the child stood near his bed clasping his arms and calling out reproachfully, "Father, don't you see that I am burning?"* The father woke and noticed a bright light coming from the adjoining room. Rushing in, he found the old man asleep, and the covers and one arm of the beloved body burned by the fallen candle.

The meaning of this affecting dream is simple enough, and the explanation given by the lecturer, as my patient reported it, was correct. The bright light coming through the open door into the eyes of the sleeper produced the same impression on him as if he had been awake; namely, that a fire had been started near the corpse by a falling candle. It is quite possible that on going to sleep he feared that the aged guardian was not equal to his task.

We can find nothing to change in this interpretation. We can add only that the contents of the dream must be overdetermined, and that the talking of the child consisted of phrases that it had uttered while still living, which recalled to the father important events. Perhaps the complaint, "I am burning," recalled the fever from which the child died, and the words quoted, "Father, don't you see?" recalled an emotional occurrence unknown to us.

But after we have recognised the dream as a senseful occurrence which can be correlated with our psychic existence, it may be surprising that a dream should have taken place under circumstances which necessitated such immediate awakening. We also notice that the dream does not lack the wish-fulfilment. The child acts as if living; it warns the father itself; it comes to his bed and clasps his arms, as it probably did on the occasion which gave origin to the first part of the speech in the dream. It was for the sake of this wish-fulfilment that the father slept a moment longer. The dream triumphed over the conscious reflection because it could show the child once more alive. If the father had awakened first, and had then drawn the conclusion which led him into the adjoining room, he would have shortened the child's life by this one moment.

The peculiar feature in this brief dream which engages our interest is quite plain. So far we have mainly endeavoured to ascertain wherein the secret meaning of the dream consists, in what way this is to be discovered, and what means the dream-work uses to conceal it. In other words, our greatest interest has hitherto centred on the problems of interpretation. We now encounter a dream, however, which can be easily explained, the sense of which is plainly presented; and we notice that in spite of this fact the dream still preserves the essential features which plainly differentiate our dreaming from our conscious thinking, and thus clearly demands an explanation. After clearing up all the problems of interpretation, we can still feel how imperfect our psychology of the dream is.

Before entering, however, into this new territory, let us stop and reflect whether we have not missed something important on our way hither. For it must be frankly admitted that we have been traversing the easy and comfortable part of our journey. Hitherto all the paths we have followed have led, if I mistake not, to light, to explication, and to full understanding, but from the moment that we wish to penetrate deeper into the psychic processes of the dream all paths lead into darkness. It is quite impossible to explain the dream as a psychic process, for to explain means to trace to the known, and as yet we do not possess any psychological knowledge under which we can range what may be inferred from our psychological investigation of dreams as their fundamental explanation. On the contrary, we shall be compelled to build a series of new assumptions concerning the structure of the psychic apparatus and its active forces; and this we shall have to be careful not to carry beyond the simplest logical concatenation, as its value may otherwise merge into uncertainty. And, even if we should make no mistake in our conclusions, and take cognisance of all the logical possibilities involved, we shall still be threatened with complete failure in our solution through the probable incompleteness of our elemental data. It will also be impossible to gain, or at least to establish, an explanation for the construction and workings of the psychic instrument even through a most careful investigation of the dream or any other single activity. On the contrary, it will be necessary for this end to bring together whatever appears decisively as constant after a comparative study of a whole series of psychic activities. Thus the psychological conceptions which we shall gain from an analysis of the dream process will have to wait, as it were, at the junction point until they can be connected with the results of other investigations which may have advanced to the nucleus of the same problem from another starting point.

(A) Forgetting in Dreams

I propose, then, first, to turn to a subject which has given rise to an objection hitherto unnoticed, threatening to undermine the foundation of our work in dream interpretation. It has been objected in more than one quarter that the dream which we wish to interpret is really unknown to us, or, to be more precise, that we have no assurance of knowing it as it has really occurred (see p. 44). What we recollect of the dream, and what we subject to our methods of interpretation, is in the first place disfigured through our treacherous memory, which seems particularly unfitted to retain the dream, and which may have omitted precisely the most important part of the dream content. For, when we pay attention to our dreams, we often find cause to complain that we have dreamed much more than we remember; that, unfortunately, we know nothing more than this one fragment, and that even this seems to us peculiarly uncertain. On the other hand, everything assures us that our memory reproduces the dream not only fragmentarily but also delusively and falsely. Just as on the one hand we may doubt whether the material dreamt was really as disconnected and confused as we remember it, so on the other hand may we doubt whether a dream was as connected as we relate it; whether in the attempt at reproduction we have not filled in the gaps existing or caused by forgetfulness with new material arbitrarily chosen; whether we have not embellished, rounded off, and prepared the dream so that all judgment as to its real content becomes impossible. Indeed, one author (Spitta) has expressed his belief that all that is orderly and connected is really first put into the dream during our attempt to recall it. Thus we are in danger of having wrested from our hands the very subject whose value we have undertaken to determine.

In our dream interpretations we have thus far ignored these warnings. Indeed, the demand for interpretation was. on the contrary, found to be no less perceptible in the smallest, most insignificant, and most uncertain ingredients of the dream content than in those containing the distinct and definite parts. In the dream of Irma's injection we read, "I quickly called in Dr. M.," and we assumed that even this small addendum would not have gotten into the dream if it had not had a special derivation. Thus we reached the history of that unfortunate patient to whose bed I "quickly" called in the older colleague. In the apparently absurd dream which treated the difference between 51 and 56 as *quantité négligé,* the number 51 was repeatedly mentioned. Instead of finding this self-evident or indifferent, we inferred from it a second train of thought in the latent content of the dream which led to the number 51. By following up this clue we came to the fears which placed 51 years as a limit of life, this being in most marked contrast to a dominant train of thought which boastfully knew no limit to life. In the dream "Non Vixit" I found, as an insignificant interposition that I at first overlooked, the sentence, "As P. does not understand him, Fl. asks me," &c. The interpretation then coming to a standstill, I returned to these words, and found through them the way to the infantile phantasy, which appeared in the dream thoughts as an intermediary point of junction. This came about by means of the poet's verses:

> Seldom have you understood me,
> Seldom have I understood you,
> But when we got into the mire,
> We at once understood each other.

Every analysis will demonstrate by examples how the most insignificant features of the dream are indispensable to the analysis, and how the finishing of the task is delayed by the fact that attention is not at first directed to them. In the same way we have in the interpretation of dreams respected, every nuance of verbal expression found in the dream; indeed, if we were confronted by a senseless or insufficient wording betraying an unsuccessful effort to translate the dream in the proper style, we have even respected these defects of expression. In brief, what the authorities have considered arbitrary improvisation, concocted hastily to suit the occasion, we have treated like a sacred text. This contradiction requires an explanation.

It is in our favour, without disparagement to the authorities. From the viewpoint of our newly-acquired understanding concerning the origin of the dream, the contradictions fall into perfect agreement. It is true that we distort the dream in our attempt to reproduce it; and herein we find another instance of what we have designated as the often misunderstood secondary elaboration of the dream through the influence of normal thinking. But this distortion is itself only a part of the elaboration to which the dream thoughts are regularly subjected by virtue of the dream censor. The authorities have here divined or observed that part of the dream distortion most obviously at work; to us this is of little importance, for we know that a more prolific work of distortion, not so easily comprehensible, has already chosen the dream from among the concealed thoughts as its object. The authorities err only in considering the modifications of the dream while it is being recalled and put in words as arbitrary and insoluble; and hence, as likely to mislead us in the interpretation of the dream. We over-estimate the determination of the psychic. There is nothing arbitrary in this field. It can quite generally be shown that a second train of thought immediately undertakes the determination of the elements which have been left undetermined by the first. I wish, *e.g.,* to think quite voluntarily of a number. This, however, is impossible. The number that occurs to me is definitely and necessarily determined by thoughts within me which may be far from my momentary intention.[1] Just as far from arbitrary are the modifications which the dream experiences through the revision of the waking state. They remain in associative connection with the content, the place of which they take, and serve to show us the way to this content, which may itself be the substitute for another.

In the analysis of dreams with patients I am accustomed to institute the following proof of this assertion, which has never proved unsuccessful. If the report of a dream appears to me at first difficult to understand, I request the dreamer to repeat it. This he rarely does in the same words. The passages wherein the expression is changed have become known to me as the weak points of the dream's disguise, which are of the same service to me as the embroidered mark on Siegfried's raiment was to Hagen. The analysis may start from these points. The narrator has been admonished by my announcement that I mean to take special pains to solve the dream, and immediately, under the impulse of resistance, he protects the weak points of the dream's disguise, replacing the treacherous expressions by remoter ones. He thus calls my attention to the expressions he has dropped. From the efforts made to guard against the solution of the dream, I can also draw conclusions as to the care with which the dream's raiment was woven.

The authors are, however, less justified in giving so much importance to the doubt which our judgment encounters in relating the dream. It is true that this doubt betrays the lack of an intellectual assurance, but our memory really knows no guarantees, and yet, much more often than is objectively justified, we yield to the pressure of lending credence to its statements. The doubt concerning the correct representation of the dream, or of its individual data, is again only an offshoot of the dream censor—that is, of the resistance against penetration to consciousness of the dream thoughts. This resistance has not entirely exhausted itself in bringing about the displacements and substitutions, and it therefore adheres as doubt to what has been allowed to pass through. We can recognise this doubt all the easier through the fact that it takes care not to attack the intensive elements of the dream, but only the weak and indistinct ones. For we already know that a transvaluation of all the psychic values has taken place between the dream thoughts and the dream. The disfigurement has been made possible only by the alteration of values; it regularly manifests itself in this way and occasionally contents itself with this. If doubt attaches to an indistinct element of the dream content, we may, following the hint, recognise in this element a direct offshoot of one of the outlawed dream thoughts. It is here just as it was after a great revolution in one of the republics of antiquity or of the Renaissance. The former noble and powerful ruling families are now banished; all high positions are filled by upstarts; in the city itself only the very poor and powerless citizens or the distant followers of the vanquished party are tolerated. Even they do not enjoy the full rights of citizenship. They are suspiciously watched. Instead of the suspicion in the comparison, we have in our case the doubt. I therefore insist that in the analysis of dreams one should emancipate one's self from the entire conception of estimating trustworthiness, and when there is the slightest possibility that this or that occurred in the dream, it should be treated as a full certainty. Until one has decided to reject these considerations in tracing the dream elements, the analysis will remain at a standstill. Antipathy toward the element concerned shows its psychic effect in the person analysed by the fact that the undesirable idea will evoke no thought in his mind. Such effect is really not self-evident. It would not be inconsistent if one would say: "Whether this or that was contained in the dream I do not know, but the following thoughts occur to me in this direction." But he never expresses himself thus; and it is just this disturbing influence of doubt in the analysis that stamps it as an offshoot and instrument of the psychic resistance. Psychoanalysis is justly suspicious. One of its rules reads: *Whatever disturbs the continuation of the work is a resistance.*

The forgetting of dreams, too, remains unfathomable as long as we do not consider the force of the psychic censor in its explanation. The feeling, indeed, that one has dreamt a great deal during the night and has retained only a little of it may have another meaning in a number of cases. It may perhaps signify that the dream-work has continued perceptibly throughout the night, and has left behind only this short dream. There is, however, no doubt of the fact that the dream is progressively forgotten on awakening. One often forgets it in spite of painful effort to remember. I believe, however, that just as one generally over-estimates the extent of one's forgetting, so also one over-estimates the deficiencies in one's knowledge, judging them by the gaps occurring in the dream. All that has been lost through forgetting in a dream content can often be brought back through analysis. At least, in a whole series of cases, it is possible to discover from one single remaining fragment, not the dream, to be sure, which is of little importance, but all the thoughts of the dream. It requires a greater expenditure of attention and self-control in the analysis; that is all. But, at the same time, this suggests that the forgetting of the dream does not lack a hostile intention.

A convincing proof of the purposeful nature of dream-forgetting, in the service of resistance, is gamed in analysis through the investigation of a preliminary stage of forgetting.[2] It often happens that in the midst of interpretation work an omitted fragment of the dream suddenly comes to the surface. This part of the dream snatched from forgetfulness is always the most important part. It lies on the shortest road toward the solution of the dream, and for that very reason it was most objectionable to the resistance. Among the examples of dreams that I have collected in connection with this treatise, it once happened that I had to interpose subsequently such a piece of dream content. It was a travelling dream, which took vengeance upon an unlovable female travelling companion; I have left it almost entirely uninterpreted on account of its being in part coarse and nasty. The part omitted read: "I said about a book by Schiller, 'It is from —' but corrected myself, for I noticed the mistake myself, 'It is by.' Upon this the man remarked to his sister, 'Indeed, he said it correctly.'"

The self-correction in dreams, which seems so wonderful to some authors, does not merit consideration by us. I shall rather show from my own memory the model for the grammatical error in the dream. I was nineteen years old when I visited England for the first time, and spent a day on the shore of the Irish Sea. I naturally amused myself by catching the sea animals left by the waves, and occupied myself in particular with a starfish (the dream begins with Hollthurn—Holothurian), when a pretty little girl came over to me and asked me, "Is it a starfish? Is it alive?" I answered, "Yes, he is alive," but was then ashamed of my mistake and repeated the sentence correctly. For the grammatical mistake which I then made, the dream substitutes another which is quite common with Germans. "Das Buch ist von Schiller" should not be translated by *the book is from,* but *the book is by.* That the dream-work produces this substitution because the word from makes possible, through consonance, a remarkable condensation with the German adjective *fromm* (pious, devout), no longer surprises us after all that we have heard about the aims of the dream-work and about its reckless selection of means of procedure. But what is the meaning of the harmless recollection of the seashore in relation to the dream? It explains by means of a very innocent example that I have used the wrong gender—*i.e.* that I have put "he," the word denoting the sex or the sexual, where it does not belong. This is surely one of the keys to the solution of dreams. Who ever has heard of the origin of the book-title *Matter and Motion* (Molière in *Malade Imaginaire:* La matière est-elle laudable?—A motion of the bowels) will readily be able to supply the missing parts.

Moreover, I can prove conclusively by a *demonstratio ad oculos* that the forgetting in dreams is in great part due to the activity of resistance. A patient tells me that he has dreamed, but that the dream has vanished without leaving a trace, as if nothing had happened. We continue to work, however; I strike a resistance which I make plain to the patient; by encouraging and urging I help him to become reconciled to some disagreeable thought; and as soon as I have succeeded he exclaims, "Now, I can recall what I have dreamed." The same resistance which that day disturbed him in the work caused him also to forget the dream. By overcoming this resistance, I brought the dream to memory.

In the same way the patient may, on reaching a certain part of the work, recall a dream which took place three, four, or more days before, and which has rested in oblivion throughout all this time.

Psychoanalytic experience has furnished us with another proof of the fact that the forgetting of dreams depends more on the resistance than on the strangeness existing between the waking and sleeping states, as the authorities have believed. It often happens to me, as well as to the other analysts and to patients under treatment, that we are awakened from sleep by a dream, as we would say, and immediately thereafter, while in full possession of our mental activity, we begin to interpret the dream. In such cases I have often not rested until I gained a full understanding of the dream, and still it would happen that after the awakening I have just as completely forgotten the interpretation work as the dream content itself, though I was aware that I had dreamed and that I had interpreted the dream. The dream has more frequently taken along into forgetfulness the result of the interpretation work than it was possible for the mental activity to retain the dream in memory. But between this interpretation work and the waking thoughts there is not that psychic gap through which alone the authorities wish to explain the forgetting of dreams. Morton Prince objects to my explanation of the forgetting of dreams on the ground that it is only a particular example of amnesia for dissociated states, and that the impossibility of harmonising my theory with other types of amnesia makes it also valueless for other purposes. He thus makes the reader suspect that in all his description of such dissociated states he has never made the attempt to find the dynamic explanation for these phenomena. For, had he done so, he surely would have discovered that the repression and the resistance produced thereby "is quite as well the cause of this dissociation as of the amnesia for its psychic content."

That the dream is as little forgotten as the other psychic acts, and that it clings to memory just as firmly as the other psychic activities was demonstrated to me by an experiment which I was able to make while compiling this manuscript. I have kept in my notes many dreams of my own which, for some reason at the time I could analyse only imperfectly or not at all. In order to get material to illustrate my assertions, I attempted to subject some of them to analysis from one to two years later. I succeeded in this attempt without any exception. Indeed, I may even state that the interpretation went more easily at this later time than at the time when the dreams were recent occurrences. As a possible explanation for this fact, I would say that I had gotten over some of the resistances which disturbed me at the time of dreaming. In such subsequent interpretations I have compared the past results in dream thoughts with the present, which have usually been more abundant, and have invariably found the past results falling under the present without change. I have, however, soon put an end to my surprise by recalling that I have long been accustomed to interpret dreams from former years which have occasionally been related to me by patients as if they were dreams of the night before, with the same method and the same success. I shall report two examples of such delayed dream interpretations in the discussion of anxiety dreams. When I instituted this experiment for the first time, I justly expected that the dream would behave in this respect like a neurotic symptom. For when I treat a neurotic, perhaps an hysteric, by psychoanalysis, I am compelled to find explanations for the first symptoms of the disease which have long been forgotten, just as for those still existing which have brought the patient to me; and I find the former problem easier to solve than the more exigent one of to-day. In the *Studien über Hysterie,* published as early as 1895, I was able to report the explanation of a first hysterical attack of anxiety which the patient, a woman over forty years of age, had experienced in her fifteenth year.[3]

I may now proceed in an informal way to some further observations on the interpretation of dreams, which will perhaps be of service to the reader who wishes to test my assertion by the analysis of his own dreams.

No one must expect that the interpretations of his dreams will come to him overnight without any exertion. Practice is required even for the perception of endoptic phenomena and other sensations usually withdrawn from attention, although this group of perceptions is not opposed by any psychic motive. It is considerably more difficult to become master of the "undesirable presentations." He who wishes to do this will have to fulfil the requirements laid down in this treatise. Obeying the rules here given, he will strive during the work to curb in himself every critique, every prejudice, and every affective or intellectual one-sidedness. We will always be mindful of the precept of Claude Bernard for the experimenter in the physiological laboratory—"Travailler comme une bête"—meaning he should be just as persistent, but also just as unconcerned about the results. He who will follow these counsels will surely no longer find the task difficult. The interpretation of a dream cannot always be accomplished in one session; you often feel, after following up a concatenation of thoughts, that your working capacity is exhausted; the dream will not tell you anything more on that day; it is then best to break off, and return to the work the following day. Another portion of the dream content then solicits your attention, and you thus find an opening to a new stratum of the dream thoughts. We may call this the "fractionary" interpretation of dreams.

It is most difficult to induce the beginner in the interpretation of dreams to recognise the fact that his task is not finished though he is in possession of a complete interpretation of the dream which is ingenious and connected, and which explains all the elements of the dream. Besides this another superimposed interpretation of the same dream may be possible which has escaped him. It is really not simple to form an idea of the abundant unconscious streams of thought striving for expression in our minds, and to believe in the skilfulness displayed by the dream-work in hitting, so to speak, with its ambiguous manner of expression, seven flies with one stroke, like the journeyman tailor in the fairy tale. The reader will constantly be inclined to reproach the author for uselessly squandering his ingenuity, but anyone who has had experience of his own will learn to know better.

The question whether every dream can be interpreted may be answered in the negative. One must not forget that in the work of interpretation one must cope with the psychic forces which are responsible for the distortion of the dream. Whether one can become master of the inner resistances through his intellectual interest, his capacity for self-control, his psychological knowledge, and his practice in dream interpretation becomes a question of the preponderance of forces. It is always possible to make some progress. One can at least go far enough to become convinced that the dream is an ingenious construction, generally far enough to gain an idea of its meaning. It happens very often that a second dream confirms and continues the interpretation assumed for the first. A whole series of dreams running for weeks or months rests on a common basis, and is therefore to be interpreted in connection. In dreams following each other, it may be often observed how one takes as its central point what is indicated only as the periphery of the next, or it is just the other way, so that the two supplement each other in interpretation. That the different dreams of the same night are quite regularly in the interpretation to be treated as a whole I have already shown by examples.

In the best interpreted dreams we must often leave one portion in obscurity because we observe in the interpretation that it represents the beginning of a tangle of dream thoughts which cannot be unravelled but which has furnished no new contribution to the dream content. This, then, is the keystone of the dream, the place at which it mounts into the unknown. For the dream thoughts which we come upon in the interpretation must generally remain without a termination, and merge in all directions into the net-like entanglement of our world of thoughts. It is from some denser portion of this texture that the dream-wish then arises like the mushroom from its mycelium.

Let us now return to the facts of dream-forgetting, as we have really neglected to draw an important conclusion from them. If the waking life shows an unmistakable intention to forget the dream formed at night, either as a whole, immediately after awakening, or in fragments during the course of the day, and if we recognise as the chief participator in this forgetting the psychic resistance against the dream which has already performed its part in opposing the dream at night—then the question arises, What has the dream formation actually accomplished against this resistance? Let us consider the most striking case in which the waking life has done away with the dream as though it had never happened. If we take into consideration the play of the psychic forces, we are forced to assert that the dream would have never come into existence had the resistance held sway during the night as during the day. We conclude then, that the resistance loses a part of its force during the night; we know that it has not been extinguished, as we have demonstrated its interest in the dream formation in the production of the distortion. We have, then, forced upon us the possibility that it abates at night, that the dream formation has become possible with this diminution of the resistance, and we thus readily understand that, having regained its full power with the awakening, it immediately sets aside what it was forced to admit as long as it was in abeyance. Descriptive psychology teaches us that the chief determinant in dream formation is the dormant state of the mind. We may now add the following elucidation: *The sleeping state makes dream formation possible by diminishing the endopsychic censor.*

We are certainly tempted to look upon this conclusion as the only one possible from the facts of dream-forgetting, and to develop from it further deductions concerning the proportions of energy in the sleeping and waking states. But we shall stop here for the present. When we have penetrated somewhat deeper into the psychology of the dream we shall find that the origin of the dream formation may be differently conceived. The resistance operating to prevent the dream thoughts coming to consciousness may perhaps be eluded without suffering diminution *per se*. It is also plausible that both the factors favourable to dream formation, the diminution as well as the eluding of the resistance, may be made possible simultaneously through the sleeping state. But we shall pause here, and continue this line of thought later.

There is another series of objections against our procedure in the dream interpretation which we must now consider. In this interpretation we proceed by dropping all the end-presentations which otherwise control reflection, we direct our attention to an individual element of the dream, and then note the unwished-for thoughts that occur to us in this connection. We then take up the next component of the dream content, and repeat the operation with it; and, without caring in what direction the thoughts take us, we allow ourselves to be led on by them until we end by rambling from one subject to another. At the same time, we harbour the confident hope that we may in the end, without effort, come upon the dream thoughts from which our dream originated. Against this the critic brings the following objection: That one can arrive somewhere, starting from a single element in the dream is nothing wonderful. Something can be associatively connected with every idea. It is remarkable only that one should succeed in hitting the dream thoughts in this aimless and arbitrary excursion of thought. It is probably a self-deception; the investigator follows the chain of association from one element until for some reason it is seen to break, when a second element is taken up; it is thus but natural that the association, originally unbounded, should now experience a narrowing. He keeps in mind the former chain of associations, and he will therefore in analysis more easily hit upon certain thoughts which have something in common with the thoughts from the first chain. He then imagines that he has found a thought which represents a point of junction between two elements of the dream. As he, moreover, allows himself every freedom of thought connection, excepting only the transitions from one idea to another which are made in normal thinking, it is not finally difficult for him to concoct something which he calls the dream thought out of a series of "intermediary thoughts"; and without any guarantee, as they are otherwise unknown, he palms these off as the psychic equivalent of the dream. But all this is accompanied by arbitrary procedure and over-ingenious exploitation of coincidence. Anyone who will go to this useless trouble can in this way work out any desired interpretation for any dream whatever.

If such objections are really advanced against us, we may refer in our defence to the agreement of our dream interpretations, to the surprising connections with other dream elements which appear in following out the different particular presentations, and to the improbability that anything which so perfectly covers and explains the dream as our dream interpretations do could be gained otherwise than by following psychic connections previously established. We can also justify ourselves by the fact that the method of dream analysis is identical with the method used in the solution of hysterical symptoms, where the correctness of the method is attested through the emergence and fading away of the symptoms—that is, where the elucidation of the text by the interposed illustrations finds corroboration. But we have no object in avoiding this problem—how one can reach to a pre-established aim by following a chain of thoughts spun out thus arbitrarily and aimlessly—for, though we are unable to solve the problem, we can get rid of it entirely.

It is in fact demonstrably incorrect to state that we abandon ourselves to an aimless course of thought when, as in the interpretation of dreams, we relinquish our reflection and allow the unwished-for idea to come to the surface. It can be shown that we can reject only those end-presentations that are familiar to us, and that as soon as these stop the unknown, or, as we say more precisely, the unconscious end-presentations, immediately come into play, which now determined the course of the unwished-for presentations. A mode of thinking without end-idea can surely not be brought about through any influence we can exert on our own mental life; nor do I know either of any state of psychic derangement in which such mode of thought establishes itself. The psychiatrists have in this field much too early rejected the solidity of the psychic structure. I have ascertained that an unregulated stream of thoughts, devoid of the end-presentation, occurs as in the realm of hysteria and paranoia as in the formation or solution of dreams. Perhaps it does not appear at all in the endogenous psychic affections, but even the deliria of confused states are senseful according to the ingenious theory of Leuret and become incomprehensible to us only through omissions. I have come to the same conviction wherever I have found opportunity for observation. The deliria are the work of a censor which no longer makes any effort to conceal its sway, which, instead of lending its support to a revision no longer obnoxious to it, cancels regardlessly that which it raises objections against, thus causing the remnant to appear disconnected. This censor behaves analogously to the Russian newspaper censor on the frontier, who allows to fall into the hands of his protected readers only those foreign journals that have passed under the black pencil.

The free play of the presentations following any associative concatenation perhaps makes its appearance in destructive organic brain lesions. What, however, is taken as such in the psychoneuroses can always be explained as the influence of the censor on a series of thoughts which have been pushed into the foreground by the concealed end-presentation.[4] It has been considered an unmistakable sign of association free from the end-presentations when the emerging presentations (or pictures) were connected with one another by means of the so-called superficial associations—that is, by assonance, word ambiguity, and causal connection without inner sense relationship; in other words, when they were connected through all those associations which we allow ourselves to make use of in wit and play upon words. This distinguishing mark proves true for the connections of thought which lead us from the elements of the dream content to the collaterals, and from these to the thoughts of the dream proper; of this we have in our dream analysis found many surprising examples. No connection was there too loose and no wit too objectionable to serve as a bridge from one thought to another. But the correct understanding of such tolerance is not remote. *Whenever one psychic element is connected with another through an obnoxious or superficial association, there also exists a correct and more profound connection between the two which succumbs to the resistance of the censor.*

The correct explanation for the predominance of the superficial associations is the pressure of the censor, and not the suppression of the end-presentations. The superficial associations supplant the deep ones in the presentation whenever the censor renders the normal connective paths impassable. It is as if in a mountainous region a general interruption of traffic, *e.g.,* an inundation, should render impassable the long and broad thoroughfares; traffic would then have to be maintained through inconvenient and steep footpaths otherwise used only by the hunter.

We can here distinguish two cases which, however, are essentially one. In the first case the censor is directed only against the connection of the two thoughts, which, having been detached from each other, escape the opposition. The two thoughts then enter successively into consciousness; their connection remains concealed; but in its place there occurs to us a superficial connection between the two which we would not otherwise have thought of, and which as a rule connects with another angle of the presentation complex instead of with the one giving rise to the suppressed but essential connection. Or, in the second case, both thoughts on account of their content succumb to the censor; both then appear not in their correct but in a modified substituted form; and both substituted thoughts are so selected that they represent, through a superficial association, the essential relation which existed between those which have been replaced by them. Under the pressure of the censor the displacement of a normal and vital association by a superficial and apparently absurd one has thus occurred in both cases.

Because we know of this displacement we unhesitatingly place reliance even upon superficial associations in the dream analysis.[5]

The psychoanalysis of neurotics makes prolific use of the two axioms, first that with the abandonment of the conscious end-presentation the domination of the train of presentation is transferred to the concealed end-presentations; and, secondly, that superficial associations are only a substitutive displacement for suppressed and more profound ones; indeed, psychoanalysis raises these two axioms to pillars of its technique. When I request a patient to dismiss all reflection, and to report to me whatever comes into his mind, I firmly cling to the presupposition that he will not be able to drop the end-idea of the treatment, and I feel justified in concluding that what he reports, even though seemingly most harmless and arbitrary, has connection with this morbid state. My own personality is another end-presentation concerning which the patient has no inkling. The full appreciation, as well as the detailed proof of both these explanations, belongs accordingly to the description of the psychoanalytic technique as a therapeutic method. We have here reached one of the allied subjects with which we propose to leave the subject of the interpretation of dreams.[6]

Of all the objections only one is correct, and still remains, namely, that we ought not to ascribe all mental occurrences of the interpretation work to the nocturnal dream-work. In the interpretation in the waking state we are making a road running from the dream elements back to the dream thoughts. The dream-work has made its way in the opposite direction, and it is not at all probable that these roads are equally passable in the opposite directions. It has, on the contrary, been shown that during the day, by means of new thought connections we make paths which strike the intermediate thoughts and the dream thoughts in different places. We can see how the recent thought material of the day takes its place in the groups of the interpretation, and probably also forces the additional resistance appearing through the night to make new and further detours. But the number and form of the collaterals which we thus spin during the day is psychologically perfectly negligible if it only leads the way to the desired dream thoughts.

(B) Regression

Now that we have guarded against objection, or at least indicated where our weapons for defence rest, we need no longer delay entering upon the psychological investigations for which we have so long prepared. Let us bring together the main results of our investigations up to this point. The dream is a momentous psychic act; its motive power is at all times to fulfil a wish; its indiscernibleness as a wish and its many peculiarities and absurdities are due to the influence of the psychic censor to which it has been subjected during its formation. Apart from the pressure to withdraw itself from this censor, the following have played a part in its formation: a strong tendency to the condensation of psychic material, a consideration for dramatisation into mental pictures, and (though not regularly) a consideration for a rational and intelligible exterior in the dream structure. From every one of these propositions the road leads further to psychological postulates and assumptions. Thus the reciprocal relation of the wish motives and the four conditions, as well as the relations of these conditions to one another will have to be investigated; and the dream will have to be brought into association with the psychic life.

At the beginning of this chapter we cited a dream in order to remind us of the riddles that are still unsolved. The interpretation of this dream of the burning child afforded us no difficulties, although it was not perfectly given in our present sense. We asked ourselves why it was necessary, after all, that the father should dream instead of awakening, and we recognised the wish to represent the child as living as the single motive of the dream. That there was still another wish playing a part in this connection, we shall be able to show after later discussions. For the present, therefore, we may say that for the sake of the wish-fulfilment the mental process of sleep was transformed into a dream.

If the wish realisation is made retrogressive, only one quality still remains which separates the two forms of psychic occurrences from each other. The dream thought might have read: "I see a glimmer coming from the room in which the corpse reposes. Perhaps a candle has been upset, and the child is burning!" The dream reports the result of this reflection unchanged, but represents it in a situation which takes place in the present, and which is conceivable by the senses like an experience in the waking state. This, however, is the most common and the most striking psychological character of the dream; a thought, usually the one wished for, is in the dream made objective and represented as a scene, or, according to our belief, as experienced.

But how are we now to explain this characteristic peculiarity of the dream-work, or, to speak more modestly, how are we to bring it into relation with the psychic processes?

On closer examination, it is plainly seen that there are two pronounced characters in the manifestations of the dream which are almost independent of each other. The one is the representation as a present situation with the omission of the "perhaps"; the other is the transformation of the thought into visual pictures and into speech.

The transformation in the dream thoughts, which shifts into the present the expectation expressed in them, is perhaps in this particular dream not so very striking. This is probably in consonance with the special or rather subsidiary rôle of the wish-fulfilment in this dream. Let us take another dream in which the dream-wish does not separate itself in sleep from a continuation of the waking thoughts, *e.g.,* the dream of Irma's injection. Here the dream thought reaching representation is in the optative, "If Otto could only be blamed for Irma's sickness!" The dream suppresses the optative, and replaces it by a simple present, "Yes, Otto is to blame for Irma's sickness." This is therefore the first of the changes which even the undistorted dream undertakes with the dream thought. But we shall not stop long at this first peculiarity of the dream. We elucidate it by a reference to the conscious phantasy, the day dream, which behaves similarly with its presentation content. When Daudet's Mr. Joyeuse wanders through the streets of Paris unemployed while his daughter is led to believe that he has a position and is in his office, he likewise dreams in the present of circumstances that might help him to obtain protection and a position. The dream therefore employs the present in the same manner and with the same right as the day dream. The present is the tense in which the wish is represented as fulfilled.

The second quality, however, is peculiar to the dream as distinguished from the day dream, namely, that the presentation content is not thought, but changed into perceptible images to which we give credence and which we believe we experience. Let us add, however, that not all dreams show this transformation of presentation into perceptible images. There are dreams which consist solely of thoughts to which we cannot, however, on that account deny the substantiality of dreams. My dream "Autodidasker— the waking phantasy with Professor N."—is of that nature; it contains hardly more perceptible elements than if I had thought its content during the day. Moreover, every long dream contains elements which have not experienced the transformation into the perceptible, and which are simply thought or known as we are wont to think or know in our waking state. We may also recall here that such transformation of ideas into perceptible images does not occur in dreams only but also in hallucinations and visions which perhaps appear spontaneously in health or as symptoms in the psychoneuroses. In brief, the relation which we are investigating here is in no way an exclusive one; the fact remains, however, that where this character of the dream occurs, it appears to us as the most noteworthy, so that we cannot think of it apart from the dream life. Its explanation, however, requires a very detailed discussion.

Among all the observations on the theory of dreams to be found in authorities on the subject, I should like to lay stress upon one as being worth mentioning. The great G. T. Fechner expresses his belief (*Psychophysik,* Part II., p. 520), in connection with some discussion devoted to the dream, that the seat of the dream is elsewhere than in the waking ideation. No other theory enables us to conceive the special qualities of the dream life.

The idea which is placed at our disposal is one of psychic locality. We shall entirely ignore the fact that the psychic apparatus with which we are here dealing is also familiar to us as an anatomical specimen, and we shall carefully avoid the temptation to determine the psychic locality in any way anatomically. We shall remain on psychological ground, and we shall think ourselves called upon only to conceive the instrument which serves the psychic activities somewhat after the manner of a compound microscope, a photographic or other similar apparatus. The psychic locality, then, corresponds to a place within such an apparatus in which one of the primary elements of the picture comes into existence. As is well known, there are in the microscope and telescope partly fanciful locations or regions in which no tangible portion of the apparatus is located. I think it superfluous to apologise for the imperfections of this and all similar figures. These comparisons are designed only to assist us in our attempt to make clear the complication of the psychic activity by breaking up this activity and referring the single activities to the single component parts of the apparatus. No one, so far as I know, has ever ventured to attempt to discover the composition of the psychic instrument through such analysis. I see no harm in such an attempt. I believe that we may give free rein to our assumptions provided we at the same time preserve our cool judgment and do not take the scaffolding for the building. As we need nothing except auxiliary ideas for the first approach to any unknown subject, we shall prefer the crudest and most tangible hypothesis to all others.

We therefore conceive the psychic apparatus as a compound instrument, the component parts of which let us call instances, or, for the sake of clearness, systems. We then entertain the expectation that these systems perhaps maintain a constant spatial relationship to each other like the different systems of lenses of the telescope, one behind another. Strictly speaking, there is no need of assuming a real spatial arrangement of the psychic system. It will serve our purpose if a firm sequence be established through the fact that in certain psychological occurrences the system will be traversed by the excitement in a definite chronological order. This sequence may experience an alteration in other processes; such possibility may be left open. For the sake of brevity, we shall henceforth speak of the component parts of the apparatus as "Psi-systems."

The first thing that strikes us is the fact that the apparatus composed of Psi-systems has a direction. All our psychic activities proceed from (inner or outer) stimuli and terminate in innervations. We thus ascribe to the apparatus a sensible and a motor end; at the sensible end we find a system which receives the perceptions, and at the motor end another which opens the locks of motility. The psychic process generally takes its course from the perception end to the motility end. The most common scheme of the psychic apparatus has therefore the following appearance: [Figure 1.] But this is only in compliance with the demand long familiar to us, that the psychic apparatus must be constructed like a reflex apparatus. The reflex act remains the model for every psychic activity.

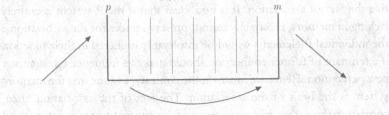

Figure 1

We have now reason to admit a first differentiation at the sensible end. The perceptions that come to us leave a trace in our psychic apparatus which we may call a "Memory trace." The function which relates to this memory trace we call the memory. If we hold seriously to our resolution to connect the psychic processes into systems, the memory trace can then consist only of lasting changes in the elements of the systems. But, as has already been shown in other places, obvious difficulties arise if one and the same system faithfully preserves changes in its elements and still remains fresh and capable of admitting new motives for change. Following the principle which directs our undertaking, we shall distribute these two activities among two different systems. We assume that a first system of the apparatus takes up the stimuli of perception, but retains nothing from them—that is, it has no memory; and that behind this there lies a second system which transforms the momentary excitement of the first into lasting traces. This would then be a diagram of our psychic apparatus: [Figure 2.]

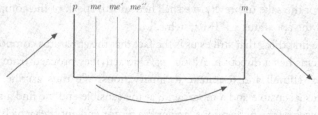

Figure 2

It is known that from the perceptions that act on the P- system we retain something else as lasting as the content itself. Our perceptions prove to be connected with one another in memory, and this is especially the case when they have once fallen together in simultaneity. We call this the fact of association. It is now clear that if the P-system is entirely lacking in memory, it certainly cannot preserve traces for the associations; the individual P-elements would be intolerably hindered in their function if a remnant of former connection should make its influence felt against a new perception. Hence we must, on the contrary, assume that the memory system is the basis of the association. The fact of the association, then, consists in this—that, in consequence of the diminutions in resistance and a smoothing of the ways from one of the Mem-elements, the excitement transmits itself to a second rather than to a third Mem-system.

On further investigation we find it necessary to assume not one but many such Mem-systems, in which the same excitement propagated by the P-elements experiences a diversified fixation. The first of these Mem-systems will contain in any case the fixation of the association through simultaneity, while in those lying further away the same exciting material will be arranged according to other forms of concurrence; so that relationships of similarity, &c., might perhaps be represented through these later systems. It would naturally be idle to attempt to report in words the psychic significance of such a system. Its characteristic would lie in the intimacy of its relations to elements of raw memory material—that is, if we wish to point to a profounder theory in the gradations of the resistances to conduction toward these elements.

We may insert here an observation of a general nature which points perhaps to something of importance. The P-system, which possesses no capability of preserving changes and hence no memory, furnishes for our consciousness the entire manifoldness of the sensible qualities. Our memories, on the other hand, are unconscious in themselves; those that are most deeply impressed form no exception. They can be made conscious, but there can be no doubt that they develop all their influences in the unconscious state. What we term our character is based, to be sure, on the memory traces of our impressions, and indeed on these impressions that have affected us most strongly, those of our early youth—those that almost never become conscious. But when memories become conscious again they show no sensible quality or a very slight one in comparison to the perceptions. If, now, it can be confirmed *that memory and quality exclude each other, as far as consciousness in the Psi-systems is concerned,* a most promising insight reveals itself to us in the determinations of the neuron excitement.

What we have so far assumed concerning the composition of the psychic apparatus at the sensible end follows regardless of the dream and the psychological explanations derived from it. The dream, however, serves as a source of proof for the knowledge of another part of the apparatus. We have seen that it became impossible to explain the dream formation unless we ventured to assume two psychic instances, one of which subjected the activity of the other to a critique as a consequence of which the exclusion from consciousness resulted.

We have seen that the criticising instance entertains closer relations with consciousness than the criticised. The former stands between the latter and consciousness like a screen. We have, moreover, found essential reasons for identifying the criticising instance with that which directs our waking life and determines our voluntary conscious actions. If we now replace these instances in the development of our theory by systems, the criticising system is then to be ascribed to the motor end because of the fact just mentioned. We now enter both systems in our scheme, and express by the names given them their relation to consciousness. [Figure 3.]

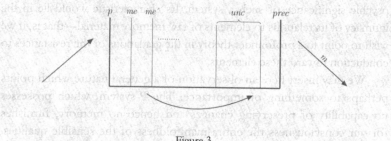

Figure 3

The last of the systems at the motor end we call the fore-conscious in order to denote that exciting processes in this system can reach consciousness without any further detention provided certain other conditions be fulfilled, *e.g.,* the attainment of a certain intensity, a certain distribution of that function which must be called attention, and the like. This is at the same time the system which possesses the keys to voluntary motility. The system behind it we call the unconscious because it has no access to consciousness except through the foreconscious, in the passage through which its excitement must submit to certain changes.

In which of these systems, now, do we localise the impulse to the dream formation? For the sake of simplicity, let us say in the system Unc. To be sure we shall find in later discussions that this is not quite correct, that the dream formation is forced to connect with dream thoughts which belong to the system of the foreconscious. But we shall learn later, when we come to deal with the dream-wish, that the motive power for the dream is furnished by the Unc., and, owing to this latter movement, we shall assume the unconscious system as the starting-point of the dream formation. This dream impulse, like all other thought structures, will now strive to continue itself in the foreconscious, and thence to gain admission to consciousness.

Experience teaches us that the road leading from the fore-conscious to consciousness is closed to the dream thoughts during the day by the resistance of the censor. At night the dream thoughts gain admission to consciousness, but the question arises, in what way and because of what change. If this admission was rendered possible to the dream thoughts through the fact that the resistance watching on the boundary between the unconscious and foreconscious sinks at night, we should then get dreams in the material of our presentations which did not show the hallucinatory character which just now interests us.

406

The sinking of the censor between the two systems, Unc. and Forec., can explain to us only such dreams as "Autodidasker," but not dreams like the one of the burning child, which we have taken as a problem at the outset in these present investigations.

What takes place in the hallucinatory dream we can describe in no other way than by saying that the excitement takes a retrogressive course. It takes its station, not at the motor end of the apparatus, but at the sensible end, and finally reaches the system of the perceptions. If we call the direction towards which the psychic process continues from the unconscious into the waking state the progressive, we may then speak of the dream as having a regressive character.

This regression is surely one of the most important peculiarities of the dream process; but we must not forget that it does not belong to the dream alone. The intentional recollection and other processes of our normal thinking also require a retrogression in the psychic apparatus from any complex presentation act to the raw material of the memory traces lying at its basis. But during the waking state this turning backward does not reach beyond the memory pictures; it is unable to produce the hallucinatory vividness of the perception pictures. Why is this different in the dream? When we spoke of the condensation work of the dream we could not avoid the assumption that the intensities adhering to the presentations are fully transferred from one to another through the dream-work. It is probably this modification of the former psychic process which makes possible the occupation of the system of P to its full sensual vividness in the opposite direction from thought.

I hope that we are far from deluding ourselves about the importance of this present discussion. We have done nothing more than give a name to an inexplicable phenomenon. We call it regression if the presentation in the dream is changed back to the perceptible image from which it once originated. But even this step demands justification. Why this naming, if it does not teach us anything new? I believe, however, that the name "Regression" will serve us to the extent of connecting a fact familiar to us with a scheme of the psychic apparatus which is supplied with a direction. At this point, for the first time, it is worth the trouble to construct such a scheme. For, with the help of this scheme, any other peculiarity of the dream formation will become clear to us without further reflection. If we look upon the dream as a process of regression in the assumed psychic apparatus, we can readily understand the empirically proven fact that all mental relation of the dream thoughts either is lost in the dream-work or can come to expression only with difficulty. According to our scheme, these mental relations are contained not in the first Mem-systems, but in those lying further to the front, and in the regression they must forfeit their expression in favour of the perception pictures. *The structure of the dream thoughts is in the regression broken up into its raw material.*

But what change renders possible this regression which is impossible during the day? Let us here be content with assumption. There must evidently be some alterations in the charge of energy belonging to the single systems causing the latter to become accessible or inaccessible to the discharge of the excitement; but in any such apparatus the same effect upon the course of excitement might be brought about through more than one form of such changes. This naturally reminds us of the state of sleep and of the many changes of energy this state produces at the sensible end of the apparatus. During the day there is a continuous coursing stream from the Psi-system of the P toward the motility; this current ceases at night, and no longer hinders a streaming of the current of excitement in the opposite direction. This would appear to be that "seclusion from the outer world" which according to the theory of some authors is supposed to explain the psychological character of the dream (*vide* p. 52). In the explanation of the regression of the dream we shall, however, have to consider those other regressions which originate during morbid waking states. In these other forms the explanation just given plainly leaves us in the lurch. Regression takes place in spite of the uninterrupted sensible current in a progressive direction.

The hallucinations of hysteria and paranoia, as well as the visions of mentally normal persons, I can explain as actually corresponding to regressions, being in fact thoughts transformed into images; and only such thoughts are subjected to this transformation as are in intimate connection with suppressed or unconscious recollections. As an example I shall cite one of my youngest hysterical patients—a boy. twelve years old, who was prevented from falling asleep by *"green faces with red eyes,"* which terrified him. The source of this manifestation was the suppressed, but once conscious, memory of a boy whom he had often seen during four years, and who offered him a deterring example of many childish bad habits, including onanism, which now formed the subject of his own reproach. His mother had noticed at the time that the complexion of the ill-bred boy was greenish and that he had *red* (*i.e. red bordered*) *eyes*. Hence the terrible vision which constantly served to remind him of his mother's warning that such boys become demented, that they are unable to make progress at school, and are doomed to an early death. A part of this prediction came true in the case of the little patient; he could not successfully pursue his high school studies, and. as appeared on examination of his involuntary fancies, he stood in great dread of the remainder of the prophecy. However. after a brief period of successful treatment, his sleep was restored, he lost his fears, and finished his scholastic year with an excellent record.

I may also add here the interpretation of a vision related to me by an hysteric forty years of age, as having occurred in her normal life. On opening her eyes one morning she beheld in the room her brother, whom she knew to be confined in an insane asylum. Her little son was asleep by her side. Lest the child should be frightened on seeing his uncle, and fall into convulsions, she pulled the sheet over the little one; this done, the phantom disappeared. This vision is the re-casting of one of her infantile reminiscences which, although conscious, is most intimately connected with all the unconscious material in her mind. Her nursemaid told her that her mother, who had died young (the patient was then only a year and a half old), had suffered from epileptic or hysterical convulsions, which dated back to a fright caused by her brother (the patient's uncle), who appeared to her disguised as a spectre with a sheet over his head. The vision contains the same elements as the reminiscence, viz. the appearance of the brother, the sheet, the fright, and its effect. These elements, however, are ranged in different relations, and are transferred to other persons. The obvious motive of the vision, which replaces the idea, is her solicitude lest her little son, who bore a striking resemblance to his uncle, should share the latter's fate. Both examples here cited are not entirely unrelated to sleep, and may therefore be unsuitable as proof for my assertion. I may therefore refer to my analysis of an hallucinatory paranoia,[7] and to the results of my hitherto unpublished studies on the psychology of the psychoneuroses in order to emphasize the fact that in these cases of regressive thought transformation one must not overlook the influence of a suppressed or unconscious reminiscence, this being in most cases of an infantile character. This recollection, so to speak, draws into the regression the thought with which it is connected, which is prevented from expression by the censor—that is, into that form of representation in which the recollection itself exists psychically. I may here mention as a result of my studies in hysteria that if we succeed in restoring infantile scenes to consciousness (whether recollections or fancies) they are seen as hallucinations, and are divested of this character only after reproduction. It is also known that the earliest infantile memories retain the character of perceptible vividness until late in life, even in persons who are otherwise not visual in memory.

If, now, we keep in mind what part is played in the dream thoughts by the infantile reminiscences or the phantasies based upon them, how often fragments of these reminiscences emerge in the dream content, and how often they even give origin to dream wishes, we cannot deny the probability that in the dream, too, the transformation of thoughts into visual images may be the result of the attraction exerted by the visually represented reminiscences, striving for reanimation, upon the thoughts severed from consciousness and struggling for expression. Following this conception, we may further describe the dream as a modified substitute for the infantile scene produced by transference to recent material. The infantile cannot enforce its renewal, and must therefore be satisfied to return as a dream.

This reference to the significance of the infantile scenes (or of their phantastic repetitions), as in a manner furnishing the pattern for the dream content, renders superfluous the assumption made by Scherner and his pupils, of an inner source of excitement. Scherner assumes a state of "visual excitation" of internal excitement in the organ of sight when the dreams manifest a particular vividness or a special abundance of visual elements. We need not object to this assumption, but may be satisfied with establishing such state of excitation for the psychic perceptive system of the organs of vision only; we shall, however, assert that this state of excitation is formed through the memory, and is merely a refreshing of the former actual visual excitation. I cannot, from my own experience, give a good example showing such an influence of infantile reminiscence; my own dreams are surely less rich in perceptible elements than I must fancy those of others; but in my most beautiful and most vivid dream of late years I can easily trace the hallucinatory distinctness of the dream contents to the sensuous nature of recently received impressions. On page 394 I mentioned a dream in which the dark blue colour of the water, the brown colour of the smoke issuing from the ship's funnels, and the sombre brown and red of the buildings which I had seen made a profound and lasting impression on my mind. This dream. if any, must be attributed to visual excitation. But what has brought my visual organ into this excitable state? It was a recent impression uniting itself with a series of former ones. The colours I beheld were those of the toy blocks with which my children erected a grand structure for my admiration on the day preceding the dream. The same sombre red colour covered the large blocks and the same blue and brown the small ones. Connected with these were the colour impression of my last journey in Italy, the charming blue of the Isonzo and the Lagoon, the brown hue of the Alpine region. The beautiful colours seen in the dream were but a repetition of those seen in the memory.

Let us review what we have learned about this peculiarity which the dream has of transforming its content of ideas into plastic images. We have neither explained this character of the dream-work nor traced it to known laws of psychology, but we have singled it out as pointing to unknown connections, and designated it by the name of the "regredient" character. Wherever this regression has occurred, we have regarded it as an effect of the resistance which opposes the progress of the thought on its normal way to consciousness, as well as a result of the simultaneous attraction exerted upon it by the vivid memories present. Regression is perhaps facilitated in the dream by the cessation of the progressive stream running from the sense organs during the day. For this auxiliary moment there must be compensation in the other forms of regression through a fortifying of the other motives of regression. We must also bear in mind that in pathological cases of regression, as in the dream, the process of transference of energy must be different from that of the regressions of normal psychic life, as it renders possible a full hallucinatory occupation of the perception systems. What we have described in the analysis of the dream-work as "Regard for Dramatic Fitness" may be referred to the selective attraction of visually recollected scenes, touched by the dream thoughts.

It is quite possible that this first part of our psychological utilisation of the dream does not entirely satisfy even us. We must, however, console ourselves with the fact that we are compelled to build in the dark. If we have not altogether strayed from the right path, we shall be sure to reach about the same ground from another starting-point and thereafter perhaps be better able to see our way.

(C) The Wish-Fulfilment

The dream of the burning child cited above affords us a welcome opportunity for appreciating the difficulties confronting the theory of wish-fulfilment. That the dream should be nothing but a wish-fulfilment surely seemed strange to us all—and that not alone because of the contradictions offered by the anxiety dream.

After learning from the first analytical explanations that the dream conceals sense and psychic validity, we could hardly expect so simple a determination of this sense. According to the correct but concise definition of Aristotle, the dream is a continuation of thinking in sleep (in so far as one sleeps). Considering that during the day our thoughts produce such a diversity of psychic acts—judgments, conclusions, contradictions, expectations, intentions, &c.—why should our sleeping thoughts be forced to confine themselves to the production of wishes? Are there not, on the contrary, many dreams that present a different psychic act in dream form, e.g., a solicitude, and is not the very transparent father's dream mentioned above of just such a nature? From the gleam of light falling into his eyes while asleep the father draws the solicitous conclusion that a candle has been upset and may have set fire to the corpse; he transforms this conclusion into a dream by investing it with a senseful situation enacted in the present tense. What part is played in this dream by the wish-fulfilment, and which are we to suspect—the predominance of the thought continued from the waking state or of the thought incited by the new sensory impression?

All these considerations are just, and force us to enter more deeply into the part played by the wish-fulfilment in the dream, and into the significance of the waking thoughts continued in sleep.

It is in fact the wish-fulfilment that has already induced us to separate dreams into two groups. We have found some dreams that were plainly wish-fulfilments; and others in which wish-fulfilment could not be recognised, and was frequently concealed by every available means. In this latter class of dreams we recognised the influence of the dream censor. The undisguised wish dreams were chiefly found in children, yet fleeting open-hearted wish dreams *seemed* (I purposely emphasize this word) to occur also in adults.

We may now ask whence the wish fulfilled in the dream originates. But to what opposition or to what diversity do we refer this "whence"? I think it is to the opposition between conscious daily life and a psychic activity remaining unconscious which can only make itself noticeable during the night. I thus find a threefold possibility for the origin of a wish. Firstly, it may have been incited during the day, and owing to external circumstances failed to find gratification, there is thus left for the night an acknowledged but unfulfilled wish. Secondly, it may come to the surface during the day but be rejected, leaving an unfulfilled but suppressed wish. Or, thirdly, it may have no relation to daily life, and belong to those wishes that originate during the night from the suppression. If we now follow our scheme of the psychic apparatus, we can localise a wish of the first order in the system Forec. We may assume that a wish of the second order has been forced back from the Forec. system into the Unc. system, where alone, if anywhere, it can maintain itself; while a wish-feeling of the third order we consider altogether incapable of leaving the Unc. system. This brings up the question whether wishes arising from these different sources possess the same value for the dream, and whether they have the same power to incite a dream.

On reviewing the dreams which we have at our disposal for answering this question, we are at once moved to add as a fourth source of the dream-wish the actual wish incitements arising during the night, such as thirst and sexual desire. It then becomes evident that the source of the dream-wish does not affect its capacity to incite a dream. This view is supported by the dream of the little girl who continued the sea trip interrupted during the day, and by the other children's dreams referred to; they are explained by an unfulfilled but not suppressed wish from the day-time. That a wish suppressed during the day asserts itself in the dream can be shown by a great many examples. I shall mention a very simple example of this class. A somewhat sarcastic young lady, whose younger friend has become engaged to be married, is asked throughout the day by her acquaintances whether she knows and what she thinks of the fiance. She answers with unqualified praise, thereby silencing her own judgment, as she would prefer to tell the truth, namely, that he is an ordinary person (Dutzendmensch).[8] The following night she dreams that the same question is put to her, and that she replies with the formula: "In case of subsequent orders it will suffice to mention the number." Finally, we have learned from numerous analyses that the wish in all dreams that have been subject to distortion has been derived from the unconscious, and has been unable to come to perception in the waking state. Thus it would appear that all wishes are of the same value and force for the dream formation.

I am at present unable to prove that the state of affairs is really different, but I am strongly inclined to assume a more stringent determination of the dream-wish. Children's dreams leave no doubt that an unfulfilled wish of the day may be the instigator of the dream. But we must not forget that it is, after all, the wish of a child, that it is a wish-feeling of infantile strength only. I have a strong doubt whether an unfulfilled wish from the day would suffice to create a dream in an adult. It would rather seem that as we learn to control our impulses by intellectual activity, we more and more reject as vain the formation or retention of such intense wishes as are natural to childhood. In this, indeed, there may be individual variations; some retain the infantile type of psychic processes longer than others. The differences are here the same as those found in the gradual decline of the originally distinct visual imagination.

In general, however, I am of the opinion that unfulfilled wishes of the day are insufficient to produce a dream in adults. I readily admit that the wish instigators originating in conscious life contribute towards the incitement of dreams, but that is probably all. The dream would not originate if the foreconscious wish were not reinforced from another source.

That source is the unconscious. I believe that *the conscious wish is a dream inciter only if it succeeds in arousing a similar unconscious wish which reinforces it.* Following the suggestions obtained through the psychoanalysis of the neuroses, I believe that these unconscious wishes are always active and ready for expression whenever they find an opportunity to unite themselves with an emotion from conscious life, and that they transfer their greater intensity to the lesser intensity of the latter.[9] It may therefore seem that the conscious wish alone has been realised in a dream; but a slight peculiarity in the formation of this dream will put us on the track of the powerful helper from the unconscious. These ever active and, as it were, immortal wishes from the unconscious recall the legendary Titans who from time immemorial have borne the ponderous mountains which were once rolled upon them by the victorious gods, and which even now quiver from time to time from the convulsions of their mighty limbs; I say that these wishes found in the repression are of themselves of an infantile origin, as we have learned from the psychological investigation of the neuroses. I should like, therefore, to withdraw the opinion previously expressed that it is unimportant whence the dream-wish originates, and replace it by another, as follows: *The wish manifested in the dream must be an infantile one.* In the adult it originates hi the Unc., while in the child, where no separation and censor as yet exist between Forec. and Unc., or where these are only in the process of formation, it is an unfulfilled and unrepressed wish from the waking state. I am aware that this conception cannot be generally demonstrated, but I maintain nevertheless that it can be frequently demonstrated, even where it was not suspected, and that it cannot be generally refuted.

The wish-feelings which remain from the conscious waking state are, therefore, relegated to the background in the dream formation. In the dream content I shall attribute to them only the part attributed to the material of actual sensations during sleep (see p. 213). If I now take into account those other psychic instigations remaining from the waking state which are not wishes, I shall only adhere to the line mapped out for me by this train of thought. We may succeed in provisionally terminating the sum of energy of our waking thoughts by deciding to go to sleep. He is a good sleeper who can do this; Napoleon I. is reputed to have been a model of this sort. But we do not always succeed in accomplishing it, or in accomplishing it perfectly. Unsolved problems, harassing cares, overwhelming impressions continue the thinking activity even during sleep, maintaining psychic processes in the system which we have termed the foreconscious. These mental processes continuing into sleep may be divided into the following groups: (1) That which has not been terminated during the day owing to casual prevention; (2) that which has been left unfinished by temporary paralysis of our mental power, *i.e.* the unsolved; (3) that which has been rejected and suppressed during the day. This unites with a powerful group (4) formed by that which has been excited in our Unc. during the day by the work of the foreconscious. Finally, we may add group (5) consisting of the indifferent and hence unsettled impressions of the day.

We should not underrate the psychic intensities introduced into sleep by these remnants of waking life, especially those emanating from the group of the unsolved. These excitations surely continue to strive for expression during the night, and we may assume with equal certainty that the sleeping state renders impossible the usual continuation of the excitement in the foreconscious and the termination of the excitement by its becoming conscious. As far as we can normally become conscious of our mental processes, even during the night, in so far we are not asleep. I shall not venture to state what change is produced in the Forec. system by the sleeping state, but there is no doubt that the psychological character of sleep is essentially due to the change of energy in this very system, which also dominates the approach to motility, which is paralysed during sleep. In contradistinction to this, there seems to be nothing in the psychology of the dream to warrant the assumption that sleep produces any but secondary changes in the conditions of the Unc. system. Hence, for the nocturnal excitation in the Forec. there remains no other path than that followed by the wish excitements from the Unc. This excitation must seek reinforcement from the Unc., and follow the detours of the unconscious excitations. But what is the relation of the foreconscious day remnants to the dream? There is no doubt that they penetrate abundantly into the dream, that they utilise the dream content to obtrude themselves upon consciousness even during the night; indeed, they occasionally even dominate the dream content, and impel it to continue the work of the day; it is also certain that the day remnants may just as well have any other character as that of wishes; but it is highly instructive and even decisive for the theory of wish-fulfilment to see what conditions they must comply with in order to be received into the dream.

Let us pick out one of the dreams cited above as examples, *e.g.,* the dream in which my friend Otto seems to show the symptoms of Basedow's disease (p. 244). My friend Otto's appearance occasioned me some concern during the day, and this worry, like everything else referring to this person, affected me. I may also assume that these feelings followed me into sleep. I was probably bent on finding out what was the matter with him. In the night my worry found expression in the dream which I have reported, the content of which was not only senseless, but failed to show any wish-fulfilment. But I began to investigate for the source of this incongruous expression of the solicitude felt during the day, and analysis revealed the connection. I identified my friend Otto with a certain Baron L. and myself with a Professor R. There was only one explanation for my being impelled to select just this substitution for the day thought. I must have always been prepared in the Unc. to identify myself with Professor R., as it meant the realisation of one of the immortal infantile wishes, viz. that of becoming great. Repulsive ideas respecting my friend, that would certainly have been repudiated in a waking state, took advantage of the opportunity to creep into the dream, but the worry of the day likewise found some form of expression through a substitution in the dream content. The day thought, which was no wish in itself but rather a worry, had in some way to find a connection with the infantile now unconscious and suppressed wish, which then allowed it, though already properly prepared, to "originate" for consciousness. The more dominating this worry, the stronger must be the connection to be established; between the contents of the wish and that of the worry there need be no connection, nor was there one in any of our examples.

We can now sharply define the significance of the unconscious wish for the dream. It may be admitted that there is a whole class of dreams in which the incitement originates preponderatingly or even exclusively from the remnants of daily life; and I believe that even my cherished desire to become at some future time a "professor extraordinarius" would have allowed me to slumber undisturbed that night had not my worry about my friend's health been still active. But this worry alone would not have produced a dream; the motive power needed by the dream had to be contributed by a wish, and it was the affair of the worriment to procure for itself such wish as a motive power of the dream. To speak figuratively, it is quite possible that a day thought plays the part of the contractor (*entrepreneur*) in the dream. But it is known that no matter what idea the contractor may have in mind, and how desirous he may be of putting it into operation, he can do nothing without capital; he must depend upon a capitalist to defray the necessary expenses, and this capitalist, who supplies the psychic expenditure for the dream is invariably and indisputably *a wish from the unconscious,* no matter what the nature of the waking thought may be.

In other cases the capitalist himself is the contractor for the dream; this, indeed, seems to be the more usual case. An unconscious wish is produced by the day's work, which in turn creates the dream. The dream processes, moreover, run parallel with all the other possibilities of the economic relationship used here as an illustration. Thus, the entrepreneur may contribute some capital himself, or several entrepreneurs may seek the aid of the same capitalist, or several capitalists may jointly supply the capital required by the entrepreneur. Thus there are dreams produced by more than one dream-wish, and many similar variations which may readily be passed over and are of no further interest to us. What we have left unfinished in this discussion of the dream-wish we shall be able to develop later.

The "tertium comparationis" in the comparisons just employed—*i.e.* the sum placed at our free disposal in proper allotment—admits of still finer application for the illustration of the dream structure. As shown on p. 313 we can recognise in most dreams a centre especially supplied with perceptible intensity. This is regularly the direct representation of the wish-fulfilment; for, if we undo the displacements of the dream-work by a process of retrogression, we find that the psychic intensity of the elements in the dream thoughts is replaced by the perceptible intensity of the elements in the dream content. The elements adjoining the wish-fulfilment have frequently nothing to do with its sense, but prove to be descendants of painful thoughts which oppose the wish. But, owing to their frequently artificial connection with the central element, they have acquired sufficient intensity to enable them to come to expression. Thus, the force of expression of the wish-fulfilment is diffused over a certain sphere of association, within which it raises to expression all elements, including those that are in themselves impotent. In dreams having several strong wishes we can readily separate from one another the spheres of the individual wish-fulfilments; the gaps in the dream likewise can often be explained as boundary zones.

Although the foregoing remarks have considerably limited the significance of the day remnants for the dream, it will nevertheless be worth our while to give them some attention. For they must be a necessary ingredient in the formation of the dream, inasmuch as experience reveals the surprising fact that every dream shows in its content a connection with some impression of a recent day, often of the most indifferent kind. So far we have failed to see any necessity for this addition to the dream mixture (p. 169). This necessity appears only when we follow closely the part played by the unconscious wish, and then seek information in the psychology of the neuroses. We thus learn that the unconscious idea, as such, is altogether incapable of entering into the foreconscious, and that it can exert an influence there only by uniting with a harmless idea already belonging to the foreconscious, to which it transfers its intensity and under which it allows itself to be concealed. This is the fact of transference which furnishes an explanation for so many surprising occurrences in the psychic life of neurotics.

The idea from the foreconscious which thus obtains an unmerited abundance of intensity may be left unchanged by the transference, or it may have forced upon it a modification from the content of the transferring idea. I trust the reader will pardon my fondness for comparisons from daily life, but I feel tempted to say i/hat the relations existing for the repressed idea are similar to the situations existing in Austria for the American dentist, who is forbidden to practise unless he gets permission from a regular physician to use his name on the public signboard and thus cover the legal requirements. Moreover, just as it is naturally not the busiest physicians who form such alliances with dental practitioners, so in the psychic life only such foreconscious or conscious ideas are chosen to cover a repressed idea as have not themselves attracted much of the attention which is operative in the foreconscious. The unconscious entangles with its connections preferentially either those impressions and ideas of the fore-conscious which have been left unnoticed as indifferent, or those that have soon been deprived of this attention through rejection. It is a familiar fact from the association studies confirmed by every experience, that ideas which have formed Intimate connections in one direction assume an almost negative attitude to whole groups of new connections. I once tried from this principle to develop a theory for hysterical paralysis.

If we assume that the same need for the transference of the repressed ideas which we have learned to know from the analysis of the neuroses makes its influence felt in the dream as well, we can at once explain two riddles of the dream, viz. that every dream analysis shows an interweaving of a recent impression, and that this recent element is frequently of the most indifferent character. We may add what we have already learned elsewhere, that these recent and indifferent elements come so frequently into the dream content as a substitute for the most deep-lying of the dream thoughts, for the further reason that they have least to fear from the resisting censor. But while this freedom from censorship explains only the preference for trivial elements, the constant presence of recent elements points to the fact that there is a need for transference. Both groups of impressions satisfy the demand of the repression for material still free from associations, the indifferent ones because they have offered no inducement for extensive associations, and the recent ones because they have had insufficient time to form such associations.

We thus see that the day remnants, among which we may now include the indifferent impressions when they participate in the dream formation, not only borrow from the Unc. the motive power at the disposal of the repressed wish, but also offer to the unconscious something indispensable, namely, the attachment necessary to the transference. If we here attempted to penetrate more deeply into the psychic processes, we should first have to throw more light on the play of emotions between the foreconscious and the unconscious, to which, indeed, we are urged by the study of the psychoneuroses, whereas the dream itself offers no assistance in this respect.

Just one further remark about the day remnants. There is no doubt that they are the actual disturbers of sleep, and not the dream, which, on the contrary, strives to guard sleep. But we shall return to this point later.

We have so far discussed the dream-wish, we have traced it to the sphere of the Unc., and analysed its relations to the day remnants, which in turn may be either wishes, psychic emotions of any other kind, or simply recent impressions. We have thus made room for any claims that may be made for the importance of conscious thought activity in dream formations in all its variations. Relying upon our thought series, it would not be at all impossible for us to explain even those extreme cases in which the dream as a continuer of the day work brings to a happy conclusion an unsolved problem of the waking state. We do not, however, possess an example, the analysis of which might reveal the infantile or repressed wish source furnishing such alliance and successful strengthening of the efforts of the foreconscious activity. But we have not come one step nearer a solution of the riddle: Why can the unconscious furnish the motive power for the wish-fulfilment only during sleep? The answer to this question must throw light on the psychic nature of wishes; and it will be given with the aid of the diagram of the psychic apparatus.

We do not doubt that even this apparatus attained its present perfection through a long course of development. Let us attempt to restore it as it existed in an early phase of its activity. From assumptions, to be confirmed elsewhere, we know that at first the apparatus strove to keep as free from excitement as possible, and in its first formation, therefore, the scheme took the form of a reflex apparatus, which enabled it promptly to discharge through the motor tracts any sensible stimulus reaching it from without. But this simple function was disturbed by the wants of life, which likewise furnish the impulse for the further development of the apparatus. The wants of life first manifested themselves to it in the form of the great physical needs. The excitement aroused by the inner want seeks an outlet in motility, which may be designated as "inner changes" or as an "expression of the emotions." The hungry child cries or fidgets helplessly, but its situation remains unchanged; for the excitation proceeding from an inner want requires, not a momentary outbreak, but a force working continuously. A change can occur only if in some way a feeling of gratification is experienced—which in the case of the child must be through outside help—in order to remove the inner excitement. An essential constituent of this experience is the appearance of a certain perception (of food in our example), the memory picture of which thereafter remains associated with the memory trace of the excitation of want.

Thanks to the established connection, there results at the next appearance of this want a psychic feeling which revives the memory picture of the former perception, and thus recalls the former perception itself, *i.e.* it actually re-establishes the situation of the first gratification. We call such a feeling a wish; the reappearance of the perception constitutes the wish-fulfilment, and the full revival of the perception by the want excitement constitutes the shortest road to the wish-fulfilment. We may assume a primitive condition of the psychic apparatus in which this road is really followed, *i.e.* where the wishing merges into an hallucination. This first psychic activity therefore aims at an identity of perception, *i.e.* it aims at a repetition of that perception which is connected with the fulfilment of the want.

This primitive mental activity must have been modified by bitter practical experience into a more expedient secondary activity. The establishment of the identity perception on the short regressive road within the apparatus does not in another respect carry with it the result which inevitably follows the revival of the same perception from without. The gratification does not take place, and the want continues. In order to equalise the internal with the external sum of energy, the former must be continually maintained, just as actually happens in the hallucinatory psychoses and in the deliriums of hunger which exhaust their psychic capacity in clinging to the object desired. In order to make more appropriate use of the psychic force, it becomes necessary to inhibit the full regression so as to prevent it from extending beyond the image of memory, whence it can select other paths leading ultimately to the establishment of the desired identity from the outer world. This inhibition and consequent deviation from the excitation becomes the task of a second system which dominates the voluntary motility, *i.e.* through whose activity the expenditure of motility is now devoted to previously recalled purposes. But this entire complicated mental activity which works its way from the memory picture to the establishment of the perception identity from the outer world merely represents a detour which has been forced upon the wish-fulfilment by experience.[10] Thinking is indeed nothing but the equivalent of the hallucinatory wish; and if the dream be called a wish-fulfilment this becomes self-evident, as nothing but a wish can impel our psychic apparatus to activity. The dream, which in fulfilling its wishes follows the short regressive path, thereby preserves for us only an example of the primary form of the psychic apparatus which has been abandoned as inexpedient. What once ruled in the waking state when the psychic life was still young and unfit seems to have been banished into the sleeping state, just as we see again in the nursery the bow and arrow, the discarded primitive weapons of grown-up humanity. *The dream is a fragment of the abandoned psychic life of the child.* In the psychoses these modes of operation of the psychic apparatus, which are normally suppressed in the waking state, reassert themselves, and then betray their inability to satisfy our wants in the outer world.

The unconscious wish-feelings evidently strive to assert themselves during the day also, and the fact of transference and the psychoses teach us that they endeavour to penetrate to consciousness and dominate motility by the road leading through the system of the foreconscious. It is, therefore, the censor lying between the Unc. and the Forec., the assumption of which is forced upon us by the dream, that we have to recognise and honour as the guardian of our psychic health. But is it not carelessness on the part of this guardian to diminish its vigilance during the night and to allow the suppressed emotions of the Unc. to come to expression, thus again making possible the hallucinatory regression? I think not. for when the critical guardian goes to rest—and we have proof that his slumber is not profound—he takes care to close the gate to motility. No matter what feelings from the otherwise inhibited Unc. may roam about on the scene, they need not be interfered with; they remain harmless because they are unable to put in motion the motor apparatus which alone can exert a modifying influence upon the outer world. Sleep guarantees the security of the fortress which is under guard. Conditions are less harmless when a displacement of forces is produced, not through a nocturnal diminution in the operation of the critical censor, but through pathological enfeeblement of the latter or through pathological reinforcement of the unconscious excitations, and this while the foreconscious is charged with energy and the avenues to motility are open. The guardian is then overpowered, the unconscious excitations subdue the Forec.; through it they dominate our speech and actions, or they enforce the hallucinatory regression, thus governing an apparatus not designed for them by virtue of the attraction exerted by the perceptions on the distribution of our psychic energy. We call this condition a psychosis.

We are now in the best position to complete our psychological construction, which has been interrupted by the introduction of the two systems, Unc. and Forec. We have still, however, ample reason for giving further consideration to the wish as the sole psychic motive power in the dream. We have explained that the reason why the dream is in every case a wish realisation is because it is a product of the Unc., which knows no other aim in its activity but the fulfilment of wishes, and which has no other forces at its disposal but wish-feelings. If we avail ourselves for a moment longer of the right to elaborate from the dream interpretation such far-reaching psychological speculations, we are in duty bound to demonstrate that we are thereby bringing the dream into a relationship which may also comprise other psychic structures. If there exists a system of the Unc.—or something sufficiently analogous to it for the purpose of our discussion— the dream cannot be its sole manifestation; every dream may be a wish-fulfilment, but there must be other forms of abnormal wish-fulfilment besides this of dreams. Indeed, the theory of all psychoneurotic symptoms culminates in the proposition *that they too must be taken as wish-fulfilments of the unconscious.* Our explanation makes the dream only the first member of a group most important for the psychiatrist, an understanding of which means the solution of the purely psychological part of the psychiatric problem. But other members of this group of wish-fulfilments, *e.g.,* the hysterical symptoms, evince one essential quality which I have so far failed to find in the dream. Thus, from the investigations frequently referred to in this treatise, I know that the formation of an hysterical symptom necessitates the combination of both streams of our psychic life. The symptom is not merely the expression of a realised unconscious wish, but it must be joined by another wish from the foreconscious which is fulfilled by the same symptom; Bo that the symptom is at least doubly determined, once by each one of the conflicting systems. Just as in the dream, there is no limit to further over-determination. The determination not derived from the Unc.

is, as far as I can see, invariably a stream of thought in reaction against the unconscious wish, *e.g.,* a self-punishment. Hence I may say, in general, that *an hysterical symptom originates only where two contrasting wish-fulfilments, having their source in different psychic systems, are able to combine in one expression.* (Compare my latest formulation of the origin of the hysterical symptoms in a treatise published by the *Zeitschrift für Sexualwissenschaft,* by Hirschfeld and others, 1908). Examples on this point would prove of little value, as nothing but a complete unveiling of the complication in question would carry conviction. I therefore content myself with the mere assertion, and will cite an example, not for conviction but for explication. The hysterical vomiting of a female patient proved, on the one hand, to be the realisation of an unconscious fancy from the time of puberty, that she might be continuously pregnant and have a multitude of children, and this was subsequently united with the wish that she might have them from as many men as possible. Against this immoderate wish there arose a powerful defensive impulse. But as the vomiting might spoil the patient's figure and beauty, so that she would not find favour in the eyes of mankind, the symptom was therefore in keeping with her punitive trend of thought, and, being thus admissible from both sides, it was allowed to become a reality. This is the same manner of consenting to a wish-fulfilment which the queen of the Parthians chose for the triumvir Crassus. Believing that he had undertaken the campaign out of greed for gold, she caused molten gold to be poured into the throat of the corpse. "Now hast thou what thou hast longed for." As yet we know of the dream only that it expresses a wish-fulfilment of the unconscious; and apparently the dominating foreconscious permits this only after it has subjected the wish to some distortions. We are really in no position to demonstrate regularly a stream of thought antagonistic to the dream-wish which is realised in the dream as in its counterpart. Only now and then have we found in the dream traces of reaction formations, as, for instance, the tenderness toward friend R. in the "uncle dream" (p. 133). But the contribution from the foreconscious, which is missing here, may be found in another place. While the dominating system has withdrawn on the wish to sleep, the dream may bring to expression with manifold distortions a wish from the Unc., and realise this wish by producing the necessary changes of energy in the psychic apparatus, and may finally retain it through the entire duration of sleep.[11]

This persistent wish to sleep on the part of the foreconscious in general facilitates the formation of the dream. Let us refer to the dream of the father who, by the gleam of light from the death chamber, was brought to the conclusion that the body has been set on fire. We have shown that one of the psychic forces decisive in causing the father to form this conclusion, instead of being awakened by the gleam of light, was the wish to prolong the life of the child seen in the dream by one moment. Other wishes proceeding from the repression probably escape us, because we are unable to analyse this dream. But as a second motive power of the dream we may mention the father's desire to sleep, for, like the life of the child, the sleep of the father is prolonged for a moment by the dream. The underlying motive is: "Let the dream go on, otherwise I must wake up." As in this dream so also in all other dreams, the wish to sleep lends its support to the unconscious wish. On page 119 we reported dreams which were apparently dreams of convenience. But, properly speaking, all dreams may claim this designation. The efficacy of the wish to continue to sleep is the most easily recognised in the waking dreams, which so transform the objective sensory stimulus as to render it compatible with the continuance of sleep; they interweave this stimulus with the dream in order to rob it of any claims it might make as a warning to the outer world. But this wish to continue to sleep must also participate in the formation of all other dreams which may disturb the sleeping state from within only. "Now, then, sleep on; why, it's but a dream"; this is in many cases the suggestion of the Force, to consciousness when the dream goes too far; and this also describes in a general way the attitude of our dominating psychic activity toward dreaming, though the thought remains tacit. I must draw the conclusion that *throughout our entire sleeping state we are just as certain that we are dreaming as we are certain that we are sleeping.* We are compelled to disregard the objection urged against this conclusion that our consciousness is never directed to a knowledge of the former, and that it is directed to a knowledge of the latter only on special occasions when the censor is unexpectedly surprised. Against this objection we may say that there are persons who are entirely conscious of their sleeping and dreaming, and who are apparently endowed with the conscious faculty of guiding their dream life. Such a dreamer, when dissatisfied with the course taken by the dream, breaks it off without awakening, and begins it anew in order to continue it with a different turn, like the popular author who, on request, gives a happier ending to his play. Or, at another time, if placed by the dream in a sexually exciting situation, he thinks in his sleep: "I do not care to continue this dream and exhaust myself by a pollution; I prefer to defer it in favour of a real situation."

(D) Waking Caused by the Dream—
The Function of the Dream—
The Anxiety Dream

Since we know that the foreconscious is suspended during the night by the wish to sleep, we can proceed to an intelligent investigation of the dream process. But let us first sum up the knowledge of this process already gained. We have shown that the waking activity leaves day remnants from which the sum of energy cannot be entirely removed; or the waking activity revives during the day one of the unconscious wishes; or both conditions occur simultaneously; we have already discovered the many variations that may take place. The unconscious wish has already made its way to the day remnants. either during the day or at any rate with the beginning of sleep, and has effected a transference to it. This produces a wish transferred to the recent material, or the suppressed recent wish comes to life again through a reinforcement from the unconscious. This wish now endeavours to make its way to consciousness on the normal path of the mental processes through the foreconscious, to which indeed it belongs through one of its constituent elements. It is confronted, however, by the censor, which is still active, and to the influence of which it now succumbs. It now takes on the distortion for which the way has already been paved by its transference to the recent material. Thus far it is in the way of becoming something resembling an obsession, delusion, or the like, *i.e.* a thought reinforced by a transference and distorted in expression by the censor. But its further progress is now checked through the dormant state of the foreconscious; this system has apparently protected itself against invasion by diminishing its excitements. The dream process, therefore, takes the regressive course, which has just been opened by the peculiarity of the sleeping state, and thereby follows the attraction exerted on it by the memory groups, which themselves exist in part only as visual energy not yet translated into terms of the later systems. On its way to regression the dream takes on the form of dramatisation. The subject of compression will be discussed later. The dream process has now terminated the second part of its repeatedly impeded course. The first part expended itself progressively from the unconscious scenes or phantasies to the foreconscious, while the second part gravitates from the advent of the censor back to the perceptions. But when the dream process becomes a content of perception it has, so to speak, eluded the obstacle set up in the Forec. by the censor and by the sleeping state. It succeeds in drawing attention to itself and in being noticed by consciousness. For consciousness, which means to us a sensory organ for the reception of psychic qualities, may receive stimuli from two sources—first, from the periphery of the entire apparatus, viz. from the perception system, and, secondly, from the pleasure and pain stimuli, which constitute the sole psychic quality produced in the transformation of energy within the apparatus. All other processes in the Psi-system, even those in the foreconscious, are devoid of any psychic quality, and are therefore not objects of consciousness inasmuch as they do not furnish pleasure or pain for perception. We shall have to assume that those liberations of pleasure and pain automatically regulate the outlet of the occupation processes. But in order to make possible more delicate functions, it was later found necessary to render the course of the presentations more independent of the manifestations of pain. To accomplish this the Forec. system needed some qualities of its own which could attract consciousness, and most probably received them through the connection of the foreconscious processes with the memory system of the signs of speech, which is not devoid of qualities. Through the qualities of this system, consciousness, which had hitherto been a sensory organ only for the perceptions, now becomes also a sensory organ for a part of our mental processes. Thus we have now, as it were, two sensory surfaces, one directed to perceptions and the other to the foreconscious mental processes.

I must assume that the sensory surface of consciousness devoted to the Forec. is rendered less excitable by sleep than that directed to the P-systems. The giving up of interest for the nocturnal mental processes is indeed purposeful. Nothing is to disturb the mind; the Forec. wants to sleep. But once the dream becomes a perception, it is then capable of exciting consciousness through the qualities thus gained. The sensory stimulus accomplishes what it was really destined for, namely, it directs a part of the energy at the disposal of the Forec. in the form of attention upon the stimulant. We must, therefore, admit that the dream invariably awakens us. that is, it puts into activity a part of the dormant force of the Forec. This force imparts to the dream that influence which we have designated as secondary elaboration for the sake of connection and comprehensibility. This means that the dream is treated by it like any other content of perception; it is subjected to the same ideas of expectation, as far at least as the material admits. As far as the direction is concerned in this third part of the dream, it may be said that here again the movement is progressive.

To avoid misunderstanding, it will not be amiss to say a few words about the temporal peculiarities of these dream processes. In a very interesting discussion, apparently suggested by Maury's puzzling guillotine dream, Goblot tries to demonstrate that the dream requires no other time than the transition period between sleeping and awakening. The awakening requires time, as the dream takes place during that period. One is inclined to believe that the final picture of the dream is so strong that it forces the dreamer to awaken; but, as a matter of fact, this picture is strong only because the dreamer is already very near awakening when it appears. "Un rêve c'est un réveil qui commence."

It has already been emphasized by Dugas that Goblot was forced to repudiate many facts in order to generalise his theory. There are, moreover, dreams from which we do not awaken, *e.g.,* some dreams in which we dream that we dream. From our knowledge of the dream-work, we can by no means admit that it extends only over the period of awakening. On the contrary, we must consider it probable that the first part of the dream-work begins during the day when we are still under the domination of the foreconscious. The second phase of the dream-work, viz. the modification through the censor, the attraction by the unconscious scenes, and the penetration to perception must continue throughout the night. And we are probably always right when we assert that we feel as though we had been dreaming the whole night, although we cannot say what. I do not, however, think it necessary to assume that, up to the time of becoming conscious, the dream processes really follow the temporal sequence which we have described, viz. that there is first the transferred dream-wish, then the distortion of the censor, and consequently the change of direction to regression, and so on. We were forced to form such a succession for the sake of *description;* in reality, however, it is much rather a matter of simultaneously trying this path and that, and of emotions fluctuating to and fro, until finally, owing to the most expedient distribution, one particular grouping is secured which remains. From certain personal experiences, I am myself inclined to believe that the dream-work often requires more than one day and one night to produce its result; if this be true, the extraordinary art manifested in the construction of the dream loses all its marvels. In my opinion, even the regard for comprehensibility as an occurrence of perception may take effect before the dream attracts consciousness to itself. To be sure, from now on the process is accelerated, as the dream is henceforth subjected to the same treatment as any other perception. It is like fireworks, which require hours of preparation and only a moment for ignition.

Through the dream-work the dream process now gains either sufficient intensity to attract consciousness to itself and arouse the foreconscious, which is quite independent of the time or profundity of sleep, or, its intensity being insufficient it must wait until it meets the attention which is set in motion immediately before awakening. Most dreams seem to operate with relatively slight psychic intensities, for they wait for the awakening. This, however, explains the fact that we regularly perceive something dreamt on being suddenly aroused from a sound sleep. Here, as well as in spontaneous awakening, the first glance strikes the perception content created by the dream-work, while the next strikes the one produced from without.

But of greater theoretical interest are those dreams which are capable of waking us in the midst of sleep. We must bear in mind the expediency elsewhere universally demonstrated, and ask ourselves why the dream or the unconscious wish has the power to disturb sleep, i.e. the fulfilment of the foreconscious wish. This is probably due to certain relations of energy into which we have no insight. If we possessed such insight we should probably find that the freedom given to the dream and the expenditure of a certain amount of detached attention represent for the dream an economy in energy, keeping in view the fact that the unconscious must be held in check at night just as during the day. We know from experience that the dream, even if it interrupts sleep, repeatedly during the same night, still remains compatible with sleep. We wake up for an instant, and immediately resume our sleep. It is like driving off a fly during sleep, we awake *ad hoc*, and when we resume our sleep we have removed the disturbance. As demonstrated by familiar examples from the sleep of wet nurses, &c., the fulfilment of the wish to sleep is quite compatible with the retention of a certain amount of attention in a given direction.

But we must here take cognisance of an objection that is based on a better knowledge of the unconscious processes. Although we have ourselves described the unconscious wishes as always active, we have, nevertheless, asserted that they are not sufficiently strong during the day to make themselves perceptible. But when we sleep, and the unconscious wish has shown its power to form a dream, and with it to awaken the foreconscious, why, then, does this power become exhausted after the dream has been taken cognisance of? Would it not seem more probable that the dream should continually renew itself, like the troublesome fly which, when driven away, takes pleasure in returning again and again? What justifies our assertion that the dream removes the disturbance of sleep?

That the unconscious wishes always remain active is quite true. They represent paths which are passable whenever a sum of excitement makes use of them. Moreover, a remarkable peculiarity of the unconscious processes is the fact that they remain indestructible. Nothing can be brought to an end in the unconscious; nothing can cease or be forgotten. This impression is most strongly gamed in the study of the neuroses, especially of hysteria. The unconscious stream of thought which leads to the discharge through an attack becomes passable again as soon as there is an accumulation of a sufficient amount of excitement. The mortification brought on thirty years ago, after having gained access to the unconscious affective source, operates during all these thirty years like a recent one. Whenever its memory is touched, it is revived and shows itself to be supplied with the excitement which is discharged in a motor attack. It is just here that the office of psychotherapy begins, its task being to bring about adjustment and forgetfulness for the unconscious processes. Indeed, the fading of memories and the flagging of affects, which we are apt to take as self-evident and to explain as a primary influence of time on the psychic memories, are in reality secondary changes brought about by painstaking work. It is the foreconscious that accomplishes this work; and the only course to be pursued by psychotherapy is to subjugate the Unc. to the domination of the Forec.

There are, therefore, two exits for the individual unconscious emotional process. It is either left to itself, in which case it ultimately breaks through somewhere and secures for once a discharge for its excitation into motility; or it succumbs to the influence of the foreconscious, and its excitation becomes confined through this influence instead of being discharged. It is the latter process that occurs in the dream. Owing to the fact that it is directed by the conscious excitement, the energy from the Forec., which confronts the dream when grown to perception, restricts the unconscious excitement of the dream and renders it harmless as a disturbing factor. When the dreamer wakes up for a moment, he has actually chased away the fly that has threatened to disturb his sleep. We can now understand that it is really more expedient and economical to give full sway to the unconscious wish, and clear its way to regression so that it may form a dream, and then restrict and adjust this dream by means of a small expenditure of fore-conscious labour, than to curb the unconscious throughout the entire period of sleep. We should, indeed, expect that the dream, even if it was not originally an expedient process, would have acquired some function in the play of forces of the psychic life. We now see what this function is. The dream has taken it upon itself to bring the liberated excitement of the Unc. back under the domination of the foreconscious; it thus affords relief for the excitement of the Unc. and acts as a safety-valve for the latter, and at the same time it insures the sleep of the foreconscious at a slight expenditure of the waking state. Like the other psychic formations of its group, the dream offers itself as a compromise serving simultaneously both systems by fulfilling both wishes in so far as they are compatible with each other. A glance at Robert's "elimination theory." referred to on page 75, will show that we must agree with this author in his main point, viz. in the determination of the function of the dream, though we differ from him in our hypotheses and in our treatment of the dream process.

The above qualification—in so far as the two wishes are compatible with each other—contains a suggestion that there may be cases in which the function of the dream suffers shipwreck. The dream process is in the first instance admitted as a wish-fulfilment of the unconscious, but if this tentative wish-fulfilment disturbs the foreconscious to such an extent that the latter can no longer maintain its rest, the dream then breaks the compromise and fails to perform the second part of its task. It is then at once broken off, and replaced by complete wakefulness. Here, too, it is not really the fault of the dream, if, while ordinarily the guardian of sleep, it is here compelled to appear as the disturber of sleep, nor should this cause us to entertain any doubts as to its efficacy. This is not the only case in the organism in which an otherwise efficacious arrangement became inefficacious and disturbing as soon as some element is changed in the conditions of its origin; the disturbance then serves at least the new purpose of announcing the change, and calling into play against it the means of adjustment of the organism. In this connection. I naturally bear in mind the case of the anxiety dream, and in order not to have the appearance of trying to exclude this testimony against the theory of wish-fulfilment wherever I encounter it, I will attempt an explanation of the anxiety dream, at least offering some suggestions.

That a psychic process developing anxiety may still be a wish-fulfilment has long ceased to impress us as a contradiction. We may explain this occurrence by the fact that the wish belongs to one system (the Unc.), while by the other system (the Forec.), this wish has been rejected and suppressed. The subjection of the Unc. by the Forec. is not complete even in perfect psychic health; the amount of this suppression shows the degree of our psychic normality. Neurotic symptoms show that there is a conflict between the two systems; the symptoms are the results of a compromise of this conflict, and they temporarily put an end to it. On the one hand, they afford the Unc. an outlet for the discharge of its excitement, and serve it as a sally port, while, on the other hand, they give the Forec. the capability of dominating the Unc. to some extent. It is highly instructive to consider, *e.g.*, the significance of any hysterical phobia or of an agoraphobia. Suppose a neurotic incapable of crossing the street alone, which we would justly call a "symptom." We attempt to remove this symptom by urging him to the action which he deems himself incapable of. The result will be an attack of anxiety, just as an attack of anxiety in the street has often been the cause of establishing an agoraphobia. We thus learn that the symptom has been constituted in order to guard against the outbreak of the anxiety. The phobia is thrown before the anxiety like a fortress on the frontier.

Unless we enter into the part played by the affects in these processes, which can be done here only imperfectly, we cannot continue our discussion. Let us therefore advance the proposition that the reason why the suppression of the unconscious becomes absolutely necessary is because, if the discharge of presentation should be left to itself, it would develop an affect in the Unc. which originally bore the character of pleasure, but which, since the appearance of the repression, bears the character of pain. The aim, as well as the result, of the suppression is to stop the development of this pain. The suppression extends over the unconscious ideation, because the liberation of pain might emanate from the ideation. The foundation is here laid for a very definite assumption concerning the nature of the affective development. It is regarded as a motor or secondary activity, the key to the innervation of which is located in the presentations of the Unc. Through the domination of the Forec. these presentations become, as it were, throttled and inhibited at the exit of the emotion-developing impulses. The danger, which is due to the fact that the Forec. ceases to occupy the energy, therefore consists in the fact that the unconscious excitations liberate such an affect as—in consequence of the repression that has previously taken place—can only be perceived as pain or anxiety.

This danger is released through the full sway of the dream process. The determinations for its realisation consist in the fact that repressions have taken place, and that the suppressed emotional wishes shall become sufficiently strong. They thus stand entirely without the psychological realm of the dream structure. Were it not for the fact that our subject is connected through just one factor, namely, the freeing of the Unc. during sleep, with the subject of the development of anxiety, I could dispense with discussion of the anxiety dream, and thus avoid all obscurities connected with it.

As I have often repeated, the theory of the anxiety belongs to the psychology of the neuroses. I would say that the anxiety in the dream is an anxiety problem and not a dream problem. We have nothing further to do with it after having once demonstrated its point of contact with the subject of the dream process. There is only one thing left for me to do. As I have asserted that the neurotic anxiety originates from sexual sources, I can subject anxiety dreams to analysis in order to demonstrate the sexual material in their dream thoughts.

For good reasons I refrain from citing here any of the numerous examples placed at my disposal by neurotic patients, but prefer to give anxiety dreams from young persons.

Personally, I have had no real anxiety dream for decades, but I recall one from my seventh or eighth year which I subjected to interpretation about thirty years later. The dream was very vivid, and showed me *my beloved mother, with peculiarly calm sleeping countenance, carried into the room and laid on the bed by two (or three) persons with bird's beaks.* I awoke crying and screaming, and disturbed my parents. The very tall figures—draped in a peculiar manner—with beaks, I had taken from the illustrations of Philippson's bible; I believe they represented deities with heads of sparrowhawks from an Egyptian tomb relief. The analysis also introduced the reminiscence of a naughty janitor's boy, who used to play with us children on the meadow in front of the house; I would add that his name was Philip. I feel that I first heard from this boy the vulgar word signifying sexual intercourse, which is replaced among the educated by the Latin "coitus," but to which the dream distinctly alludes by the selection of the bird's heads.[12] I must have suspected the sexual significance of the word from the facial expression of my worldly-wise teacher. My mother's features in the dream were copied from the countenance of my grandfather, whom I had seen a few days before his death snoring in the state of coma. The interpretation of the secondary elaboration in the dream must therefore have been that my mother was dying; the tomb relief, too, agrees with this. In this anxiety I awoke, and could not calm myself until I had awakened my parents. I remember that I suddenly became calm on coming face to face with my mother, as if I needed the assurance that my mother was not dead. But this secondary interpretation of the dream had been effected only under the influence of the developed anxiety. I was not frightened because I dreamed that my mother was dying, but I interpreted the dream in this manner in the fore-conscious elaboration because I was already under the domination of the anxiety. The latter, however, could be traced by means of the repression to an obscure obviously sexual desire, which had found its satisfying expression in the visual content of the dream.

A man twenty-seven years old who had been severely ill for a year had had many terrifying dreams between the ages of eleven and thirteen. He thought that a man with an axe was running after him; he wished to run, but felt paralysed and could not move from the spot. This may be taken as a good example of a very common, and apparently sexually indifferent, anxiety dream. In the analysis the dreamer first thought of a story told him by his uncle, which chronologically was later than the dream, viz. that he was attacked at night by a suspicious-looking individual. This occurrence led him to believe that he himself might have already heard of a similar episode at the time of the dream. In connection with the axe he recalled that during that period of his life he once hurt his hand with an axe while chopping wood. This immediately led to his relations with his younger brother, whom he used to maltreat and knock down. In particular, he recalled an occasion when he struck his brother on the head with his boot until he bled, whereupon his mother remarked: "I fear he will kill him some day." While he was seemingly thinking of the subject of violence, a reminiscence from his ninth year suddenly occurred to him. His parents came home late and went to bed while he was feigning sleep. He soon heard panting and other noises that appeared strange to him, and he could also make out the position of his parents in bed. His further associations showed that he had established an analogy between this relation between his parents and his own relation toward his younger brother. He subsumed what occurred between his parents under the conception "violence and wrestling," and thus reached a sadistic conception of the coitus act, as often happens among children. The fact that he often noticed blood on his mother's bed corroborated his conception.

That the sexual intercourse of adults appears strange to children who observe it, and arouses fear in them, I dare say is a fact of daily experience. I have explained this fear by the fact that sexual excitement is not mastered by their understanding, and is probably also inacceptable to them because their parents are involved in it. For the same reason this excitement is converted into fear. At a still earlier period of life sexual emotion directed toward the parent of opposite sex does not meet with repression but finds free expression, as we have seen above (pp. 235).

For the night terrors with hallucinations (*pavor nocturnus*) frequently found in children, I would unhesitatingly give the same explanation. Here, too, we are certainly dealing with the incomprehensible and rejected sexual feelings, which, if noted, would probably show a temporal periodicity, for an enhancement of the sexual *libido* may just as well be produced accidentally through emotional impressions as through the spontaneous and gradual processes of development.

I lack the necessary material to sustain these explanations from observation. On the other hand, the podiatrists seem to lack the point of view which alone makes comprehensible the whole series of phenomena, on the somatic as well as on the psychic side. To illustrate by a comical example how one wearing the blinders of medical mythology may miss the understanding of such cases I will relate a case which I found in a thesis on *pavor nocturnus* by Debacker, 1881 (p. 66). A thirteen-year-old boy of delicate health began to become anxious and dreamy; his sleep became restless, and about once a week it was interrupted by an acute attack of anxiety with hallucinations. The memory of these dreams was invariably very distinct. Thus, he related that the devil shouted at him: "Now we have you, now we have you," and this was followed by an odour of sulphur; the fire burned his skin. This dream aroused him, terror-stricken. He was unable to scream at first; then his voice returned, and he was heard to say distinctly: "No, no, not me; why, I have done nothing," or, "Please don't, I shall never do it again." Occasionally, also, he said: "Albert has not done that." Later he avoided undressing, because, as he said, the fire attacked him only when he was undressed. From amid these evil dreams, which menaced his health, he was sent into the country, where he recovered within a year and a half, but at the age of fifteen he once confessed: "Je n'osais pas l'avouer, mais j'éprouvais continuellement des picotements et des surexcitations *parties*,[13] à la fin, cela m'énervait tant que plusieurs fois, j'ai pensé me jeter par la fenêtre au dortoir."

It is certainly not difficult to suspect: 1. that the boy had practised masturbation in former years, that he probably denied it, and was threatened with severe punishment for his wrongdoing (his confession: Je ne le ferai plus; his denial: Albert n'a jamais fait ça). 2. That under the pressure of puberty the temptation to self-abuse through the tickling of the genitals was reawakened. 3. That now, however, a struggle of repression arose in him, suppressing the *libido* and changing it into fear, which subsequently took the form of the punishments with which he was then threatened.

Let us, however, quote the conclusions drawn by our author (p. 69). This observation shows:

1. That the influence of puberty may produce in a boy of delicate health a condition of extreme weakness, and that it may lead to a *very marked cerebral anaemia*.[14]

2. This cerebral anaemia produces a transformation of character, demonomaniacal hallucinations, and very violent nocturnal, perhaps also diurnal, states of anxiety.

3. Demonomania and the self-reproaches of the day can be traced to the influences of religious education which the subject underwent as a child.

4. All manifestations disappeared as a result of a lengthy sojourn in the country, bodily exercise, and the return of physical strength after the termination of the period of puberty.

5. A predisposing influence for the origin of the cerebral condition of the boy may be attributed to heredity and to the father's chronic syphilitic state.

The concluding remarks of the author read: "Nous avons fait entrer cette observation dans le cadre des délires apyrétiques d'inanition, car c'est à l'ischémie cérébrale que nous rattachons cet état particulier."

(E) The Primary and Secondary Processes—Regression

In venturing to attempt to penetrate more deeply into the psychology of the dream processes, I have undertaken a difficult task, to which, indeed, my power of description is hardly equal. To reproduce in description by a succession of words the simultaneousness of so complex a chain of events, and in doing so to appear unbiassed throughout the exposition, goes fairly beyond my powers. I have now to atone for the fact that I have been unable in my description of the dream psychology to follow the historic development of my views. The view-points for my conception of the dream were reached through earlier investigations in the psychology of the neuroses, to which I am not supposed to refer here, but to which I am repeatedly forced to refer, whereas I should prefer to proceed in the opposite direction, and, starting from the dream, to establish a connection with the psychology of the neuroses. I am well aware of all the inconveniences arising for the reader from this difficulty, but I know of no way to avoid them.

As I am dissatisfied with this state of affairs, I am glad to dwell upon another view-point which seems to raise the value of my efforts. As has been shown in the introduction to the first chapter, I found myself confronted with a theme which had been marked by the sharpest contradictions on the part of the authorities. After our elaboration of the dream problems we found room for most of these contradictions. We have been forced, however, to take decided exception to two of the views pronounced, viz. that the dream is a senseless and that it is a somatic process; apart from these cases we have had to accept all the contradictory views in one place or another of the complicated argument, and we have been able to demonstrate that they had discovered something that was correct. That the dream continues the impulses and interests of the waking state has been quite generally confirmed through the discovery of the latent thoughts of the dream. These thoughts concern themselves only with things that seem important and of momentous interest to us. The dream never occupies itself with trifles. But we have also concurred with the contrary view, viz. that the dream gathers up the indifferent remnants from the day, and that not until it has in some measure withdrawn itself from the waking activity can an important event of the day be taken up by the dream. We found this holding true for the dream content, which gives the dream thought its changed expression by means of disfigurement. We have said that from the nature of the association mechanism the dream process more easily takes possession of recent or indifferent material which has not yet been seized by the waking mental activity; and by reason of the censor it transfers the psychic intensity from the important but also disagreeable to the indifferent material. The hypermnesia of the dream and the resort to infantile material have become main supports in our theory. In our theory of the dream we have attributed to the wish originating from the infantile the part of an indispensable motor for the formation of the dream. We naturally could not think of doubting the experimentally demonstrated significance of the objective sensory stimuli during sleep; but we have brought this material into the same relation to the dream-wish as the thought remnants from the waking activity. There was no need of disputing the fact that the dream interprets the objective sensory stimuli after the manner of an illusion; but we have supplied the motive for this interpretation which has been left undecided by the authorities. The interpretation follows in such a manner that the perceived object is rendered harmless as a sleep disturber and becomes available for the wish-fulfilment. Though we do not admit as special sources of the dream the subjective state of excitement of the sensory organs during sleep, which seems to have been demonstrated by Trumbull Ladd, we are nevertheless able to explain this excitement through the regressive revival of active memories behind the dream. A modest part in our conception has also been assigned to the inner organic sensations which are wont to be taken as the cardinal point in the explanation of the dream. These—the sensation of falling, flying, or inhibition—stand as an ever ready material to be used by the dream-work to express the dream thought as often as need arises.

That the dream process is a rapid and momentary one seems to be true for the perception through consciousness of the already prepared dream content; the preceding parts of the dream process probably take a slow, fluctuating course. We have solved the riddle of the superabundant dream content compressed within the briefest moment by explaining that this is due to the appropriation of almost fully formed structures from the psychic life. That the dream is disfigured and distorted by memory we found to be correct, but not troublesome, as this is only the last manifest operation in the work of disfigurement which has been active from the beginning of the dream-work. In the bitter and seemingly irreconcilable controversy as to whether the psychic life sleeps at night or can make the same use of all its capabilities as during the day, we have been able to agree with both sides, though not fully with either. We have found proof that the dream thoughts represent a most complicated intellectual activity, employing almost every means furnished by the psychic apparatus; still it cannot be denied that these dream thoughts have originated during the day, and it is indispensable to assume that there is a sleeping state of the psychic life. Thus, even the theory of partial sleep has come into play; but the characteristics of the sleeping state have been found not in the dilapidation of the psychic connections but in the cessation of the psychic system dominating the day, arising from its desire to sleep. The withdrawal from the outer world retains its significance also for our conception; though not the only factor, it nevertheless helps the regression to make possible the representation of the dream. That we should reject the voluntary guidance of the presentation course is uncontestable; but the psychic life does not thereby become aimless, for we have seen that after the abandonment of the desired end-presentation undesired ones gain the mastery. The loose associative connection in the dream we have not only recognised, but we have placed under its control a far greater territory than could have been supposed; we have, however, found it merely the feigned substitute for another correct and senseful one. To be sure we, too, have called the dream absurd; but we have been able to learn from examples how wise the dream really is when it simulates absurdity. We do not deny any of the functions that have been attributed to the dream. That the dream relieves the mind like a valve, and that, according to Robert's assertion, all kinds of harmful material are rendered harmless through representation in the dream, not only exactly coincides with our theory of the twofold wish-fulfilment in the dream, but, in his own wording, becomes even more comprehensible for us than for Robert himself. The free indulgence of the psychic in the play of its faculties finds expression with us in the non-interference with the dream on the part of the foreconscious activity. The "return to the embryonal state of psychic life in the dream" and the observation of Havelock Ellis, "an archaic world of vast emotions and imperfect thoughts," appear to us as happy anticipations of our deductions to the effect that primitive modes of work suppressed during the day participate in the formation of the dream; and with us, as with Delage, the *suppressed* material becomes the mainspring of the dreaming.

We have fully recognised the rôle which Scherner ascribes to the dream phantasy, and even his interpretation; but we have been obliged, so to speak, to conduct them to another department in the problem. It is not the dream that produces the phantasy but the unconscious phantasy that takes the greatest part in the formation of the dream thoughts. We are indebted to Scherner for his clue to the source of the dream thoughts, but almost everything that he ascribes to the dream-work is attributable to the activity of the unconscious, which is at work during the day, and which supplies incitements not only for dreams but for neurotic symptoms as well. We have had to separate the dream-work from this activity as being something entirely different and far more restricted. Finally, we have by no means abandoned the relation of the dream to mental disturbances, but, on the contrary, we have given it a more solid foundation on new ground.

Thus held together by the new material of our theory as by a superior unity, we find the most varied and most contradictory conclusions of the authorities fitting into our structure; some of them are differently disposed, only a few of them are entirely rejected. But our own structure is still unfinished. For, disregarding the many obscurities which we have necessarily encountered in our advance into the darkness of psychology, we are now apparently embarrassed by a new contradiction. On the one hand, we have allowed the dream thoughts to proceed from perfectly normal mental operations, while, on the other hand, we have found among the dream thoughts a number of entirely abnormal mental processes which extend likewise to the dream contents. These, consequently, we have repeated in the interpretation of the dream. All that we have termed the "dream-work" seems so remote from the psychic processes recognised by us as correct, that the severest judgments of the authors as to the low psychic activity of dreaming seem to us well founded.

Perhaps only through still further advance can enlightenment and improvement be brought about. I shall pick out one of the constellations leading to the formation of dreams.

We have learned that the dream replaces a number of thoughts derived from daily life which are perfectly formed logically. We cannot therefore doubt that these thoughts originate from our normal mental life. All the qualities which we esteem in our mental operations, and which distinguish these as complicated activities of a high order, we find repeated in the dream thoughts. There is, however, no need of assuming that this mental work is performed during sleep, as this would materially impair the conception of the psychic state of sleep we have hitherto adhered to. These thoughts may just as well have originated from the day, and, unnoticed by our consciousness from their inception, they may have continued to develop until they stood complete at the onset of sleep. If we are to conclude anything from this state of affairs, it will at most prove *that the most complex mental operations are possible without the co-operation of consciousness,* which we have already learned independently from every psychoanalysis of persons suffering from hysteria or obsessions. These dream thoughts are in themselves surely not incapable of consciousness; if they have not become conscious to us during the day, this may have various reasons. The state of becoming conscious depends on the exercise of a certain psychic function, viz. attention, which seems to be extended only in a definite quantity, and which may have been withdrawn from the stream of thought in question by other aims. Another way in which such mental streams are kept from consciousness is the following:—Our conscious reflection teaches us that when exercising attention we pursue a definite course. But if that course leads us to an idea which does not hold its own with the critic, we discontinue and cease to apply our attention. Now, apparently, the stream of thought thus started and abandoned may spin on without regaining attention unless it reaches a spot of especially marked intensity which forces the return of attention. An initial rejection, perhaps consciously brought about by the judgment on the ground of incorrectness or unfitness for the actual purpose of the mental act, may therefore account for the fact that a mental process continues until the onset of sleep unnoticed by consciousness.

Let us recapitulate by saying that we call such a stream of thought a foreconscious one, that we believe it to be perfectly correct, and that it may just as well be a more neglected one or an interrupted and suppressed one. Let us also state frankly in what manner we conceive this presentation course. We believe that a certain sum of excitement, which we call occupation energy, is displaced from an end-presentation along the association paths selected by that end-presentation. A "neglected" stream of thought has received no such occupation, and from a "suppressed" or "rejected" one this occupation has been withdrawn; both have thus been left to their own emotions. The end-stream of thought stocked with energy is under certain conditions able to draw to itself the attention of consciousness, through which means it then receives a "surplus of energy."

We shall be obliged somewhat later to elucidate our assumption concerning the nature and activity of consciousness.

A train of thought thus incited in the Forec. may either disappear spontaneously or continue. The former issue we conceive as follows: It diffuses its energy through all the association paths emanating from it, and throws the entire chain of ideas into a state of excitement which, after lasting for a while, subsides through the transformation of the excitement requiring an outlet into dormant energy.[15] If this first issue is brought about the process has no further significance for the dream formation. But other end-presentations are lurking in our foreconscious that originate from the sources of our unconscious and from the ever active wishes. These may take possession of the excitations in the circle of thought thus left to itself, establish a connection between it and the unconscious wish, and transfer to it the energy inherent in the unconscious wish. Henceforth the neglected or suppressed train of thought is in a position to maintain itself, although this reinforcement does not help it to gain access to consciousness. We may say that the hitherto foreconscious train of thought has been drawn into the unconscious.

Other constellations for the dream formation would result if the foreconscious train of thought had from the beginning been connected with the unconscious wish, and for that reason met with rejection by the dominating end-occupation; or if an unconscious wish were made active for other—possibly somatic—reasons and of its own accord sought a transference to the psychic remnants not occupied by the Forec. All three cases finally combine in one issue, so that there is established in the foreconscious a stream of thought which, having been abandoned by the foreconscious occupation, receives occupation from the unconscious wish.

The stream of thought is henceforth subjected to a series of transformations which we no longer recognise as normal psychic processes and which give us a surprising result, viz. a psychopathological formation. Let us emphasize and group the same.

1. The intensities of the individual ideas become capable of discharge in their entirety, and, proceeding from one conception to the other, they thus form single presentations endowed with marked intensity. Through the repeated recurrence of this process the intensity of an entire train of ideas may ultimately be gathered in a single presentation element. This is the principle of *compression or condensation* with which we became

acquainted in the chapter on "The Dream-Work." It is condensation that is mainly responsible for the strange impression of the dream, for we know of nothing analogous to it in the normal psychic life accessible to consciousness. We find here, also, presentations which possess great psychic significance as junctions or as end-results of whole chains of thought; but this validity does not manifest itself in any character conspicuous enough for internal perception; hence, what has been presented in it does not become in any way more intensive. In the process of condensation the entire psychic connection becomes transformed into the intensity of the presentation content. It is the same as in a book where we space or print in heavy type any word upon which particular stress is laid for the understanding of the text. In speech the same word would be pronounced loudly and deliberately and with emphasis. The first comparison leads us at once to an example taken from the chapter on "The Dream-Work" (trimethylamine in the dream of Irma's injection). Historians of art call our attention to the fact that the most ancient historical sculptures follow a similar principle in expressing the rank of the persons represented by the size of the statue. The king is made two or three times as large as his retinue or the vanquished enemy. A piece of art, however, from the Roman period makes use of more subtle means to accomplish the same purpose. The figure of the emperor is placed in the centre in a firmly erect posture; special care is bestowed on the proper modelling of his figure; his enemies are seen cowering at his feet; but he is no longer represented a giant among dwarfs. However, the bowing of the subordinate to his superior in our own days is only an echo of that ancient principle of representation.

The direction taken by the condensations of the dream is prescribed on the one hand by the true foreconscious relations of the dream thoughts, on the other hand by the attraction of the visual reminiscences in the unconscious. The success of the condensation work produces those intensities which are required for penetration into the perception systems.

2. Through this free transferability of the intensities, moreover, and in the service of condensation, *intermediary presentations*—compromises, as it were—are formed (*cf.* the numerous examples). This, likewise, is something unheard of in the normal presentation course, where it is above all a question of selection and retention of the "proper" presentation element. On the other hand, composite and compromise formations occur with

extraordinary frequency when we are trying to find the linguistic expression for foreconscious thoughts; these are considered "slips of the tongue."

3. The presentations which transfer their intensities to one another are *very loosely connected,* and are joined together by such forms of association as are spurned in our serious thought and are utilised in the production of the effect of wit only. Among these we particularly find associations of the sound and consonance types.

4. Contradictory thoughts do not strive to eliminate one another, but remain side by side. They often unite to produce condensation *as if no contradiction* existed, or they form compromises for which we should never forgive our thoughts, but which we frequently approve of in our actions.

These are some of the most conspicuous abnormal processes to which the thoughts which have previously been rationally formed are subjected in the course of the dream-work. As the main feature of these processes we recognise the high importance attached to the fact of rendering the occupation energy mobile and capable of discharge; the content and the actual significance of the psychic elements, to which these energies adhere, become a matter of secondary importance. One might possibly think that the condensation and compromise formation is effected only in the service of regression, when occasion arises for changing thoughts into pictures. But the analysis and—still more distinctly—the synthesis of dreams which lack regression toward pictures, *e.g.* the dream "Autodidasker—Conversation with Court-Councillor N.," present the same processes of displacement and condensation as the others.

Hence we cannot refuse to acknowledge that the two kinds of essentially different psychic processes participate in the formation of the dream; one forms perfectly correct dream thoughts which are equivalent to normal thoughts, while the other treats these ideas in a highly surprising and incorrect manner. The latter process we have already set apart in Chapter VI as the dream-work proper. What have we now to advance concerning this latter psychic process?

We should be unable to answer this question here if we had not penetrated considerably into the psychology of the neuroses and especially of hysteria. From this we learn that the same incorrect psychic processes—as well as others that have not been enumerated—control the formation of hysterical symptoms. In hysteria, too, we at once find a series of perfectly correct thoughts equivalent to our conscious thoughts, of whose existence, however, in this form we can learn nothing and which we can only subsequently reconstruct. If they have forced their way anywhere to our perception, we discover from the analysis of the symptom formed that these normal thoughts have been subjected to abnormal treatment and *have been transformed into the symptom by means of condensation and compromise formation, through superficial associations, under cover of contradictions, and eventually over the road of regression.* In view of the complete identity found between the peculiarities of the dream-work and of the psychic activity forming the psychoneurotic symptoms, we shall feel justified in transferring to the dream the conclusions urged upon us by hysteria.

From the theory of hysteria we borrow the proposition that *such an abnormal psychic elaboration of a normal train of thought takes place only when the latter has been used for the transference of an unconscious wish which dates from the infantile life and is in a state of repression.* In accordance with this proposition we have construed the theory of the dream on the assumption that the actuating dream-wish invariably originates in the unconscious, which, as we ourselves have admitted, cannot be universally demonstrated though it cannot be refuted. But in order to explain the real meaning of the term *repression,* which we have employed so freely, we shall be obliged to make some further addition to our psychological construction.

We have above elaborated the fiction of a primitive psychic apparatus, whose work is regulated by the efforts to avoid accumulation of excitement and as far as possible to maintain itself free from excitement. For this reason it was constructed after the plan of a reflex apparatus; the motility, originally the path for the inner bodily change, formed a discharging path standing at its disposal. We subsequently discussed the psychic results of a feeling of gratification, and we might at the same time have introduced the second assumption, viz. that accumulation of excitement—following certain modalities that do not concern us—is perceived as pain and sets the apparatus in motion in order to reproduce a feeling of gratification in which the diminution of the excitement is perceived as pleasure. Such a current in the apparatus which emanates from pain and strives for pleasure we call a wish. We have said that nothing but a wish is capable of setting the apparatus in motion, and that the discharge of excitement in the apparatus is regulated automatically by the perception of pleasure and pain. The first wish must have been an hallucinatory occupation of the memory for gratification. But this hallucination, unless it were maintained to the point of exhaustion, proved incapable of bringing about a cessation of the desire and consequently of securing the pleasure connected with gratification.

Thus there was required a second activity—in our terminology the activity of a second system—which should not permit the memory occupation to advance to perception and therefrom to restrict the psychic forces, but should lead the excitement emanating from the craving stimulus by a devious path over the spontaneous motility which ultimately should so change the outer world as to allow the real perception of the object of gratification to take place. Thus far we have elaborated the plan of the psychic apparatus; these two systems are the germ of the Unc. and Forec. which we include in the fully developed apparatus.

In order to be in a position successfully to change the outer world through the motility, there is required the accumulation of a large sum of experiences in the memory systems as well as a manifold fixation of the relations which are evoked in this memory material by different end-presentations. We now proceed further with our assumption. The manifold activity of the second system, tentatively sending forth and retracting energy, must on the one hand have full command over all memory material, but on the other hand it would be a superfluous expenditure for it to send to the individual mental paths large quantities of energy which would thus flow off to no purpose, diminishing the quantity available for the transformation of the outer world. In the interests of expediency I therefore postulate that the second system succeeds in maintaining the greater part of the occupation energy in a dormant state and in using but a small portion for the purposes of displacement. The mechanism of these processes is entirely unknown to me; anyone who wishes to follow up these ideas must try to find the physical analogies and prepare the way for a demonstration of the process of motion in the stimulation of the neuron. I merely hold to the idea that the activity of the first Psi-system is directed *to the free outflow of the quantities of excitement,* and that the second system brings about an inhibition of this outflow through the energies emanating from it, *i.e.* it produces a *transformation into dormant energy, probably by raising the level.* I therefore assume that under the control of the second system as compared with the first, the course of the excitement is bound to entirely different mechanical conditions. After the second system has finished its tentative mental work, it removes the inhibition and congestion of the excitements and allows these excitements to flow off to the motility.

An interesting train of thought now presents itself if we consider the relations of this inhibition of discharge by the second system to the regulation through the principle of pain. Let us now seek the counterpart of the primary feeling of gratification, namely, the objective feeling of fear. A perceptive stimulus acts on the primitive apparatus, becoming the source of a painful emotion. This will then be followed by irregular motor manifestations until one of these withdraws the apparatus from perception and at the same time from pain, but on the reappearance of the perception this manifestation will immediately repeat itself (perhaps as a movement of flight) until the perception has again disappeared. But there will here remain no tendency again to occupy the perception of the source of pain in the form of an hallucination or in any other form. On the contrary, there will be a tendency in the primary apparatus to abandon the painful memory picture as soon as it is in any way awakened, as the overflow of its excitement would surely produce (more precisely, begin to produce) pain. The deviation from memory, which is but a repetition of the former flight from perception, is facilitated also by the fact that, unlike perception, memory does not possess sufficient quality to excite consciousness and thereby to attract to itself new energy. This easy and regularly occurring deviation of the psychic process from the former painful memory presents to us the model and the first example of *psychic repression*. As is generally known, much of this deviation from the painful, much of the behaviour of the ostrich, can be readily demonstrated even in the normal psychic life of adults.

By virtue of the principle of pain the first system is therefore altogether incapable of introducing anything unpleasant into the mental associations. The system cannot do anything but wish. If this remained so the mental activity of the second system, which should have at its disposal all the memories stored up by experiences, would be hindered. But two ways are now opened: the work of the second system either frees itself completely from the principle of pain and continues its course, paying no heed to the painful reminiscence, or it contrives to occupy the painful memory in such a manner as to preclude the liberation of pain. We may reject the first possibility, as the principle of pain also manifests itself as a regulator for the emotional discharge of the second system; we are, therefore, directed to the second possibility, namely, that this system occupies a reminiscence in such a manner as to inhibit its discharge and hence, also, to inhibit the discharge comparable to a motor innervation for the development of pain. Thus from two starting points we are led to the hypothesis that occupation through the second system is at the same time an inhibition for the emotional discharge, viz. from a consideration of the principle of pain and from the principle of the smallest expenditure of innervation. Let us, however, keep to the fact—this is the key to the theory of repression—that the second system is capable of occupying an idea only when it is in position to check the development of pain emanating from it. Whatever withdraws itself from this inhibition also remains inaccessible for the second system and would soon be abandoned by virtue of the principle of pain. The inhibition of pain, however, need not be complete; it must be permitted to begin, as it indicates to the second system the nature of the memory and possibly its defective adaptation for the purpose sought by the mind.

The psychic process which is admitted by the first system only I shall now call the *primary* process; and the one resulting from the inhibition of the second system I shall call the *secondary* process. I show by another point for what purpose the second system is obliged to correct the primary process. The primary process strives for a discharge of the excitement in order to establish a *perception* identity with the sum of excitement thus gathered; the secondary process has abandoned this intention and undertaken instead the task of bringing about a *thought identity*. All thinking is only a circuitous path from the memory of gratification taken as an end-presentation to the identical occupation of the same memory, which is again to be attained on the track of the motor experiences. The state of thinking must take an interest in the connecting paths between the presentations without allowing itself to be misled by their intensities. But it is obvious that condensations and intermediate or compromise formations occurring in the presentations impede the attainment of this end-identity; by substituting one idea for the other they deviate from the path which otherwise would have been continued from the original idea. Such processes are therefore carefully avoided in the secondary thinking. Nor is it difficult to understand that the principle of pain also impedes the progress of the mental stream in its pursuit of the thought identity, though, indeed, it offers to the mental stream the most important points of departure. Hence the tendency of the thinking process must be to free itself more and more from exclusive adjustment by the principle of pain, and through the working of the mind to restrict the affective development to that minimum which is necessary as a signal. This refinement of the activity must have been attained through a recent over-occupation of energy brought about by consciousness. But we are aware that this refinement is seldom completely successful even in the most normal psychic life and that our thoughts ever remain accessible to falsification through the interference of the principle of pain.

This, however, is not the breach in the functional efficiency of our psychic apparatus through which the thoughts forming the material of the secondary mental work are enabled to make their way into the primary psychic process—with which formula we may now describe the work leading to the dream and to the hysterical symptoms. This case of insufficiency results from the union of the two factors from the history of our evolution; one of which belongs solely to the psychic apparatus and has exerted a determining influence on the relation of the two systems, while the other operates fluctuatingly and introduces motive forces of organic origin into the psychic life. Both originate in the infantile life and result from the transformation which our psychic and somatic organism has undergone since the infantile period.

When I termed one of the psychic processes in the psychic apparatus the primary process, I did so not only in consideration of the order of precedence and capability, but also as admitting the temporal relations to a share in the nomenclature. As far as our knowledge goes there is no psychic apparatus possessing only the primary process, and in so far it is a theoretic fiction; but so much is based on fact that the primary processes are present in the apparatus from the beginning, while the secondary processes develop gradually in the course of life, inhibiting and covering the primary ones, and gaining complete mastery over them perhaps only at the height of life. Owing to this retarded appearance of the secondary processes, the essence of our being, consisting in unconscious wish feelings, can neither be seized nor inhibited by the foreconscious, whose part is once for all restricted to the indication of the most suitable paths for the wish feelings originating in the unconscious. These unconscious wishes establish for all subsequent psychic efforts a compulsion to which they have to submit and which they must strive if possible to divert from its course and direct to higher aims. In consequence of this retardation of the foreconscious occupation a large sphere of the memory material remains inaccessible.

Among these indestructible and unincumbered wish feelings originating from the infantile life, there are also some, the fulfilments of which have entered into a relation of contradiction to the end-presentation of the secondary thinking. The fulfilment of these wishes would no longer produce an affect of pleasure but one of pain; *and it is just this transformation of affect that constitutes the nature of what we designate as "repression," in which we recognise the infantile first step of passing adverse sentence or of rejecting through reason.* To investigate in what way and through what motive forces such a transformation can be produced constitutes the problem of repression, which we need here only skim over. It will suffice to remark that such a transformation of affect occurs in the course of development (one may think of the appearance in infantile life of disgust which was originally absent), and that it is connected with the activity of the secondary system. The memories from which the unconscious wish brings about the emotional discharge have never been accessible to the Force., and for that reason their emotional discharge cannot be inhibited. It is just on account of this affective development that these ideas are not even now accessible to the foreconscious thoughts to which they have transferred their wishing power. On the contrary, the principle of pain comes into play, and causes the Forec. to deviate from these thoughts of transference. The latter, left to themselves, are "repressed," and thus the existence of a store of infantile memories, from the very beginning withdrawn from the Force., becomes the preliminary condition of repression.

In the most favourable case the development of pain terminates as soon as the energy has been withdrawn from the thoughts of transference in the Forec., and this effect characterises the intervention of the principle of pain as expedient. It is different, however, if the repressed unconscious wish receives an organic enforcement which it can lend to its thoughts of transference and through which it can enable them to make an effort towards penetration with their excitement, even after they have been abandoned by the occupation of the Forec. A defensive struggle then ensues, inasmuch as the Forec. reinforces the antagonism against the repressed ideas, and subsequently this leads to a penetration by the thoughts of transference (the carriers of the unconscious wish) in some form of compromise through symptom formation. But from the moment that the suppressed thoughts are powerfully occupied by the unconscious wish-feeling and abandoned by the fore-conscious occupation, they succumb to the primary psychic process and strive only for motor discharge; or, if the path be free, for hallucinatory revival of the desired perception identity. We have previously found, empirically, that the incorrect processes described are enacted only with thoughts that exist in the repression. We now grasp another part of the connection. These incorrect processes are those that are primary in the psychic apparatus; *they appear wherever thoughts abandoned by the foreconscious occupation are left to themselves, and can fill themselves with the uninhibited energy, striving for discharge from the unconscious.* We may add a few further observations to support the view that these processes designated "incorrect" are really not falsifications of the normal defective thinking, but the modes of activity of the psychic apparatus when freed from inhibition. Thus we see that the transference of the foreconscious excitement to the motility takes place according to the same processes, and that the connection of the foreconscious presentations with words readily manifest the same displacements and mixtures which are described to inattention. Finally, I should like to adduce proof that an increase of work necessarily results from the inhibition of these primary courses from the fact that we gain a *comical effect,* a surplus to be discharged through laughter, *if we allow these streams of thought to come to consciousness.*

The theory of the psychoneuroses asserts with complete certainty that only sexual wish-feelings from the infantile life experience repression (emotional transformation) during the developmental period of childhood. These are capable of returning to activity at a later period of development, and then have the faculty of being revived, either as a consequence of the sexual constitution, which is really formed from the original bisexuality, or in consequence of unfavourable influences of the sexual life; and they thus supply the motive power for all psychoneurotic symptom formations. It is only by the introduction of these sexual forces that the gaps still demonstrable in the theory of repression can be filled. I will leave it undecided whether the postulate of the sexual and infantile may also be asserted for the theory of the dream; I leave this here unfinished because I have already passed a step beyond the demonstrable in assuming that the dream-wish invariably originates from the unconscious.[16] Nor will I further investigate the difference in the play of the psychic forces in the dream formation and in the formation of the hysterical symptoms, for to do this we ought to possess a more explicit knowledge of one of the members to be compared. But I regard another point as important, and will here confess that it was on account of this very point that I have just undertaken this entire discussion concerning the two psychic systems, their modes of operation, and the repression. For it is now immaterial whether I have conceived the psychological relations in question with approximate correctness, or, as is easily possible in such a difficult matter, in an erroneous and fragmentary manner. Whatever changes may be made in the interpretation of the psychic censor and of the correct and of the abnormal elaboration of the dream content, the fact nevertheless remains that such processes are active in dream formation, and that essentially they show the closest analogy to the processes observed in the formation of the hysterical symptoms. The dream is not a pathological phenomenon, and it does not leave behind an enfeeblement of the mental faculties. The objection that no deduction can be drawn regarding the dreams of healthy persons from my own dreams and from those of neurotic patients may be rejected without comment. Hence, when we draw conclusions from the phenomena as to their motive forces, we recognise that the psychic mechanism made use of by the neuroses is not created by a morbid disturbance of the psychic life, but is found ready in the normal structure of the psychic apparatus. The two psychic systems, the censor crossing between them, the inhibition and the covering of the one activity by the other, the relations of both to consciousness—or whatever may offer a more correct interpretation of the actual conditions in their stead—all these belong to the normal structure of our psychic instrument, and the dream points out for us one of the roads leading to a knowledge of this structure. If, in addition to our knowledge, we wish to be contented with a minimum perfectly established. we shall say that the dream gives us proof that the *suppressed material continues to exist even in the normal person and remains capable of psychic activity.* The dream itself is one of the manifestations of this suppressed material; theoretically, this is true in *all* cases; according to substantial experience it is true in at least a great number of such as most conspicuously display the prominent characteristics of dream life. The suppressed psychic material, which in the waking state has been prevented from expression and cut off from internal perception *by the antagonistic adjustment of the contradictions,* finds ways and means of obtruding itself on consciousness during the night under the domination of the compromise formations.

"Flectere si nequeo superos, Acheronta movebo."

At any rate the interpretation of dreams is the via regia to a knowledge of the unconscious in the psychic life.

In following the analysis of the dream we have made some progress toward an understanding of the composition of this most marvellous and most mysterious of instruments; to be sure, we have not gone very far, but enough of a beginning has been made to allow us to advance from other so-called pathological formations further into the analysis of the unconscious. Disease—at least that which is justly termed functional—is not due to the destruction of this apparatus, and the establishment of new splittings in its interior; it is rather to be explained dynamically through the strengthening and weakening of the components in the play of forces by which so many activities are concealed during the normal function. We have been able to show in another place how the composition of the apparatus from the two systems permits a subtilisation even of the normal activity which would be impossible for a single system.[17]

(F) The Unconscious and Consciousness—Reality

On closer inspection we find that it is not the existence of two systems near the motor end of the apparatus but of two kinds of processes or modes of emotional discharge, the assumption of which was explained in the psychological discussions of the previous chapter. This can make no difference for us, for we must always be ready to drop our auxiliary ideas whenever we deem ourselves in position to replace them by something else approaching more closely to the unknown reality. Let us now try to correct some views which might be erroneously formed as long as we regarded the two systems in the crudest and most obvious sense as two localities within the psychic apparatus, views which have left their traces in the terms "repression" and "penetration." Thus, when we say that an unconscious idea strives for transference into the foreconscious in order later to penetrate consciousness, we do not mean that a second idea is to be formed situated in a new locality like an interlineation near which the original continues to remain; also, when we speak of penetration into consciousness, we wish carefully to avoid any idea of change of locality. When we say that a foreconscious idea is repressed and subsequently taken up by the unconscious, we might be tempted by these figures, borrowed from the idea of a struggle over a territory, to assume that an arrangement is really broken up in one psychic locality and replaced by a new one in the other locality. For these comparisons we substitute what would seem to correspond better with the real state of affairs by saying that an energy occupation is displaced to or withdrawn from a certain arrangement so that the psychic formation falls under the domination of a system or is withdrawn from the same. Here again we replace a topical mode of presentation by a dynamic; it is not the psychic formation that appears to us as the moving factor but the innervation of the same.

I deem it appropriate and justifiable, however, to apply ourselves still further to the illustrative conception of the two systems. We shall avoid any misapplication of this manner of representation if we remember that presentations, thoughts, and psychic formations should generally not be localised in the organic elements of the nervous system, but, so to speak, between them, where resistances and paths form the correlate corresponding to them. Everything that can become an object of our internal perception is virtual, like the image in the telescope produced by the passage of the rays of light. But we are justified in assuming the existence of the systems, which have nothing psychic in themselves and which never become accessible to our psychic perception, corresponding to the lenses of the telescope which design the image. If we continue this comparison, we may say that the censor between two systems corresponds to the refraction of rays during their passage into a new medium.

Thus far we have made psychology on our own responsibility; it is now time to examine the theoretical opinions governing present-day psychology and to test their relation to our theories. The question of the unconscious in psychology is, according to the authoritative words of Lipps,[18] less a psychological question than the question of psychology. As long as psychology settled this question with the verbal explanation that the "psychic" is the "conscious" and that "unconscious psychic occurrences" are an obvious contradiction, a psychological estimate of the observations gained by the physician from abnormal mental states was precluded. The physician and the philosopher agree only when both acknowledge that unconscious psychic processes are "the appropriate and well-justified expression for an established fact." The physician cannot but reject with a shrug of his shoulders the assertion that "consciousness is the indispensable quality of the psychic"; he may assume, if his respect for the utterings of the philosophers still be strong enough, that he and they do not treat the same subject and do not pursue the same science. For a single intelligent observation of the. psychic life of a neurotic, a single analysis of a dream must force upon him the unalterable conviction that the most complicated and correct mental operations, to which no one will refuse the name of psychic occurrences, may take place without exciting the consciousness of the person. It is true that the physician does not learn of these unconscious processes until they have exerted such an effect on consciousness as to admit communication or observation. But this effect of consciousness may show a psychic character widely differing from the unconscious process, so that the internal perception cannot possibly recognise the one as a substitute for the other. The physician must reserve for himself the right to penetrate, by a process of deduction, from the effect on consciousness to the unconscious psychic process; he learns in this way that the effect on consciousness is only a remote psychic product of the unconscious process and that the latter has not become conscious as such; that it has been in existence and operative without betraying itself in any way to consciousness.

A reaction from the over-estimation of the quality of consciousness becomes the indispensable preliminary condition for any correct insight into the behaviour of the psychic. In the words of Lipps, the unconscious must be accepted as the general basis of the psychic life. The unconscious is the larger circle which includes within itself the smaller circle of the conscious; everything conscious has its preliminary step in the unconscious, whereas the unconscious may stop with this step and still claim full value as a psychic activity. Properly speaking, the unconscious is the real psychic; *its inner nature is just as unknown to us as the reality of the external world, and it is just as imperfectly reported to us through the data of consciousness as is the external world through the indications of our sensory organs.*

A series of dream problems which have intensely occupied older authors will be laid aside when the old opposition between conscious life and dream life is abandoned and the unconscious psychic assigned to its proper place. Thus many of the activities whose performances in the dream have excited our admiration are now no longer to be attributed to the dream but to unconscious thinking, which is also active during the day. If, according to Scherner, the dream seems to play with a symbolising representation of the body, we know that this is the work of certain unconscious phantasies which have probably given in to sexual emotions, and that these phantasies come to expression not only in dreams but also in hysterical phobias and in other symptoms. If the dream continues and settles activities of the day and even brings to light valuable inspirations, we have only to subtract from it the dream disguise as a feat of dream-work and a mark of assistance from obscure forces in the depth of the mind (*cf.* the devil in Tartini's sonata dream). The intellectual task as such must be attributed to the same psychic forces which perform all such tasks during the day. We are probably far too much inclined to over-estimate the conscious character even of intellectual and artistic productions. From the communications of some of the most highly productive persona, such as Goethe and Helmholtz, we learn, indeed, that the most essential and original parts in their creations came to them in the form of inspirations and reached their perceptions almost finished. There is nothing strange about the assistance of the conscious activity in other cases where there was a concerted effort of all the psychic forces. But it is a much abused privilege of the conscious activity that it is allowed to hide from us all other activities wherever it participates.

It will hardly be worth while to take up the historical significance of dreams as a special subject. Where, for instance, a chieftain has been urged through a dream to engage in a bold undertaking the success of which has had the effect of changing history, a new problem results only so long as the dream, regarded as a strange power, is contrasted with other more familiar psychic forces; the problem, however, disappears when we regard the dream as a form of expression for feelings which are burdened with resistance during the day and which can receive reinforcements at night from deep emotional sources.[19] But the great respect shown by the ancients for the dream is based on a correct psychological surmise. It is a homage paid to the unsubdued and indestructible in the human mind, and to the demoniacal which furnishes the dream-wish and which we find again in our unconscious.

Not inadvisedly do I use the expression "in our unconscious," for what we so designate does not coincide with the unconscious of the philosophers, nor with the unconscious of Lipps. In the latter uses it is intended to designate only the opposite of conscious. That there are also unconscious psychic processes beside the conscious ones is the hotly contested and energetically defended issue. Lipps gives us the more far-reaching theory that everything psychic exists as unconscious, but that some of it may exist also as conscious. But it was not to prove this theory that we have adduced the phenomena of the dream and of the hysterical symptom formation; the observation of normal life alone suffices to establish its correctness beyond any doubt. The new fact that we have learned from the analysis of the psychopathological formations, and indeed from their first member, viz. dreams, is that the unconscious—hence the psychic—occurs as a function of two separate systems and that it occurs as such even in normal psychic life. Consequently there are two kinds of unconscious, which we do not as yet find distinguished by the psychologists. Both are unconscious in the psychological sense; but in our sense the first, which we call Unc., is likewise incapable of consciousness, whereas the second we term "Forec." because its emotions, after the observance of certain rules, can reach consciousness, perhaps not before they have again undergone censorship, but still regardless of the Unc. system. The fact that in order to attain consciousness the emotions must traverse an unalterable series of events or succession of instances, as is betrayed through their alteration by the censor, has helped us to draw a comparison from spatiality. We described the relations of the two systems to each other and to consciousness by saying that the system Forec. is like a screen between the system Unc. and consciousness. The system Forec. not only bars access to consciousness, but also controls the entrance to voluntary motility and is capable of sending out a sum of mobile energy, a portion of which is familiar to us as attention.

We must also steer clear of the distinctions superconscious and subconscious which have found so much favour in the more recent literature on the psychoneuroses, for just such a distinction seems to emphasize the equivalence of the psychic and the conscious.

What part now remains in our description of the once all-powerful and all-overshadowing consciousness? None other than that of a sensory organ for the perception of psychic qualities. According to the fundamental idea of schematic undertaking we can conceive the conscious perception only as the particular activity of an independent system for which the abbreviated designation "Cons." commends itself. This system we conceive to be similar in its mechanical characteristics to the perception system P, hence excitable by qualities and incapable of retaining the trace of changes, i.e. it is devoid of memory. The psychic apparatus which, with the sensory organs of the P-systems, is turned to the outer world, is itself the outer world for the sensory organ of Cons.; the teleological justification of which rests on this relationship. We are here once more confronted with the principle of the succession of instances which seems to dominate the structure of the apparatus. The material under excitement flows to the Cons. sensory organ from two sides, firstly from the P-system whose excitement, qualitatively determined, probably experiences a new elaboration until it comes to conscious perception; and, secondly, from the interior of the apparatus itself, the quantitative processes of which are perceived as a qualitative series of pleasure and pain as soon as they have undergone certain changes.

The philosophers, who have learned that correct and highly complicated thought structures are possible even without the co-operation of consciousness, have found it difficult to attribute any function to consciousness; it has appeared to them a superfluous mirroring of the perfected psychic process. The analogy of our Cons. system with the systems of perception relieves us of this embarrassment. We see that perception through our sensory organs results in directing the occupation of attention to those paths on which the incoming sensory excitement is diffused; the qualitative excitement of the P-system serves the mobile quantity of the psychic apparatus as a regulator for its discharge. We may claim the same function for the overlying sensory organ of the Cons. system. By assuming new qualities, it furnishes a new contribution toward the guidance and suitable distribution of the mobile occupation quantities. By means of the perceptions of pleasure and pain, it influences the course of the occupations within the psychic apparatus, which normally operates unconsciously and through the displacement of quantities. It is probable that the principle of pain first regulates the displacements of occupation automatically, but it is quite possible that the consciousness of these qualities adds a second and more subtle regulation which may even oppose the first and perfect the working capacity of the apparatus by placing it in a position contrary to its original design for occupying and developing even that which is connected with the liberation of pain. We learn from neuropsychology that an important part in the functional activity of the apparatus is attributed to such regulations through the qualitative excitation of the sensory organs. The automatic control of the primary principle of pain and the restriction of mental capacity connected with it are broken by the sensible regulations, which in their turn are again automatisms. We learn that the repression which, though originally expedient, terminates nevertheless in a harmful rejection of inhibition and of psychic domination, is so much more easily accomplished with reminiscences than with perceptions, because in the former there is no increase in occupation through the excitement of the psychic sensory organs. When an idea to be rejected has once failed to become conscious because it has succumbed to repression, it can be repressed on other occasions only because it has been withdrawn from conscious perception on other grounds. These are hints employed by therapy in order to bring about a retrogression of accomplished repressions.

The value of the over-occupation which is produced by the regulating influence of the Cons. sensory organ on the mobile quantity, is demonstrated in the teleological connection by nothing more clearly than by the creation of a new series of qualities and consequently a new regulation which constitutes the precedence of man over the animals. For the mental processes are in themselves devoid of quality except for the excitements of pleasure and pain accompanying them, which, as we know, are to be held in check as possible disturbances of thought. In order to endow them with a quality, they are associated in man with verbal memories, the qualitative remnants of which suffice to draw upon them the attention of consciousness which in turn endows thought with a new mobile energy.

The manifold problems of consciousness in their entirety can be examined only through an analysis of the hysterical mental process. From this analysis we receive the impression that the transition from the foreconscious to the occupation of consciousness is also connected with a censorship similar to the one between the Unc. and the Forec. This censorship, too, begins to act only with the reaching of a certain quantitative degree, so that few intense thought formations escape it. Every possible case of detention from consciousness, as well as of penetration to consciousness, under restriction is found included within the picture of the psychoneurotic phenomena; every case points to the intimate and twofold connection between the censor and consciousness. I shall conclude these psychological discussions with the report of two such occurrences.

On the occasion of a consultation a few years ago the subject was an intelligent and innocent-looking girl. Her attire was strange; whereas a woman's garb is usually groomed to the last fold, she had one of her stockings hanging down and two of her waist buttons opened. She complained of pains in one of her legs, and exposed her leg unrequested. Her chief complaint, however, was in her own words as follows: She had a feeling in her body as if something was stuck into it which moved to and fro and made her tremble through and through. This sometimes made her whole body stiff. On hearing this, my colleague in consultation looked at me; the complaint was quite plain to him. To both of us it seemed peculiar that the patient's mother thought nothing of the matter; of course she herself must have been repeatedly in the situation described by her child. As for the girl, she had no idea of the import of her words or she would never have allowed them to pass her lips. Here the censor had been deceived so successfully that under the mask of an innocent complaint a phantasy was admitted to consciousness which otherwise would have remained in the foreconscious.

Another example: I began the psychoanalytic treatment of a boy of fourteen years who was suffering from *tic convulsif,* hysterical vomiting, headache, &c., by assuring him that, after closing his eyes, he would see pictures or have ideas, which I requested him to communicate to me. He answered by describing pictures. The last impression he had received before coming to me was visually revived in his memory. He had played a game of checkers with his uncle, and now saw the checker-board before him. He commented on various positions that were favourable or unfavourable, on moves that were not safe to make. He then saw a dagger lying on the checkerboard, an object belonging to his father, but transferred to the checker-board by his phantasy. Then a sickle was lying on the board; next a scythe was added; and, finally, he beheld the likeness of an old peasant mowing the grass in front of the boy's distant parental home. A few days later I discovered the meaning of this series of pictures. Disagreeable family relations had made the boy nervous. It was the case of a strict and crabbed father who lived unhappily with his mother, and whose educational methods consisted in threats; of the separation of his father from his tender and delicate mother, and the remarrying of his father, who one day brought home a young woman as his new mamma. The illness of the fourteen-year-old boy broke out a few days later. It was the suppressed anger against his father that had composed these pictures into intelligible allusions. The material was furnished by a reminiscence from mythology. The sickle was the one with which Zeus castrated his father; the scythe and the likeness of the peasant represented Kronos, the violent old man who eats his children and upon whom Zeus wreaks vengeance in so unfilial a manner. The marriage of the father gave the boy an opportunity to return the reproaches and threats of his father—which had previously been made because the child played with his genitals (the checker-board; the prohibitive moves; the dagger with which a person may be killed). We have here long repressed memories and their unconscious remnants which, under the guise of senseless pictures have slipped into consciousness by devious paths left open to them.

I should then expect to find the theoretical value of the study of dreams in its contribution to psychological knowledge and in its preparation for an understanding of neuroses. Who can foresee the importance of a thorough knowledge of the structure and activities of the psychic apparatus when even our present state of knowledge produces a happy therapeutic influence in the curable forms of the psychoneuroses? What about the practical value of such study someone may ask, for psychic knowledge and for the discovering of the secret peculiarities of individual character? Have not the unconscious feelings revealed by the dream the value of real forces in the psychic life? Should we take lightly the ethical significance of the suppressed wishes which, as they now create dreams, may some day create other things?

I do not feel justified in answering these questions. I have not thought further upon this side of the dream problem. I believe, however, that at all events the Roman Emperor was in the wrong who ordered one of his subjects executed because the latter dreamt that he had killed the Emperor. He should first have endeavoured to discover the significance of the dream; most probably it was not what it seemed to be. And even if a dream of different content had the significance of this offence against majesty, it would still have been in place to remember the words of Plato, that the virtuous man contents himself with dreaming that which the wicked man does in actual life. I am therefore of the opinion that it is best to accord freedom to dreams. Whether any reality is to be attributed to the unconscious wishes, and in what sense, I am not prepared to say offhand. Reality must naturally be denied to all transition—and intermediate thoughts. If we had before us the unconscious wishes, brought to their last and truest expression, we should still do well to remember that more than one single form of existence must be ascribed to the psychic reality. Action and the conscious expression of thought mostly suffice for the practical need of judging a man's character. Action, above all, merits to be placed in the first rank; for many of the impulses penetrating consciousness are neutralised by real forces of the psychic life before they are converted into action; indeed, the reason why they frequently do not encounter any psychic obstacle on their way is because the unconscious is certain of their meeting with resistances later. In any case it is instructive to become familiar with the much raked-up soil from which our virtues proudly arise. For the complication of human character moving dynamically in all directions very rarely accommodates itself to adjustment through a simple alternative, as our antiquated moral philosophy would have it.

And how about the value of the dream for a knowledge of the future? That, of course, we cannot consider.[20] One feels inclined to substitute: "for a knowledge of the past." For the dream originates from the past in every sense. To be sure the ancient belief that the dream reveals the future is not entirely devoid of truth. By representing to us a wish as fulfilled the dream certainly leads us into the future; but this future, taken by the dreamer as present, has been formed into the likeness of that past by the indestructible wish.

Notes:

1. See the *Psychopathology of Everyday Life,* 4th ed., 1912. (English translation in preparation.)

2. Concerning the object of forgetting in general, see the *Psychopathology of Everyday Life.*

3. Translated by A. A. Brill, appearing under the title *Selected Papers on Hysteria.*

4. Jung has brilliantly corroborated this statement by analyses of Dementia Praecox. (*The Psychology of Dementia Praecox,* translated by F. Peterson and A. A. Brill.)

5. The same considerations naturally hold true also for the case where superficial associations are exposed in the dream, as, *e.g.,* in both dreams reported by Maury (p. 58, *pélerinage—pelletier—pelle, kilometer—kilogram—gilolo, Lobelia—Lopez—Lotto*). I know from my work with neurotics what kind of reminiscence preferentially represents itself in this manner. It is the consultation of encyclopaedias by which most people pacify their desire for explanation of the sexual riddle during the period of curiosity in puberty.

6. The above sentences, which when written sounded very improbable, have since been justified experimentally by Jung and his pupils in the *Diagnostiche Assoziationsstudien.*

7. *Selected Papers on Hysteria and Other Psychoneuroses,* p. 165, translated by A. A. Brill (*Journal Mental and Nervous Disease* Publishing Co.).

8. The German word "Dutzendmensch" (a man of dozens) which the young lady wished to use in order to express her real opinion of her friend's fiancé, denotes a person with whom figures are everything. (Translator.)

9. They share this character of indestructibility with all psychic acts that are really unconscious—that is, with psychic acts belonging to the system of the unconscious only. These paths are constantly open and never fall into disuse; they conduct the discharge of the exciting process as often as it becomes endowed with unconscious excitement. To speak metaphorically they suffer the same form of annihilation as the shades of the lower region in the *Odyssey,* who awoke to new life the moment they drank blood. The processes depending on the foreconscious system are destructible in a different way. The psychotherapy of the neuroses is based on this difference.

10. Le Lorrain justly extols the wish-fulfilment of the dream: "Sans fatigue sérieuse, sans être obligé de recourir à cette lutte opinâtre et longue qui use et corrode les jouissances poursuivies."

11. This idea has been borrowed from *The Theory of Sleep* by Liébault, who revived hypnotic investigation in our days. (*Du Sommeil provoqué*, etc.; Paris, 1889.)

12. The German of the word *bird* is "Vogel," which gives origin to the vulgar expression "vöglen," denoting sexual intercourse. (Trans. note.)

13. The italics are my own, though the meaning is plain enough without them.

14. The italics are mine.

15. *Cf.* the significant observations by J. Breuer in our *Studies on Hysteria,* 1895, and 2nd ed. 1909.

16. Here, as in other places, there are gaps in the treatment of the subject, which I have left intentionally, because to fill them up would require on the one hand too great effort, and on the other hand an extensive reference to material that is foreign to the dream. Thus I have avoided stating whether I connect with the word "suppressed" another sense than with the word "repressed." It has been made clear only that the latter emphasizes more than the former the relation to the unconscious. I have not entered into the cognate problem why the dream thoughts also experience distortion by the censor when they abandon the progressive continuation to consciousness and choose the path of

regression. I have been above all anxious to awaken an interest in the problems to which the further analysis of the dream-work leads and to indicate the other themes which meet these on the way. It was not always easy to decide just where the pursuit should be discontinued. That I have not treated exhaustively the part played in the dream by the psychosexual life and have avoided the interpretation of dreams of an obvious sexual content is due to a special reason which may not come up to the reader's expectation. To be sure, it is very far from my ideas and the principles expressed by me in neuropathology to regard the sexual life as a "pudendum" which should be left unconsidered by the physician and the scientific investigator. I also consider ludicrous the moral indignation which prompted the translator of Artemidoros of Daldis to keep from the reader's knowledge the chapter on sexual dreams contained in the *Symbolism of the Dreams*. As for myself, I have been actuated solely by the conviction that in the explanation of sexual dreams I should be bound to entangle myself deeply in the still unexplained problems of perversion and bisexuality; and for that reason I have reserved this material for another connection.

17. The dream is not the only phenomenon tending to base psychopathology on psychology. In a short series of unfinished articles ("Monatsschrift für Psychiatrie und Neurologie" entitled *Über den psychischen Mechanismus der Vergeslichkeit*, 1898, and *Über Deckerinnerungen*, 1899) I attempt to interpret a number of psychic manifestations from everyday life in support of the same conception. These and other articles on "Forgetting," "Lapse of Speech," &c., have since been published collectively under the title of *Psychopathology of Everyday Life*, 1904 and 1907, of which an English translation will shortly appear.

18. "The Conception of the Unconscious in Psychology": Lecture delivered at the Third International Congress of Psychology at Munich, 1897.

19. *Cf.* here (p. 116) the dream ([Greek]) of Alexander the Great at the siege of Tyrus.

20. Professor Ernst Oppenheim (Vienna) has shown me from folk-lore material that there is a class of dreams for which even the people drop the expectation of future interpretation, and which they trace in a perfectly correct manner to wish feelings and wants arising during sleep. He will in the near future fully report upon these dreams, which for the most part are in the form of "funny stories."

Bibliography

Aristoteles. *Über Träume und Traumdeutungen.* Translated by Bender.

Artemidoros aus Daldis. *Symbolik der Traüme.* Translated by Friedrich. S. Krauss. Wien, 1881.

Benini, V. "La Memoria e la Durata dei Sogni." *Rivista Italiana de Filosofia,*Marz–April 1898.

Binz, C. *Über den Traum.* Bonn, 1878.

Borner, J. *Das Alpdrücken, seine Begründung und Verhütung.* Würzburg, 1855.

Bradley, J. H. "On the Failure of Movement in Dream." *Mind,* July 1894.

Brander, R. *Der Schlaf und das Traumleben.* 1884.

Burdach. *Die Physiologie als Erfahrungswissenschaft,* 3 Bd. 1830.

Büchsenschütz, B. *Traum und Traumdeutung in Altertum.* Berlin, 1868.

Chaslin, Ph. *Du Röle du Rêve dans l'Evolution du Délire.* Thèse de Paris. 1887.

Chabaneix. *Le Subconscient chez les Artistes, les Savants et les Ecrivains.* Paris, 1897.

Calkins, Mary Whiton. "Statistics of Dreams." *Amer. J. of Psychology,* V., 1893.

Clavière. "La Rapidité de la Pensée dans le Rêve." *Revue philosophique,* XLIII., 1897.

Dandolo, G. *La Coscienza nel Sonno.* Padova, 1889.

Delage, Yves. "Une Théorie de Rêve." *Revue scientifique,* II., Juli 1891.

Delboeuf, J. *Le Sommeil et les Rêves.* Paris, 1885.

Debacker. *Terreurs nocturnes des Enfants.* Thèses de Paris. 1881.

Dugas. "Le Souvenir du Rêve." *Revue philosophique,* XLIV., 1897.

Dugas. "Le Sommeil et la Cérébration inconsciente durant le Sommeil." *Revue philosophique,* XLIII., 1897.

Egger, V. "La Durée apparente des Rêves." *Revue philosophique,* Juli 1895.

Egger. "Le Souvenir dans le Rêve." *Revue philosophique,* XLVI., 1898.

Ellis Havelock. "On Dreaming of the Dead." *The Psychological Review,* II., Nr. 5, September 1895.

Ellis Havelock. "The Stuff that Dreams are made of." *Appleton's Popular Science Monthly,* April 1899.

Ellis Havelock. "A Note on Hypnogogic Paramnesia." *Mind,* April 1897.

Fechner, G. Th. *Elemente der Psychophysik.* 2 Aufl., 1889.

Fichte, J. H. "Psychologie." *Die Lehre vom bewussten Geiste des Menschen.* I. Teil. Leipzig, 1864.

Giessler, M. Aus den Tiefen des Traumlebens. Halle, 1890.

Giessler, M. *Die physiologischen Beziehungen der Traumvorgänge.* Halle, 1896.

Goblot. "Sur le Souvenir des Rêves." *Revue philosophique,* XLII., 1896.

Graffunder. *Traum und Traumdeutung.* 1894.

Griesinger. *Pathologie und Therapie der psychischen Krankheiten.* 3 Aufl. 1871.

Haffner, P. "Schlafen und Träumen. 1884." *Frankfurter zeitgemässe Broschüren,* 5 Bd., Heft. 10.

Hallam, Fl., and Sarah Weed. "A Study of the Dream Consciousness." *Amer. J. of Psychology,* VII., Nr. 3, April 1896.

D'Hervey. *Les Rêves et les Moyens de les Diriger.* Paris, 1867 (anonym.).

Hildebrandt, F. W. *Der Traum und seine Verwertung für Leben.* Leipzig, 1875.

Jessen. *Versuch einer Wissenschaftlichen Begründung der Psychologie.* Berlin, 1856.

Jodl. *Lehrbuch der Psychologie.* Stuttgart, 1896.

Kant, J. *Anthropologie in pragmatischer Hinsicht.* Kirchmannsche Ausgabe. Leipzig, 1880.

Krauss, A. "Der Sinn im Wahnsinn." *Allgemeine Zeitschrift für Psychologie,* XV. u. XVI., 1858–1859.

Ladd. "Contribution to the Psychology of Visual Dreams." *Mind,* April 1892.

Leidesdorf, M. *Das Traumleben.* Wien, 1880. Sammlung der "Alma Mater."

Lémoine. *Du Sommeil au Point de Vue physiologique et psychologique.* Paris, 1885.

Lièbeault, A. *Le Sommeil provogué et les Etats analogues.* Paris, 1889.

Lipps, Th. *Grundtatsachen des Sellenlebens.* Bonn, 1883.

Le Lorrain. "Le Rêve." *Revue philosophique.* July 1895.

Mandsley. *The Pathology of Mind.* 1879.

Maury, A. "Analogies des Phénomènes du Rêve et de l'Aliénation Mentale."*Annales med. psych.,* 1854, p. 404.

Maury, A. *Le Sommeil et les Rêves.* Paris, 1878.

Moreau, J. "De l'Identité de l'Etat de Rêve et de Folie." *Annales med. psych.,*1855, p. 361.

Nelson, J. "A Study of Dreams." *Amer. J. of Psychology,* I., 1888.

Pilcz. "Über eine gewisse Gesetzmassigkeit in den Träumen." Autorreferat in *Monatsschrift für Psychologie und Neurologie.* März 1899.

Pfaff, E. R. *Das Traumleben und seine Deutung nach den Prinzipien der Araber, Perser, Griechen, Indier und Ägypter.* Leipzig, 1868.

Purkinje. Artikel. *Wachen, Schlaf, Traum und verwandte Zustände in Wagners Handwörterbuch der Physiologie.* 1846.

Radestock, P. *Schlaf und Traum.* Leipzig, 1878.

Robert, W. *Der Traum als Naturnotwendigkeit erklärt.* 1886.

Sante de Sanctis. *Les Maladies mentales et les Rêves.* 1897. Extrait des *Annales de la Société de Médecine de Gand.*

Sante de Sanctis. "Sui rapporti d'Identità, di Somiglianza, di Analogia e di Equivalenza fra Sogno e Pazzia." *Rivista quindicinale di Psicologia, Psichiatria, Neuropatologia.* 15, Nov. 1897.

Scherner, R. A. *Das Leben des Traümes.* Berlin. 1861.

Scholz, Fr. *Schlaf und Traum.* Leipzig, 1887.

Schopenhauer. "Versuch über das Geistersehen und was damit zusammenhängt." *Parerga und Paralipomena,* 1 Bd., 1857.

Schleiermacher, Fr. *Psychologie.* Edited by L. George. Berlin, 1862.

Siebek, A. *Das Traumleben der Seele.* 1877. *Sammlung Virchow-Holtzendorf.* Nr. 279.

Simon, M. "Le Monde des Rêves." Paris, 1888. *Bibliothèque scientifique contemporaine.*

Spitta, W. *Die Schlaf- und Traumzustände der menschlichen Seele.* 2 Aufl. Freiburg, I. B. 1892.

Stumpf, E. J. G. *Der Traum und seine Deutung.* Leipzig, 1899.

Strümpell, L. *Die Natur und Entstehung der Träume.* Leipzig, 1877.

Tannery. "Sur la Mémoire dans le Rêve." *Revue philosophique,* XLV., 1898.

Tissié, Ph. "Les Rêves, Physiologie et Pathologie." 1898. *Bibliothèque de Philosophie contemporaine.*

Titchener. "Taste Dreams." *Amer. Jour. of Psychology,* VI., 1893.

Thomayer. "Sur la Signification de quelques Rêves." *Revue neurologique.* Nr. 4, 1897.

Vignoli. "Von den Träumen, Illusionen und Halluzinationen." *Internationale wissenschaftliche Bibliothek,* Bd. 47.

Volkelt, J. *Die Traumphantasie.* Stuttgart, 1875.

Vold, J. Mourly. "Expériences sur les Rêves et en particulier sur ceux d'Origine musculaire et optique." Christiania, 1896. Abstract in the *Revue philosophique, XLII., 1896.*

Vold, J. Mourly. "Einige Experimente über Gesichtsbilder im Traume." *Dritter internationaler Kongress für Psychologie in München.* 1897.

(Vold, J. Mourly. "Über den Traum." *Experimentellpsychologische Untersuchungen.* Herausgegeben von O. Klemm. Erster Band. Leipzig, 1910.)

Weygandt, W. *Entstehung der Träume.* Leipzig, 1893.

Wundt. *Grundzüge der physiologischen Psychologie.* II. Bd., 2 Aufl. 1880.

Stricker. *Studien Über das Bewusstsein.* Wien, 1879.

Stricker. *Studien über die Assoziation der Vorstellungen.* Wien, 1883.

Psychoanalytic Literature of Dreams

Abraham, Karl (Berlin): *Traum und Mythos: Eine Studie zur Volker-psychologie.* Schriften z. angew. Seelenkunde, Heft 4, Wien und Leipzig, 1909.

Abraham, Karl (Berlin): "Über hysterische Traumzustände." (*Jahrbuch f. psychoanalyt. und psychopatholog.* Forschungen, Vol. II., 1910.)

Adler, Alfred (Wien): "Zwei Träume einer Prostituierten." (*Zeitschrift f. Sexualwissenschaft,* 1908, Nr. 2.)

Adler, Alfred (Wien): "Ein erlogener Traum." (*Zentralbl. f. Psychoanalyse,* 1 Jahrg. 1910, Heft 3.)

Bleuler, E. (Zürich): "Die Psychanalyse Freuds." (*Jahrb. f. psychoanalyt. u. psychopatholog.* Forschungen, Bd. II., 1910.)

Brill, A. A. (New York): "Dreams and their Relation to the Neuroses." (*New York Medical Journal,* April 23, 1910.)

Brill. Hysterical Dreamy States. Ebenda, May 25, 1912.

Ellis, Havelock: "The Symbolism of Dreams." (*The Popular Science Monthly,* July 1910.)

Ellis, Havelock: *The World of Dreams.* London, 1911.

Ferenczi, S. (Budapest): "Die psychologische Analyse der Traüme." (*Psychiatrisch-neurologische Wochenschrift,* XII., Jahrg., Nr. 11–13, Juni 1910.

English translation under the title: *The Psychological Analysis of Dreams* in the *American Journal of Psychology,* April 1910.)

Freud, S. (Wien): "Über den Traum. (*Grenzfragen des Nerven- und Seelenlebens.* Edited by Löwenfeld und Kurella, Heft 8. Wiesbaden, Bergmann, 1901, 2 Aufl. 1911.)

Freud, S. (Wien): "Bruchstük einer Hysterieanalyse. (*Monatsschr. f. Psychiatrie und Neurologie,* Bd. 18, Heft 4 und 5, 1905. Reprinted in Sammlung kleiner Schriften zur Neurosenlehre, 2 Folge. Leipzig u. Wien, 1909.)

Freud, S (Wien): "Der Wahn und die Träume in W. Jensen's *Gradiva.* (*Schriften zur angewandten Seelenkunde,* Heft 1, Wien und Leipzig, 1907.)

Freud, S. (Wien): "Über den Gegensinn der Urworte." A review of the brochure of the same name by Karl Abel, 1884. (*Jahrbuch für psychoanalyt. und psychopatholog.* Forschungen, Bd. II., 1910.)

"Typisches Beispiel eines verkappten Ödipustraumes." (*Zentralbl. für Psychoanalyse,* I. Jahrg. 1910, Heft. 1.)

Freud, S. (Wien): *Nachträge zur Traumdeutung.* (Ebenda, Heft 5.)

Hitschmann, Ed. (Wien): *Freud's Neurosenlehre. Nach ihrem gegenwartigen Stande zusammenfassend dargestellt.* Wien und Leipzig, 1911. (Kap. V., "Der Traum.")

Jones, Ernest (Toronto): "Freud's Theory of Dreams." (*American Journal of Psychology,* April 1910.)

Jones, Ernest (Toronto): "Some Instances of the Influence of Dreams on Waking Life." (*The Journ. of Abnormal Psychology, April–May 1911.)*

Jung, C. G. (Zürich): "L'Analyse des Rêves." (L'Année psychologique, tome XV.)

Jung, C. G. (Zürich): "Assoziation, Traum und hysterisches Symptom." (*Diagnostische Assoziationsstudien.* Beiträge zur experimentellen Psychopathologie, hrg. von Doz. C. G. Jung, II. Bd., Leipzig 1910. Nr. VTII., S. 31–66.)

Jung, C. G. (Zürich): "Ein Beitrag zur Psychologie des Gerüchtes." (*Zentralblatt für Psychoanalyse,* I. Jahrg. 1910, Heft 3.)

Maeder, Alphonse (Zürich): "Essai d'Interprétation de quelques Rêves." (*Archives de Psychologie,* t. VI., Nr. 24, April 1907.)

Maeder, Alphonse (Zürich): "Die Symbolik in den Legenden, Märchen, Gebrauchen und Träumen." (*Psychiatrisch-Neurolog.* Wochenschr. X. Jahrg.)

Meisl, Alfred (Wien): *Der Traum. Analytische Studien über die Elemente der Psychischen Funktion* V. (Wr. klin. Rdsch., 1907, Nr. 3–6.)

Onuf, B. (New York): "Dreams and their Interpretations as Diagnostic and Therapeutic Aids in Psychology." (*The Journal of Abnormal Psychology,* Feb.–Mar. 1910.)

Pfister, Oskar (Zürich): *Wahnvorstellung und Schülerselbstmord. Auf Grund einer Traumanalyse beleuchtet.* (Schweiz. Blätter für Schulgesundheitspflege, 1909, Nr. 1.)

Prince, Morton (Boston): "The Mechanism and Interpretation of Dreams." (*The Journal of Abnormal Psychology,* Oct.–Nov. 1910.)

Rank, Otto (Wien): "Ein Traum, der sich selbst deutet." (*Jahrbuch für psychoanalyt. und psychopatholog.* Forschungen, Bd. II., 1910.)

Rank, Otto (Wien): *Ein Beitrag zum Narzissismus.* (Ebenda, Bd. III., 1.).

Rank, Otto (Wien): "Beispiel eines verkappten Ödipustraumes." (*Zentralblatt für Psychoanalyse,* I. Jahrg., 1910.)

Rank, Otto (Wien): *Zum Thema der Zahnreiztraume.* (Ebenda.)

Rank, Otto (Wien): *Das Verlienen als Symptomhandlung. Zugleich ein Beitrag zum Verstandnis der Beziehungen des Traumlebens zu den Fehlleistungen des Alltagslebens.* (Ebenda.)

Robitsek, Alfred (Wien): "Die Analyse von Egmonts Traum." (*Jahrb. f. psychoanalyt. u. psychopathol.* Forschungen, Bd. II. 1910.)

Silberer, Herbert (Wien): "Bericht über eine Methode, gewisse symbolische Halluzinationserscheinungen hervorzurufen und zu beobachten." (*Jahr. Bleuler-Freud,* Bd. I., 1909.)

Silberer, Herbert (Wien): *Phantasie und Mythos.* (Ebenda, Bd. II., 1910.)

Stekel, Wilhelm (Wien): "Beiträge zur Traumdeutung." (*Jahrbuch für psychoanalytische und psychopatholog.* Forschungen, Bd. I., 1909.)

Stekel, Wilhelm (Wien): *Nervöse Angstzustände und ihre Behandlung.* (Wien und Berlin, 1908.)

Stekel, Wilhelm (Wien): *Die Sprache des Traumes.* A description of the symbolism and interpretation of the Dream and its relation to the normal and abnormal mind for physicians and psychologists. (Wiesbaden, 1911.)

Swoboda Hermann. *Die Perioden des menschlichen Organismus.* (Wien und Leipzig, 1904.)

Waterman, George A. (Boston): "Dreams as a Cause of Symptoms." (*The Journal of Abnormal Psychol.,* Oct.–Nov. 1910.)